NUMBER THEORY

An Introduction to Mathematics: Part B

NUMBER THEORY

An Introduction to Mathematics: Part B

By

WILLIAM A. COPPEL

 Springer

Library of Congress Control Number: 2005934653

PART A
   ISBN-10: 0-387-29851-7   e-ISBN: 0-387-29852-5
   ISBN-13: 978-0387-29851-1

PART B
   ISBN-10: 0-387-29853-3   e-ISBN: 0-387-29854-1
   ISBN-13: 978-0387-29853-5

2-VOLUME SET
   ISBN-10: 0-387-30019-8   e-ISBN: 0-387-30529-7
   ISBN-13: 978-0387-30019-1

Printed on acid-free paper.

AMS Subject Classifications: 11-xx, 05B20, 33E05

Contents

Part B

VII
The arithmetic of quadratic forms

We have already determined the integers which can be represented as a sum of two squares. Similarly, one may ask which integers can be represented in the form $x^2 + 2y^2$ or, more generally, in the form $ax^2 + 2bxy + cy^2$, where a,b,c are given integers. The arithmetic theory of binary quadratic forms, which had its origins in the work of Fermat, was extensively developed during the 18th century by Euler, Lagrange, Legendre and Gauss. The extension to quadratic forms in more than two variables, which was begun by them and is exemplified by Lagrange's theorem that every positive integer is a sum of four squares, was continued during the 19th century by Dirichlet, Hermite, H.J.S. Smith, Minkowski and others. In the 20th century Hasse and Siegel made notable contributions. With Hasse's work especially it became apparent that the theory is more perspicuous if one allows the variables to be rational numbers, rather than integers. This opened the way to the study of quadratic forms over arbitrary fields, with pioneering contributions by Witt (1937) and Pfister (1965-67).

From this vast theory we focus attention on one central result, the *Hasse–Minkowski theorem*. However, we first study quadratic forms over an arbitrary field in the geometric formulation of Witt. Then, following an interesting approach due to Fröhlich (1967), we study quadratic forms over a *Hilbert field*.

1 Quadratic spaces

The theory of quadratic spaces is simply another name for the theory of quadratic forms. The advantage of the change in terminology lies in its appeal to geometric intuition. It has in fact led to new results even at quite an elementary level. The new approach had its debut in a paper by Witt (1937) on the arithmetic theory of quadratic forms, but it is appropriate also if one is interested in quadratic forms over the real field or any other field.

For the remainder of this chapter *we will restrict attention to fields for which* $1 + 1 \neq 0$. Thus the phrase 'an arbitrary field' will mean 'an arbitrary field of characteristic $\neq 2$'. The proofs of many results make essential use of this restriction on the characteristic. For any field F, we will denote by $F^\times$ the multiplicative group of all nonzero elements of F. The squares in $F^\times$ form a subgroup $F^{\times 2}$ and any coset of this subgroup is called a *square class*.

Let V be a finite-dimensional vector space over such a field F. We say that V is a *quadratic space* if with each ordered pair u,v of elements of V there is associated an element (u,v) of F such that

(i) $(u_1 + u_2, v) = (u_1, v) + (u_2, v)$ for all $u_1, u_2, v \in V$;

(ii) $(\alpha u, v) = \alpha(u, v)$ for every $\alpha \in F$ and all $u, v \in V$;

(iii) $(u, v) = (v, u)$ for all $u, v \in V$.

It follows that

(i)' $(u, v_1 + v_2) = (u, v_1) + (u, v_2)$ for all $u, v_1, v_2 \in V$;

(ii)' $(u, \alpha v) = \alpha(u, v)$ for every $\alpha \in F$ and all $u, v \in V$.

Let $e_1, \ldots, e_n$ be a basis for the vector space V. Then any $u, v \in V$ can be uniquely expressed in the form

$$u = \sum_{j=1}^{n} \xi_j e_j, \quad v = \sum_{j=1}^{n} \eta_j e_j,$$

where $\xi_j, \eta_j \in F$ ($j = 1, \ldots, n$), and

$$(u, v) = \sum_{j,k=1}^{n} \alpha_{jk} \xi_j \eta_k,$$

where $\alpha_{jk} = (e_j, e_k) = \alpha_{kj}$. Thus

$$(u, u) = \sum_{j,k=1}^{n} \alpha_{jk} \xi_j \xi_k$$

is a *quadratic form* with coefficients in F. The quadratic space is completely determined by the quadratic form, since

$$(u, v) = \{(u + v, u + v) - (u, u) - (v, v)\}/2. \tag{1}$$

Conversely, for a given basis $e_1, \ldots, e_n$ of V, any $n \times n$ symmetric matrix $A = (\alpha_{jk})$ with elements from F, or the associated quadratic form $f(x) = x^t A x$, may be used in this way to give V the structure of a quadratic space.

Let $e_1', \ldots, e_n'$ be any other basis for V. Then

$$e_i = \sum_{j=1}^{n} \tau_{ji} e_j',$$

where $T = (\tau_{ij})$ is an invertible $n \times n$ matrix with elements from F. Conversely, any such matrix T defines in this way a new basis $e_1', \ldots, e_n'$. Since

$$(e_i, e_k) = \sum_{j,h=1}^{n} \tau_{ji} \beta_{jh} \tau_{hk},$$

where $\beta_{jh} = (e_j', e_h')$, it follows that

$$A = T^t B T. \tag{2}$$

Two symmetric matrices A,B with elements from F are said to be *congruent* if (2) holds for some invertible matrix T with elements from F. Thus congruence of symmetric matrices corresponds to a change of basis in the quadratic space. Evidently congruence is an equivalence relation, i.e. it is reflexive, symmetric and transitive. Two quadratic forms are said to be *equivalent over* F if their coefficient matrices are congruent. Equivalence over F of the quadratic forms f and g will be denoted by $f \sim_F g$ or simply $f \sim g$.

It follows from (2) that

$$\det A = (\det T)^2 \det B.$$

Thus, although $\det A$ is not uniquely determined by the quadratic space, if it is nonzero, its *square class* is uniquely determined. By abuse of language, we will call any representative of this square class the *determinant* of the quadratic space V and denote it by $\det V$.

Although quadratic spaces are better adapted for proving theorems, quadratic forms and symmetric matrices are useful for computational purposes. Thus a familiarity with both languages is desirable. However, we do not feel obliged to give two versions of each definition or result, and a version in one language may later be used in the other without explicit comment.

A vector v is said to be *orthogonal* to a vector u if $(u,v) = 0$. Then also u is orthogonal to v. The *orthogonal complement* $U^\perp$ of a subspace U of V is defined to be the set of all $v \in V$ such that $(u,v) = 0$ for every $u \in U$. Evidently $U^\perp$ is again a subspace. A subspace U will be said to be *non-singular* if $U \cap U^\perp = \{0\}$.

The whole space V is itself non-singular if and only if $V^\perp = \{0\}$. Thus V is non-singular if and only if some, and hence every, symmetric matrix describing it is non-singular, i.e. if and only if $\det V \neq 0$.

We say that a quadratic space V is the *orthogonal sum* of two subspaces V_1 and V_2, and we write $V = V_1 \perp V_2$, if $V = V_1 + V_2$, $V_1 \cap V_2 = \{0\}$ and $(v_1, v_2) = 0$ for all $v_1 \in V_1$, $v_2 \in V_2$.

If A_1 is a coefficient matrix for V_1 and A_2 a coefficient matrix for V_2, then

$$A = \begin{pmatrix} A_1 & 0 \\ 0 & A_2 \end{pmatrix}$$

is a coefficient matrix for $V = V_1 \perp V_2$. Thus $\det V = (\det V_1)(\det V_2)$. Evidently V is non-singular if and only if both V_1 and V_2 are non-singular.

If W is any subspace supplementary to the orthogonal complement $V^\perp$ of the whole space V, then $V = V^\perp \perp W$ and W is non-singular. Many problems for arbitrary quadratic spaces may be reduced in this way to non-singular quadratic spaces.

PROPOSITION 1 *If a quadratic space V contains a vector u such that $(u,u) \neq 0$, then*

$$V = U \perp U^\perp,$$

where $U = \langle u \rangle$ is the one-dimensional subspace spanned by u.

Proof For any vector $v \in V$, put $v' = v - \alpha u$, where $\alpha = (v,u)/(u,u)$. Then $(v',u) = 0$ and hence $v' \in U^\perp$. Since $U \cap U^\perp = \{0\}$, the result follows. $\square$

A vector space basis $u_1,...,u_n$ of a quadratic space V is said to be an *orthogonal basis* if $(u_j,u_k) = 0$ whenever $j \neq k$.

PROPOSITION 2 *Any quadratic space V has an orthogonal basis.*

Proof If V has dimension 1, there is nothing to prove. Suppose V has dimension $n > 1$ and the result holds for quadratic spaces of lower dimension. If $(v,v) = 0$ for all $v \in V$, then any basis is an orthogonal basis, by (1). Hence we may assume that V contains a vector u_1 such that $(u_1,u_1) \neq 0$. If U_1 is the 1-dimensional subspace spanned by u_1 then, by Proposition 1,

$$V = U_1 \perp U_1^\perp.$$

By the induction hypothesis $U_1^\perp$ has an orthogonal basis $u_2,...,u_n$, and $u_1,u_2,...,u_n$ is then an orthogonal basis for V. $\square$

Proposition 2 says that any symmetric matrix A is congruent to a diagonal matrix, or that the corresponding quadratic form f is equivalent over F to a diagonal form $\delta_1\xi_1^2 + ... + \delta_n\xi_n^2$. Evidently $\det f = \delta_1 \cdots \delta_n$ and f is non-singular if and only if $\delta_j \neq 0$ $(1 \leq j \leq n)$. If $A \neq 0$ then, by Propositions 1 and 2, we can take δ_1 to be any element of $F^\times$ which is represented by f.

Here $\gamma \in F^\times$ is said to be *represented* by a quadratic space V over the field F if there exists a vector $v \in V$ such that $(v,v) = \gamma$.

As an application of Proposition 2 we prove

PROPOSITION 3 *If U is a non-singular subspace of the quadratic space V, then $V = U \perp U^\perp$.*

Proof Let $u_1,...,u_m$ be an orthogonal basis for U. Then $(u_j,u_j) \neq 0$ $(1 \leq j \leq m)$, since U is non-singular. For any vector $v \in V$, let $u = \alpha_1 u_1 + ... + \alpha_m u_m$, where $\alpha_j = (v,u_j)/(u_j,u_j)$ for each j. Then $u \in U$ and $(u,u_j) = (v,u_j)$ $(1 \le j \le m)$. Hence $v - u \in U^\perp$. Since $U \cap U^\perp = \{0\}$, the result follows. $\square$

It may be noted that if U is a non-singular subspace and $V = U \perp W$ for some subspace W, then necessarily $W = U^\perp$. For it is obvious that $W \subseteq U^\perp$ and $\dim W = \dim V - \dim U = \dim U^\perp$, by Proposition 3.

PROPOSITION 4 *Let V be a non-singular quadratic space. If $v_1,...,v_m$ are linearly independent vectors in V then, for arbitrary $\eta_1,...,\eta_m \in F$, there exists a vector $v \in V$ such that $(v_j,v) = \eta_j$ $(1 \le j \le m)$.*

Moreover, if U is any subspace of V, then

(i) $\dim U + \dim U^\perp = \dim V$;

(ii) $U^{\perp\perp} = U$;

(iii) $U^\perp$ *is non-singular if and only if U is non-singular.*

Proof   There exist vectors $v_{m+1},...,v_n \in V$ such that $v_1,...,v_n$ form a basis for V. If we put $\alpha_{jk} = (v_j,v_k)$ then, since V is non-singular, the $n \times n$ symmetric matrix $A = (\alpha_{jk})$ is non-singular. Hence, for any $\eta_1,...,\eta_n \in F$, there exist *unique* $\xi_1,...,\xi_n \in F$ such that $v = \xi_1 v_1 + ... + \xi_n v_n$ satisfies

$$(v_1,v) = \eta_1, \ ... \ , (v_n,v) = \eta_n.$$

This proves the first part of the proposition.

By taking $U = <v_1,...,v_m>$ and $\eta_1 = ... = \eta_m = 0$, we see that $\dim U^\perp = n - m$. Replacing U by $U^\perp$, we obtain $\dim U^{\perp\perp} = \dim U$. Since it is obvious that $U \subseteq U^{\perp\perp}$, this implies $U = U^{\perp\perp}$. Since U non-singular means $U \cap U^\perp = \{0\}$, (iii) follows at once from (ii). $\square$

We now introduce some further definitions. A vector u is said to be *isotropic* if $u \ne 0$ and $(u,u) = 0$. A subspace U of V is said to be *isotropic* if it contains an isotropic vector and *anisotropic* otherwise. A subspace U of V is said to be *totally isotropic* if every nonzero vector in U is isotropic, i.e. if $U \subseteq U^\perp$. According to these definitions, the trivial subspace $\{0\}$ is both anisotropic and totally isotropic.

A quadratic space V over a field F is said to be *universal* if it represents every $\gamma \in F^\times$, i.e. if for each $\gamma \in F^\times$ there is a vector $v \in V$ such that $(v,v) = \gamma$.

PROPOSITION 5 *If a non-singular quadratic space V is isotropic, then it is universal.*

Proof   Since V is isotropic, it contains a vector $u \ne 0$ such that $(u,u) = 0$. Since V is non-singular, it contains a vector w such that $(u,w) \ne 0$. Then w is linearly independent of u and by replacing w by a scalar multiple we may assume $(u,w) = 1$. If $v = \alpha u + w$, then $(v,v) = \gamma$ for $\alpha = \{\gamma - (w,w)\}/2$. $\square$

On the other hand, a non-singular universal quadratic space need not be isotropic. As an example, take F to be the finite field with three elements and V the 2-dimensional quadratic space corresponding to the quadratic form $\xi_1^2 + \xi_2^2$.

PROPOSITION 6 *A non-singular quadratic form* $f(\xi_1,...,\xi_n)$ *with coefficients from a field* F *represents* $\gamma \in F^\times$ *if and only if the quadratic form*

$$g(\xi_0,\xi_1,...,\xi_n) = -\gamma\xi_0^2 + f(\xi_1,...,\xi_n)$$

is isotropic.

Proof Obviously if $f(x_1,...,x_n) = \gamma$ and $x_0 = 1$, then $g(x_0,x_1,...,x_n) = 0$. Suppose on the other hand that $g(x_0,x_1,...,x_n) = 0$ for some $x_j \in F$, not all zero. If $x_0 \neq 0$, then f certainly represents γ. If $x_0 = 0$, then f is isotropic and hence, by Proposition 5, it still represents γ. $\square$

PROPOSITION 7 *Let V be a non-singular isotropic quadratic space. If $V = U \perp W$, then there exists* $\gamma \in F^\times$ *such that, for some $u \in U$ and $w \in W$,*

$$(u,u) = \gamma, \quad (w,w) = -\gamma.$$

Proof Since V is non-singular, so also are U and W, and since V contains an isotropic vector v', there exist $u' \in U$, $w' \in W$, not both zero, such that

$$(u',u') = -(w',w').$$

If this common value is nonzero, we are finished. Otherwise either U or W is isotropic. Without loss of generality, suppose U is isotropic. Since W is non-singular, it contains a vector w such that $(w,w) \neq 0$, and U contains a vector u such that $(u,u) = -(w,w)$, by Proposition 5. $\square$

We now show that the totally isotropic subspaces of a quadratic space are important for an understanding of its structure, even though they are themselves trivial as quadratic spaces.

PROPOSITION 8 *All maximal totally isotropic subspaces of a quadratic space have the same dimension.*

Proof Let U_1 be a maximal totally isotropic subspace of the quadratic space V. Then $U_1 \subseteq U_1^\perp$ and $U_1^\perp \setminus U_1$ contains no isotropic vector. Since $V^\perp \subseteq U_1^\perp$, it follows that $V^\perp \subseteq U_1$. If V' is a subspace of V supplementary to $V^{\perp}$, then V' is non-singular and $U_1 = V^{\perp} + U_1'$, where $U_1' \subseteq V'$. Since U_1' is a maximal totally isotropic subspace of V', this shows that it is sufficient to establish the result when V itself is non-singular.

Let U_2 be another maximal totally isotropic subspace of V. Put $W = U_1 \cap U_2$ and let W_1, W_2 be subspaces supplementary to W in U_1, U_2 respectively. We are going to show that $W_2 \cap W_1^{\perp} = \{0\}$.

Let $v \in W_2 \cap W_1^{\perp}$. Since $W_2 \subseteq U_2$, v is isotropic and $v \in U_2^{\perp} \subseteq W^{\perp}$. Hence $v \in U_1^{\perp}$ and actually $v \in U_1$, since v is isotropic. Since $W_2 \subseteq U_2$ this implies $v \in W$, and since $W \cap W_2 = \{0\}$ this implies $v = 0$.

It follows that $\dim W_2 + \dim W_1^{\perp} \leq \dim V$. But, since V is now assumed non-singular, $\dim W_1 = \dim V - \dim W_1^{\perp}$, by Proposition 4. Hence $\dim W_2 \leq \dim W_1$ and, for the same reason, $\dim W_1 \leq \dim W_2$. Thus $\dim W_2 = \dim W_1$, and hence $\dim U_2 = \dim U_1$. $\quad\blacksquare$

We define the *index*, ind V, of a quadratic space V to be the dimension of any maximal totally isotropic subspace. Thus V is anisotropic if and only if ind $V = 0$.

A field F is said to be *ordered* if it contains a subset P of *positive* elements, which is closed under addition and multiplication, such that F is the disjoint union of the sets $\{0\}$, P and $-P = \{-x : x \in P\}$. The rational field $\mathbb{Q}$ and the real field $\mathbb{R}$ are ordered fields, with the usual interpretation of 'positive'. For quadratic spaces over an ordered field there are other useful notions of index.

A subspace U of a quadratic space V over an ordered field F is said to be *positive definite* if $(u,u) > 0$ for all nonzero $u \in U$ and *negative definite* if $(u,u) < 0$ for all nonzero $u \in U$. Evidently positive definite and negative definite subspaces are anisotropic.

PROPOSITION 9 *All maximal positive definite subspaces of a quadratic space V over an ordered field F have the same dimension.*

Proof Let U_+ be a maximal positive definite subspace of the quadratic space V. Since U_+ is certainly non-singular, we have $V = U_+ \perp W$, where $W = U_+^{\perp}$, and since U_+ is maximal, $(w,w) \leq 0$ for all $w \in W$. Since $U_+ \subseteq V$, we have $V^{\perp} \subseteq W$. If U_- is a maximal negative definite subspace of W, then in the same way $W = U_- \perp U_0$, where $U_0 = U_-^{\perp} \cap W$. Evidently U_0 is totally isotropic and $U_0 \subseteq V^{\perp}$. In fact $U_0 = V^{\perp}$, since $U_- \cap V^{\perp} = \{0\}$. Since $(v,v) \geq 0$ for all $v \in U_+ \perp V^{\perp}$, it follows that U_- is a maximal negative definite subspace of V.

If U_+' is another maximal positive definite subspace of V, then $U_+' \cap W = \{0\}$ and hence

$$\dim U_+' + \dim W = \dim (U_+' + W) \leq \dim V.$$

Thus $\dim U_+' \leq \dim V - \dim W = \dim U_+$. But U_+ and U_+' can be interchanged. ◻

If V is a quadratic space over an ordered field F, we define the *positive index* $\mathrm{ind}^+ V$ to be the dimension of any maximal positive definite subspace. Similarly all maximal negative definite subspaces have the same dimension, which we call the *negative index* of V and denote by $\mathrm{ind}^- V$. The proof of Proposition 9 shows that

$$\mathrm{ind}^+ V + \mathrm{ind}^- V + \dim V^\perp = \dim V.$$

PROPOSITION 10 *Let F denote the real field $\mathbb{R}$ or, more generally, an ordered field in which every positive element is a square. Then any non-singular quadratic form f in n variables with coefficients from F is equivalent over F to a quadratic form*

$$g = \xi_1^2 + \dots + \xi_p^2 - \xi_{p+1}^2 - \dots - \xi_n^2,$$

where $p \in \{0,1,\dots,n\}$ is uniquely determined by f. In fact,

$$\mathrm{ind}^+ f = p, \quad \mathrm{ind}^- f = n - p, \quad \mathrm{ind}\, f = \min (p, n - p).$$

Proof By Proposition 2, f is equivalent over F to a diagonal form $\delta_1 \eta_1^2 + \dots + \delta_n \eta_n^2$, where $\delta_j \neq 0$ $(1 \leq j \leq n)$. We may choose the notation so that $\delta_j > 0$ for $j \leq p$ and $\delta_j < 0$ for $j > p$. The change of variables $\xi_j = \delta_j^{1/2} \eta_j$ $(j \leq p)$, $\xi_j = (- \delta_j)^{1/2} \eta_j$ $(j > p)$ now brings f to the form g. Since the corresponding quadratic space has a p-dimensional maximal positive definite subspace, $p = \mathrm{ind}^+ f$ is uniquely determined. Similarly $n - p = \mathrm{ind}^- f$, and the formula for $\mathrm{ind}\, f$ follows readily. ◻

It follows that, for quadratic spaces over a field of the type considered in Proposition 10, a subspace is anisotropic if and only if it is either positive definite or negative definite.

Proposition 10 completely solves the problem of equivalence for real quadratic forms. (The uniqueness of p is known as *Sylvester's law of inertia*.) It will now be shown that the problem of equivalence for quadratic forms over a finite field can also be completely solved.

LEMMA 11 *If V is a non-singular 2-dimensional quadratic space over a finite field $\mathbb{F}_q$, of (odd) cardinality q, then V is universal.*

Proof By choosing an orthogonal basis for V we are reduced to showing that if $\alpha, \beta, \gamma \in \mathbb{F}_q^\times$, then there exist $\xi, \eta \in \mathbb{F}_q$ such that $\alpha \xi^2 + \beta \eta^2 = \gamma$. As ξ runs through $\mathbb{F}_q$, $\alpha \xi^2$ takes $(q + 1)/2$ $= 1 + (q - 1)/2$ distinct values. Similarly, as η runs through $\mathbb{F}_q$, $\gamma - \beta \eta^2$ takes $(q + 1)/2$ distinct values. Since $(q + 1)/2 + (q + 1)/2 > q$, there exist $\xi, \eta \in \mathbb{F}_q$ for which $\alpha \xi^2$ and $\gamma - \beta \eta^2$ take the same value. ◻

PROPOSITION 12 *Any non-singular quadratic form f in n variables over a finite field* $\mathbb{F}_q$ *is equivalent over* $\mathbb{F}_q$ *to the quadratic form*

$$\xi_1^2 + \dots + \xi_{n-1}^2 + \delta\xi_n^2,$$

where $\delta = \det f$ *is the determinant of f.*

    There are exactly two equivalence classes of non-singular quadratic forms in n variables over $\mathbb{F}_q$, *one consisting of those forms f whose determinant* $\det f$ *is a square in* $\mathbb{F}_q^\times$, *and the other those for which* $\det f$ *is not a square in* $\mathbb{F}_q^\times$.

Proof Since the first statement of the proposition is trivial for $n = 1$, we assume that $n > 1$ and it holds for all smaller values of n. It follows from Lemma 11 that f represents 1 and hence, by the remark after the proof of Proposition 2, f is equivalent over $\mathbb{F}_q$ to a quadratic form $\xi_1^2 + g(\xi_2,...,\xi_n)$. Since f and g have the same determinant, the first statement of the proposition now follows from the induction hypothesis.

    Since $\mathbb{F}_q^\times$ contains $(q - 1)/2$ distinct squares, every element of $\mathbb{F}_q^\times$ is either a square or a square times a fixed non-square. The second statement of the proposition now follows from the first. $\quad\square$

    We now return to quadratic spaces over an arbitrary field. A 2-dimensional quadratic space is said to be a *hyperbolic plane* if it is non-singular and isotropic.

PROPOSITION 13 *For a 2-dimensional quadratic space V, the following statements are equivalent:*

(i)   *V is a hyperbolic plane;*
(ii)   *V has a basis* u_1,u_2 *such that* $(u_1,u_1) = (u_2,u_2) = 0$, $(u_1,u_2) = 1$;
(iii) *V has a basis* v_1,v_2 *such that* $(v_1,v_1) = 1$, $(v_2,v_2) = -1$, $(v_1,v_2) = 0$;
(iv) $-\det V$ *is a square in* $F^\times$.

Proof Suppose first that V is a hyperbolic plane and let u_1 be *any* isotropic vector in V. If v is any linearly independent vector, then $(u_1,v) \neq 0$, since V is non-singular. By replacing v by a scalar multiple we may assume $(u_1,v) = 1$. If we put $u_2 = v + \alpha u_1$, where $\alpha = -(v,v)/2$, then

$$(u_2,u_2) = (v,v) + 2\alpha = 0, \quad (u_1,u_2) = (u_1,v) = 1,$$

and u_1,u_2 is a basis for V.

    If u_1,u_2 are isotropic vectors in V such that $(u_1,u_2) = 1$, then the vectors $v_1 = u_1 + u_2/2$ and $v_2 = u_1 - u_2/2$ satisfy (iii), and if v_1,v_2 satisfy (iii) then $\det V = -1$.

Finally, if (iv) holds then V is certainly non-singular. Let w_1, w_2 be an orthogonal basis for V and put $\delta_j = (w_j, w_j)$ ($j = 1,2$). By hypothesis, $\delta_1 \delta_2 = -\gamma^2$, where $\gamma \in F^\times$. Since $\gamma w_1 + \delta_1 w_2$ is an isotropic vector, this proves that (iv) implies (i). $\square$

PROPOSITION 14 *Let V be a non-singular quadratic space. If U is a totally isotropic subspace with basis $u_1, \ldots, u_m$, then there exists a totally isotropic subspace U' with basis $u_1', \ldots, u_m'$ such that*

$$(u_j, u_k') = 1 \text{ or } 0 \text{ according as } j = k \text{ or } j \neq k.$$

Hence $U \cap U' = \{0\}$ and

$$U + U' = H_1 \perp \ldots \perp H_m,$$

where H_j is the hyperbolic plane with basis u_j, u_j' ($1 \leq j \leq m$).

Proof Suppose first that $m = 1$. Since V is non-singular, there exists a vector $v \in V$ such that $(u_1, v) \neq 0$. The subspace H_1 spanned by u_1, v is a hyperbolic plane and hence, by Proposition 13, it contains a vector u_1' such that $(u_1', u_1') = 0$, $(u_1, u_1') = 1$. This proves the proposition for $m = 1$.

Suppose now that $m > 1$ and the result holds for all smaller values of m. Let W be the totally isotropic subspace with basis $u_2, \ldots, u_m$. By Proposition 4, there exists a vector $v \in W^\perp$ such that $(u_1, v) \neq 0$. The subspace H_1 spanned by u_1, v is a hyperbolic plane and hence it contains a vector u_1' such that $(u_1', u_1') = 0$, $(u_1, u_1') = 1$. Since H_1 is non-singular, $H_1^\perp$ is also non-singular and $V = H_1 \perp H_1^\perp$. Since $W \subseteq H_1^\perp$, the result now follows by applying the induction hypothesis to the subspace W of the quadratic space $H_1^\perp$. $\square$

PROPOSITION 15 *Any quadratic space V can be represented as an orthogonal sum*

$$V = V^\perp \perp H_1 \perp \ldots \perp H_m \perp V_0,$$

where $H_1, \ldots, H_m$ are hyperbolic planes and the subspace V_0 is anisotropic.

Proof Let V_1 be any subspace supplementary to $V^\perp$. Then V_1 is non-singular, by the definition of $V^\perp$. If V_1 is anisotropic, we can take $m = 0$ and $V_0 = V_1$. Otherwise V_1 contains an isotropic vector and hence also a hyperbolic plane H_1, by Proposition 14. By Proposition 3,

$$V_1 = H_1 \perp V_2,$$

where $V_2 = H_1^\perp \cap V_1$ is non-singular. If V_2 is anisotropic, we can take $V_0 = V_2$. Otherwise we repeat the process. After finitely many steps we must obtain a representation of the required form, possibly with $V_0 = \{0\}$. $\square$

Let V and V' be quadratic spaces over the same field F. The quadratic spaces V,V' are said to be *isometric* if there exists a linear map $\varphi\colon V \to V'$ which is an *isometry*, i.e. it is bijective and

$$(\varphi v,\varphi v) = (v,v) \quad \text{for all } v \in V.$$

By (1), this implies

$$(\varphi u,\varphi v) = (u,v) \quad \text{for all } u,v \in V.$$

The concept of isometry is only another way of looking at equivalence. For if $\varphi\colon V \to V'$ is an isometry, then V and V' have the same dimension. If $u_1,\dots,u_n$ is a basis for V and $u_1',\dots,u_n'$ a basis for V' then, since $(u_j,u_k) = (\varphi u_j,\varphi u_k)$, the isometry is completely determined by the change of basis in V' from $\varphi u_1,\dots,\varphi u_n$ to $u_1',\dots,u_n'$.

A particularly simple type of isometry is defined in the following way. Let V be a quadratic space and w a vector such that $(w,w) \neq 0$. The map $\tau\colon V \to V$ defined by

$$\tau v = v - \{2(v,w)/(w,w)\}w$$

is obviously linear. If W is the non-singular one-dimensional subspace spanned by w, then $V = W \perp W^{\perp}$. Since $\tau v = v$ if $v \in W^{\perp}$ and $\tau v = -v$ if $v \in W$, it follows that τ is bijective. Writing $\alpha = -2(v,w)/(w,w)$, we have

$$(\tau v,\tau v) = (v,v) + 2\alpha(v,w) + \alpha^2(w,w) = (v,v).$$

Thus τ is an isometry. Geometrically, τ is a *reflection* in the hyperplane orthogonal to w. We will refer to $\tau = \tau_w$ as the reflection corresponding to the non-isotropic vector w.

PROPOSITION 16 *If u,u' are vectors of a quadratic space V such that $(u,u) = (u',u') \neq 0$, then there exists an isometry $\varphi\colon V \to V$ such that $\varphi u = u'$.*

Proof Since

$$(u + u',u + u') + (u - u',u - u') = 2(u,u) + 2(u',u') = 4(u,u),$$

at least one of the vectors $u + u', u - u'$ is not isotropic. If $u - u'$ is not isotropic, the reflection τ corresponding to $w = u - u'$ has the property $\tau u = u'$, since $(u - u',u - u') = 2(u,u - u')$. If $u + u'$ is not isotropic, the reflection τ corresponding to $w = u + u'$ has the property $\tau u = -u'$. Since u' is not isotropic, the corresponding reflection σ maps u' onto $-u'$, and hence the isometry $\sigma\tau$ maps u onto u'. $\quad\blacksquare$

The proof of Proposition 16 has the following interesting consequence:

PROPOSITION 17 *Any isometry* $\varphi\colon V \to V$ *of a non-singular quadratic space* V *is a product of reflections.*

Proof Let $u_1,\dots,u_n$ be an orthogonal basis for V. By Proposition 16 and its proof, there exists an isometry ψ, which is either a reflection or a product of two reflections, such that $\psi u_1 = \varphi u_1$. If U is the subspace with basis u_1 and W the subspace with basis $u_2,\dots,u_n$, then $V = U \perp W$ and $W = U^\perp$ is non-singular. Since the isometry $\varphi_1 = \psi^{-1}\varphi$ fixes u_1, we have also $\varphi_1 W = W$. But if $\sigma\colon W \to W$ is a reflection, the extension $\tau\colon V \to V$ defined by $\tau u = u$ if $u \in U$, $\tau w = \sigma w$ if $w \in W$, is also a reflection. By using induction on the dimension n, it follows that φ_1 is a product of reflections, and hence so also is φ. $\square$

By a more elaborate argument E. Cartan (1938) showed that any isometry of an n-dimensional non-singular quadratic space is a product of at most n reflections.

PROPOSITION 18 *Let* V *be a quadratic space with two orthogonal sum representations*

$$V = U \perp W = U' \perp W'.$$

If there exists an isometry $\varphi\colon U \to U'$, *then there exists an isometry* $\psi\colon V \to V$ *such that* $\psi u = \varphi u$ *for all* $u \in U$ *and* $\psi W = W'$. *Thus if* U *is isometric to* U', *then* W *is isometric to* W'.

Proof Let $u_1,\dots,u_m$ and $u_{m+1},\dots,u_n$ be bases for U and W respectively. If $u_j' = \varphi u_j$ $(1 \le j \le m)$, then $u_1',\dots,u_m'$ is a basis for U'. Let $u_{m+1}',\dots,u_n'$ be a basis for W'. The symmetric matrices associated with the bases $u_1,\dots,u_n$ and $u_1',\dots,u_n'$ of V have the form

$$\begin{pmatrix} A & 0 \\ 0 & B \end{pmatrix}, \begin{pmatrix} A & 0 \\ 0 & C \end{pmatrix},$$

which we will write as $A \oplus B$, $A \oplus C$. Thus the two matrices $A \oplus B$, $A \oplus C$ are congruent. It is enough to show that this implies that B and C are congruent. For suppose $C = S^t B S$ for some invertible matrix $S = (\sigma_{ij})$. If we define $u_{m+1}'',\dots,u_n''$ by

$$u_i' = \sum_{j=m+1}^{n} \sigma_{ji} u_j'' \quad (m+1 \le i \le n),$$

then $(u_j'',u_k'') = (u_j,u_k)$ $(m+1 \le j,k \le n)$ and the linear map $\psi\colon V \to V$ defined by

$$\psi u_j = u_j' \quad (1 \le j \le m), \quad \psi u_j = u_j'' \quad (m+1 \le j \le n),$$

is the required isometry.

By taking the bases for U,W,W' to be orthogonal bases we are reduced to the case in which

A,B,C are diagonal matrices. We may choose the notation so that $A = \text{diag } [a_1,...,a_m]$, where $a_j \neq 0$ for $j \leq r$ and $a_j = 0$ for $j > r$. If $a_1 \neq 0$, i.e. if $r > 0$, and if we write $A' = \text{diag } [a_2,...,a_m]$, then it follows from Propositions 1 and 16 that the matrices $A' \oplus B$ and $A' \oplus C$ are congruent. Proceeding in this way, we are reduced to the case $A = O$.

Thus we now suppose $A = O$. We may assume $B \neq O$, $C \neq O$, since otherwise the result is obvious. We may choose the notation also so that $B = O_s \oplus B'$ and $C = O_s \oplus C'$, where B' is non-singular and $0 \leq s < n - m$. If $T^t(O_{m+s} \oplus C')T = O_{m+s} \oplus B'$, where

$$T = \begin{pmatrix} T_1 & T_2 \\ T_3 & T_4 \end{pmatrix},$$

then $T_4^t C T_4 = B'$. Since B' is non-singular, so also is T_4 and thus B' and C' are congruent. It follows that B and C are also congruent. $\square$

COROLLARY 19 *If a non-singular subspace U of a quadratic space V is isometric to another subspace U', then $U^\perp$ is isometric to $U'^\perp$.*

Proof This follows at once from Proposition 18, since U' is also non-singular and

$$V = U \perp U^\perp = U' \perp U'^\perp. \quad \square$$

The first statement of Proposition 18 is known as *Witt's extension theorem* and the second statement as *Witt's cancellation theorem*. It was Corollary 19 which was actually proved by Witt (1937).

There is also another version of the extension theorem, which says that if $\varphi: U \to U'$ is an isometry between two subspaces U,U' of a *non-singular* quadratic space V, then there exists an isometry $\psi: V \to V$ such that $\psi u = \varphi u$ for all $u \in U$. For non-singular U this has just been proved, and the singular case can be reduced to the non-singular by applying (several times, if necessary) the following lemma.

LEMMA 20 *Let V be a non-singular quadratic space. If U,U' are singular subspaces of V and if there exists an isometry $\varphi: U \to U'$, then there exist subspaces $\overline{U},\overline{U}'$ properly containing U,U' respectively and an isometry $\overline{\varphi}: \overline{U} \to \overline{U}'$ such that $\overline{\varphi} u = \varphi u$ for all $u \in U$.*

Proof By hypothesis there exists a nonzero vector $u_1 \in U \cap U^\perp$. Then U has a basis $u_1,...,u_m$ with u_1 as first vector. By Proposition 4, there exists a vector $w \in V$ such that

$$(u_1,w) = 1, \quad (u_j,w) = 0 \text{ for } 1 < j \leq m.$$

Moreover we may assume $(w,w) = 0$, by replacing w by $w - \alpha u_1$, where $\alpha = (w,w)/2$. If W is the 1-dimensional subspace spanned by w, then $U \cap W = \{0\}$ and $\overline{U} = U + W$ contains U properly.

The same construction can be applied to U', with the basis $\varphi u_1,...,\varphi u_m$, to obtain an isotropic vector w' and a subspace $\overline{U}' = U' + W'$. The linear map $\overline{\varphi} : \overline{U} \to \overline{U}'$ defined by

$$\overline{\varphi} u_j = \varphi u_j \ (1 \le j \le m), \quad \overline{\varphi} w = w',$$

is easily seen to have the required properties. $\square$

As an application of Proposition 18, we will consider the uniqueness of the representation obtained in Proposition 15.

PROPOSITION 21 *Suppose the quadratic space V can be represented as an orthogonal sum*

$$V = U \perp H \perp V_0,$$

where U is totally isotropic, H is the orthogonal sum of m hyperbolic planes, and the subspace V_0 is anisotropic.

Then $U = V^\perp$, $m = \text{ind } V - \dim V^\perp$, and V_0 is uniquely determined up to an isometry.

Proof Since H and V_0 are non-singular, so also is $W = H \perp V_0$. Hence, by the remark after the proof of Proposition 3, $U = W^\perp$. Since $U \subseteq U^\perp$, it follows that $U \subseteq V^\perp$. In fact $U = V^\perp$, since $W \cap V^\perp = \{0\}$.

The subspace H has two m-dimensional totally isotropic subspaces U_1, U_1' such that

$$H = U_1 + U_1', \quad U_1 \cap U_1' = \{0\}.$$

Evidently $V_1 := V^\perp + U_1$ is a totally isotropic subspace of V. In fact V_1 is maximal, since any isotropic vector in $U_1' \perp V_0$ is already contained in U_1'. Thus $m = \text{ind } V - \dim V^\perp$ is uniquely determined and H is uniquely determined up to an isometry. If also

$$V = V^\perp \perp H' \perp V_0',$$

where H' is the orthogonal sum of m hyperbolic planes and V_0' is anisotropic then, by Proposition 18, V_0 is isometric to V_0'. $\square$

Proposition 21 reduces the problem of equivalence for quadratic forms over an arbitrary field to the case of anisotropic forms. As we will see, this can still be a difficult problem, even for the field of rational numbers.

Two quadratic spaces V, V' over the same field F may be said to be *Witt-equivalent*, in symbols $V \approx V'$, if their anisotropic components V_0, V_0' are isometric. This is certainly an equivalence relation. The cancellation law makes it possible to define various algebraic operations on the set $\mathcal{W}(F)$ of all quadratic spaces over the field F, with equality replaced by Witt-equivalence. If we define $-V$ to be the quadratic space with the same underlying vector space as V but with (v_1, v_2) replaced by $-(v_1, v_2)$, then

$$V \perp (-V) \approx \{O\}.$$

If we define the *sum* of two quadratic spaces V and W to be $V \perp W$, then

$$V \approx V', W \approx W' \implies V \perp W \approx V' \perp W'.$$

Similarly, if we define the *product* of V and W to be the tensor product $V \otimes W$ of the underlying vector spaces with the quadratic space structure defined by

$$(\{v_1, w_1\}, \{v_2, w_2\}) = (v_1, v_2)(w_1, w_2),$$

then

$$V \approx V', W \approx W' \implies V \otimes W \approx V' \otimes W'.$$

It is readily seen that in this way $\mathcal{W}(F)$ acquires the structure of a commutative ring, the *Witt ring* of the field F.

2 The Hilbert symbol

Again let F be any field of characteristic $\neq 2$ and $F^\times$ the multiplicative group of all nonzero elements of F. We define the *Hilbert symbol* $(a,b)_F$, where $a,b \in F^\times$, by

$$(a,b)_F = 1 \text{ if there exist } x,y \in F \text{ such that } ax^2 + by^2 = 1,$$

$$= -1 \text{ otherwise.}$$

By Proposition 6, $(a,b)_F = 1$ *if and only if the ternary quadratic form* $a\xi^2 + b\eta^2 - \zeta^2$ *is isotropic*.

The following lemma shows that the Hilbert symbol can also be defined in an asymmetric way:

LEMMA 22 *For any field F and any $a,b \in F^\times$, $(a,b)_F = 1$ if and only if the binary quadratic form $f_a = \xi^2 - a\eta^2$ represents b. Morover, for any $a \in F^\times$, the set G_a of all $b \in F^\times$ which are represented by f_a is a subgroup of $F^\times$.*

Proof Suppose first that $ax^2 + by^2 = 1$ for some $x,y \in F$. If a is a square, the quadratic form f_a is isotropic and hence universal. If a is not a square, then $y \neq 0$ and $(y^{-1})^2 - a(xy^{-1})^2 = b$.

Suppose next that $u^2 - av^2 = b$ for some $u,v \in F$. If $-ba^{-1}$ is a square, the quadratic form $a\xi^2 + b\eta^2$ is isotropic and hence universal. If $-ba^{-1}$ is not a square, then $u \neq 0$ and $a(vu^{-1})^2 + b(u^{-1})^2 = 1$.

It is obvious that if $b \in G_a$, then also $b^{-1} \in G_a$, and it is easily verified that if

$$\zeta_1 = \xi_1\eta_1 + a\xi_2\eta_2, \quad \zeta_2 = \xi_1\eta_2 + \xi_2\eta_1,$$

then

$$\zeta_1^2 - a\zeta_2^2 = (\xi_1^2 - a\xi_2^2)(\eta_1^2 - a\eta_2^2).$$

(In fact this is just Brahmagupta's identity, already encountered in §4 of Chapter IV.) It follows that G_a is a subgroup of $F^\times$. ◻

PROPOSITION 23 *For any field F, the Hilbert symbol has the following properties*:

(i) $(a,b)_F = (b,a)_F$,

(ii) $(a,bc^2)_F = (a,b)_F$ *for any* $c \in F^\times$,

(iii) $(a,1)_F = 1$,

(iv) $(a,-ab)_F = (a,b)_F$,

(v)   *if* $(a,b)_F = 1$, *then* $(a,bc)_F = (a,c)_F$ *for any* $c \in F^\times$.

Proof The first three properties follow immediately from the definition. The fourth property follows from Lemma 22. For, since G_a is a group and f_a represents $-a$, f_a represents $-ab$ if and only if it represents b. The proof of (v) is similar: if f_a represents b, then it represents bc if and only if it represents c. ◻

The Hilbert symbol will now be evaluated for the real field $\mathbb{R} = \mathbb{Q}_\infty$ and the p-adic fields $\mathbb{Q}_p$ studied in Chapter VI. In these cases it will be denoted simply by $(a,b)_\infty$, resp. $(a,b)_p$. For the real field, we obtain at once from the definition of the Hilbert symbol

PROPOSITION 24 *Let $a,b \in \mathbb{R}^\times$. Then $(a,b)_\infty = -1$ if and only if both $a < 0$ and $b < 0$.* ◻

To evaluate $(a,b)_p$, we first note that we can write $a = p^\alpha a'$, $b = p^\beta b'$, where $\alpha,\beta \in \mathbb{Z}$ and $|a'|_p = |b'|_p = 1$. It follows from (i),(ii) of Proposition 23 that we may assume $\alpha,\beta \in \{0,1\}$.

Furthermore, by (ii),(iv) of Proposition 23 we may assume that α and β are not both 1. Thus we are reduced to the case where a is a p-adic unit and either b is a p-adic unit or $b = pb'$, where b' is a p-adic unit. To evaluate $(a,b)_p$ under these assumptions we will use the conditions for a p-adic unit to be a square which were derived in Chapter VI. It is convenient to treat the case $p = 2$ separately.

PROPOSITION 25 *Let p be an odd prime and $a,b \in \mathbb{Q}_p$ with $|a|_p = |b|_p = 1$. Then*

(i) $(a,b)_p = 1$,

(ii) $(a,pb)_p = 1$ *if and only if $a = c^2$ for some $c \in \mathbb{Q}_p$.*

   In particular, for any integers a,b not divisible by p, $(a,b)_p = 1$ and $(a,pb)_p = (a/p)$, where (a/p) is the Legendre symbol.

Proof Let $S \subseteq \mathbb{Z}_p$ be a set of representatives, with $0 \in S$, of the finite residue field $\mathbb{F}_p = \mathbb{Z}_p/p\mathbb{Z}_p$. There exist non-zero $a_0,b_0 \in S$ such that

$$|a - a_0|_p < 1, \quad |b - b_0|_p < 1.$$

But Lemma 11 implies that there exist $x_0,y_0 \in S$ such that

$$|a_0 x_0^2 + b_0 y_0^2 - 1|_p < 1.$$

Since $|x_0|_p \leq 1, |y_0|_p \leq 1$, it follows that

$$|ax_0^2 + by_0^2 - 1|_p < 1.$$

Hence, by Proposition VI.16, $ax_0^2 + by_0^2 = z^2$ for some $z \in \mathbb{Q}_p$. Since $z \neq 0$, this implies $(a,b)_p = 1$. This proves (i).

   If $a = c^2$ for some $c \in \mathbb{Q}_p$, then $(a,pb)_p = 1$, by Proposition 23. Conversely, suppose there exist $x,y \in \mathbb{Q}_p$ such that $ax^2 + pby^2 = 1$. Then $|ax^2|_p \neq |pby^2|_p$, since $|a|_p = |b|_p = 1$. It follows that $|x|_p = 1, |y|_p \leq 1$. Thus $|ax^2 - 1|_p < 1$ and hence $ax^2 = z^2$ for some $z \in \mathbb{Q}_p^\times$. This proves (ii).

   The special case where a and b are integers now follows from Corollary VI.17. $\square$

COROLLARY 26   *If p is an odd prime and if $a,b,c \in \mathbb{Q}_p$ are p-adic units, then the quadratic form $a\xi^2 + b\eta^2 + c\zeta^2$ is isotropic.*

Proof The quadratic form $-c^{-1}a\xi^2 - c^{-1}b\eta^2 - \zeta^2$ is isotropic, since $(-c^{-1}a, -c^{-1}b)_p = 1$, by Proposition 25. $\square$

PROPOSITION 27 *Let $a,b \in \mathbb{Q}_2$ with $|a|_2 = |b|_2 = 1$. Then*

(i) $(a,b)_2 = 1$ *if and only if at least one of $a,b,a-4,b-4$ is a square in $\mathbb{Q}_2$;*

(ii) $(a,2b)_2 = 1$ *if and only if either a or $a+2b$ is a square in $\mathbb{Q}_2$.*

In particular, for any odd integers a,b, $(a,b)_2 = 1$ if and only if $a \equiv 1$ or $b \equiv 1$ mod 4, and $(a,2b)_2 = 1$ if and only if $a \equiv 1$ or $a+2b \equiv 1$ mod 8.

Proof Suppose there exist $x,y \in \mathbb{Q}_2$ such that $ax^2 + by^2 = 1$ and assume, for definiteness, that $|x|_2 \geq |y|_2$. Then $|x|_2 \geq 1$ and $|x|_2 = 2^\alpha$, where $\alpha \geq 0$. By Corollary VI.14,

$$x = 2^\alpha(x_0 + 4x'), \quad y = 2^\alpha(y_0 + 4y'),$$

where $x_0 \in \{1,3\}$, $y_0 \in \{0,1,2,3\}$ and $x',y' \in \mathbb{Z}_2$. If a and b are not squares in $\mathbb{Q}_2$ then, by Proposition VI.16, $|a-1|_2 > 2^{-3}$ and $|b-1|_2 > 2^{-3}$. Thus

$$a = a_0 + 8a', \quad b = b_0 + 8b',$$

where $a_0,b_0 \in \{3,5,7\}$ and $a',b' \in \mathbb{Z}_2$. Hence

$$1 = ax^2 + by^2 = 2^{2\alpha}(a_0 + b_0y_0^2 + 8z'),$$

where $z' \in \mathbb{Z}_2$. Since a_0,b_0 are odd and $y_0^2 \equiv 0,1$ or 4 mod 8, we must have $\alpha = 0$, $y_0^2 \equiv 1$ mod 8 and $a_0 = 5$. Thus, by Proposition VI.16 again, $a-4$ is a square in $\mathbb{Q}_2$. This proves that the condition in (i) is necessary.

If a is a square in $\mathbb{Q}_2$, then certainly $(a,b)_2 = 1$. If $a-4$ is a square, then $a = 5 + 8a'$, where $a' \in \mathbb{Z}_2$, and $a + 4b = 1 + 8c'$, where $c' \in \mathbb{Z}_2$. Hence $a+4b$ is a square in $\mathbb{Q}_2$ and the quadratic form $a\xi^2 + b\eta^2$ represents 1. This proves that the condition in (i) is sufficient.

Suppose next that there exist $x,y \in \mathbb{Q}_2$ such that $ax^2 + 2by^2 = 1$. By the same argument as for odd p in Proposition 25, we must have $|x|_2 = 1$, $|y|_2 \leq 1$. Thus $x = x_0 + 4x'$, $y = y_0 + 4y'$, where $x_0 \in \{1,3\}$, $y_0 \in \{0,1,2,3\}$ and $x',y' \in \mathbb{Z}_2$. Writing $a = a_0 + 8a'$, $b = b_0 + 8b'$, where $a_0,b_0 \in \{1,3,5,7\}$ and $a',b' \in \mathbb{Z}_2$, we obtain $a_0x_0^2 + 2b_0y_0^2 \equiv 1$ mod 8. Since $2y_0^2 \equiv 0$ or 2 mod 8, this implies either $a_0 \equiv 1$ or $a_0 + 2b_0 \equiv 1$ mod 8. Hence either a or $a+2b$ is a square in $\mathbb{Q}_2$. It is obvious that, conversely, $(a,2b)_2 = 1$ if either a or $a+2b$ is a square in $\mathbb{Q}_2$.

The special case where a and b are integers again follows from Corollary VI.17. $\quad\blacksquare$

For $F = \mathbb{R}$, the factor group $F^\times/F^{\times 2}$ is of order 2, with 1 and -1 as representatives of the two square classes. For $F = \mathbb{Q}_p$, with p odd, it follows from Corollary VI.17 that the factor group $F^\times/F^{\times 2}$ is of order 4. Moreover, if r is an integer such that $(r/p) = -1$, then $1,r,p,rp$ are representatives of the four square classes. Similarly for $F = \mathbb{Q}_2$, the factor group $F^\times/F^{\times 2}$ is of order 8 and 1,3,5,7,2,6,10,14 are representatives of the eight square classes. The Hilbert symbol $(a,b)_F$ for these representatives, and hence for all $a,b \in F^\times$, may be determined directly from Propositions 24, 25 and 27. The values obtained are listed in Table 1, where $\varepsilon = (-1/p)$ and thus

$$\varepsilon = \pm 1 \text{ according as } p \equiv \pm 1 \bmod 4.$$

$\mathbb{Q}_\infty = \mathbb{R}$

$\mathbb{Q}_p: p$ odd

$a \backslash b$	1	-1
1	+	+
-1	+	$-$

$a \backslash b$	1	p	rp	r
1	+	+	+	+
p	+	ε	$-\varepsilon$	$-$
rp	+	$-\varepsilon$	ε	$-$
r	+	$-$	$-$	+

$\mathbb{Q}_2$

$a \backslash b$	1	3	6	2	14	10	5	7
1	+	+	+	+	+	+	+	+
3	+	$-$	+	$-$	+	$-$	+	$-$
6	+	+	$-$	$-$	+	+	$-$	$-$
2	+	$-$	$-$	+	+	$-$	$-$	+
14	+	+	+	+	$-$	$-$	$-$	$-$
10	+	$-$	+	$-$	$-$	+	$-$	+
5	+	+	$-$	$-$	$-$	$-$	+	+
7	+	$-$	$-$	+	$-$	+	+	$-$

Table 1: Values of the Hilbert symbol $(a,b)_F$ for $F = \mathbb{Q}_v$

It will be observed that each of the three symmetric matrices in Table 1 is a Hadamard matrix! In particular, in each row after the first row of +'s there are equally many + and $-$ signs. This property turns out to be of basic importance and prompts the following definition:

A field F is a *Hilbert field* if some $a \in F^\times$ is not a square and if, for every such a, the subgroup G_a has index 2 in $F^\times$.

Thus the real field $\mathbb{R} = \mathbb{Q}_\infty$ and the p-adic fields $\mathbb{Q}_p$ are all Hilbert fields. We now show that in Hilbert fields further properties of the Hilbert symbol may be derived.

PROPOSITION 28 *In any Hilbert field F, the Hilbert symbol has the following additional properties*:

(i)   *if* $(a,b)_F = 1$ *for every* $b \in F^\times$, *then a is a square in* $F^\times$;

(ii) $(a,bc)_F = (a,b)_F \, (a,c)_F$ *for all* $a,b,c \in F^\times$.

Proof The first property is immediate, since $G_a \neq F^\times$ if a is not a square. If $(a,b)_F = 1$ or $(a,c)_F = 1$, then (ii) follows from Proposition 23(v). Suppose now that $(a,b)_F = (a,c)_F = -1$. Then a is not a square and f_a does not represent b or c. Since F is a Hilbert field and $b,c \notin G_a$, it follows that $bc \in G_a$. Thus $(a,bc)_F = 1$. ☐

The definition of a Hilbert field can be reformulated in terms of quadratic forms. If f is an anisotropic binary quadratic form with determinant d, then $-d$ is not a square and f is equivalent to a diagonal form $a(\xi^2 + d\eta^2)$. It follows that F is a Hilbert field if and only if there exists an anisotropic binary quadratic form and for each such form there is, apart from equivalent forms, exactly one other whose determinant is in the same square class. We are going to show that Hilbert fields can also be characterized by means of quadratic forms in 4 variables.

LEMMA 29 *Let F be an arbitrary field and a,b elements of* $F^\times$ *with* $(a,b)_F = -1$. *Then the quadratic form*

$$f_{a,b} = \xi_1^2 - a\xi_2^2 - b(\xi_3^2 - a\xi_4^2)$$

is anisotropic. Morover, the set $G_{a,b}$ *of all elements of* $F^\times$ *which are represented by* $f_{a,b}$ *is a subgroup of* $F^\times$.

Proof Since $(a,b)_F = -1$, a is not a square and hence the binary form f_a is anisotropic. If $f_{a,b}$ were isotropic, some $c \in F^\times$ would be represented by both f_a and bf_a. But then $(a,c)_F = 1$ and $(a,bc)_F = 1$. Since $(a,b)_F = -1$, this contradicts Proposition 23.

Clearly if $c \in G_{a,b}$, then also $c^{-1} \in G_{a,b}$, and it is easily verified that if

$$\zeta_1 = \xi_1\eta_1 + a\xi_2\eta_2 + b\xi_3\eta_3 - ab\xi_4\eta_4, \quad \zeta_2 = \xi_1\eta_2 + \xi_2\eta_1 - b\xi_3\eta_4 + b\xi_4\eta_3,$$
$$\zeta_3 = \xi_1\eta_3 + \xi_3\eta_1 + a\xi_2\eta_4 - a\xi_4\eta_2, \quad \zeta_4 = \xi_1\eta_4 + \xi_4\eta_1 + \xi_2\eta_3 - \xi_3\eta_2,$$

then

$$\zeta_1^2 - a\zeta_2^2 - b\zeta_3^2 + ab\zeta_4^2 = (\xi_1^2 - a\xi_2^2 - b\xi_3^2 + ab\xi_4^2)(\eta_1^2 - a\eta_2^2 - b\eta_3^2 + ab\eta_4^2).$$

It follows that $G_{a,b}$ is a subgroup of $F^\times$. ☐

PROPOSITION 30 *A field F is a Hilbert field if and only if one of the following mutually exclusive conditions is satisfied:*

(A) *F is an ordered field and every positive element of F is a square;*
(B) *there exists, up to equivalence, one and only one anisotropic quaternary quadratic form over F.*

Proof Suppose first that the field F is of type (A). Then -1 is not a square, since $-1+1=0$ and any nonzero square is positive. By Proposition 10, any anisotropic binary quadratic form is equivalent over F to exactly one of the forms $\xi^2 + \eta^2$, $-\xi^2 - \eta^2$ and thus F is a Hilbert field. Since the quadratic forms $\xi_1^2 + \xi_2^2 + \xi_3^2 + \xi_4^2$ and $-\xi_1^2 - \xi_2^2 - \xi_3^2 - \xi_4^2$ are anisotropic and inequivalent, the field F is not of type (B).

Suppose next that the field F is of type (B). The anisotropic quaternary quadratic form must be universal, since it is equivalent to any nonzero scalar multiple. Hence, for any $a \in F^\times$ there exists an anisotropic diagonal form

$$-a\xi_1^2 - b'\xi_2^2 - c'\xi_3^2 - d'\xi_4^2,$$

where $b',c',d' \in F^\times$. In particular, for $a = -1$, this shows that not every element of $F^\times$ is a square. The ternary quadratic form $h = -b'\xi_2^2 - c'\xi_3^2 - d'\xi_4^2$ is certainly anisotropic. If h does not represent 1, the quaternary quadratic form $-\xi_1^2 + h$ is also anisotropic and hence, by Witt's cancellation theorem, a must be a square. Consequently, if $a \in F^\times$ is not a square, then there exists an anisotropic form

$$-a\xi_1^2 + \xi_2^2 - b\xi_3^2 - c\xi_4^2.$$

Thus for any $a \in F^\times$ which is not a square, there exists $b \in F^\times$ such that $(a,b)_F = -1$. If $(a,b)_F = (a,b')_F = -1$ then, by Lemma 29, the forms

$$\xi_1^2 - a\xi_2^2 - b(\xi_3^2 - a\xi_4^2), \quad \xi_1^2 - a\xi_2^2 - b'(\xi_3^2 - a\xi_4^2)$$

are anisotropic and thus equivalent. It follows from Witt's cancellation theorem that the binary forms $b(\xi_3^2 - a\xi_4^2)$ and $b'(\xi_3^2 - a\xi_4^2)$ are equivalent. Consequently $\xi_3^2 - a\xi_4^2$ represents bb' and $(a,bb')_F = 1$. Thus G_a has index 2 in $F^\times$ for any $a \in F^\times$ which is not a square, and F is a Hilbert field.

Suppose now that F is a Hilbert field. Then there exists $a \in F^\times$ which is not a square and, for any such a, there exists $b \in F^\times$ such that $(a,b)_F = -1$. Consequently, by Lemma 29, the quaternary quadratic form $f_{a,b}$ is anisotropic and represents 1. Conversely, any anisotropic quaternary quadratic form which represents 1 is equivalent to some form

$$g = \xi_1^2 - a\xi_2^2 - b(\xi_3^2 - c\xi_4^2)$$

with $a,b,c \in F^\times$. Evidently a and c are not squares, and if d is represented by $\xi_3^2 - c\xi_4^2$, then bd is not represented by $\xi_1^2 - a\xi_2^2$. Thus $(c,d)_F = 1$ implies $(a,bd)_F = -1$. In particular, $(a,b)_F = -1$ and hence $(c,d)_F = 1$ implies $(a,d)_F = 1$. Interchanging the roles of $\xi_1^2 - a\xi_2^2$ and $\xi_3^2 - c\xi_4^2$, we see that $(a,d)_F = 1$ also implies $(c,d)_F = 1$. Hence $(ac,d)_F = 1$ for all $d \in F^\times$. Thus ac is a square and g is equivalent to

$$f_{a,b} = \xi_1^2 - a\xi_2^2 - b(\xi_3^2 - a\xi_4^2).$$

We now show that $f_{a,b}$ and $f_{a',b'}$ are equivalent if $(a,b)_F = (a',b')_F = -1$. Suppose first that $(a,b')_F = -1$. Then $(a,bb')_F = 1$ and there exist $x_3, x_4 \in F$ such that $b' = b(x_3^2 - ax_4^2)$. Since

$$(x_3^2 - ax_4^2)(\xi_3^2 - a\xi_4^2) = \eta_3^2 - a\eta_4^2,$$

where $\eta_3 = x_3\xi_3 + ax_4\xi_4$, $\eta_4 = x_4\xi_3 + x_3\xi_4$, it follows that $f_{a,b'}$ is equivalent to $f_{a,b}$. For the same reason $f_{a,b'}$ is equivalent to $f_{a',b'}$ and thus $f_{a,b}$ is equivalent to $f_{a',b'}$. By symmetry, the same conclusion holds if $(a',b)_F = -1$. Thus we now suppose

$$(a,b')_F = (a',b)_F = 1.$$

But then $(a,bb')_F = (a',bb')_F = -1$ and so, by what we have already proved,

$$f_{a,b} \sim f_{a,bb'} \sim f_{a',bb'} \sim f_{a',b'}.$$

Together, the last two paragraphs show that if F is a Hilbert field, then all anisotropic quaternary quadratic forms which represent 1 are equivalent. Hence the Hilbert field F is of type (B) if every anisotropic quaternary quadratic form represents 1.

Suppose now that some anisotropic quaternary quadratic form does not represent 1. Then some scalar multiple of this form represents 1, but is not universal. Thus $f_{a,b}$ is not universal for some $a,b \in F^\times$ with $(a,b)_F = -1$. By Lemma 29, the set $G_{a,b}$ of all $c \in F^\times$ which are represented by $f_{a,b}$ is a subgroup of $F^\times$. In fact $G_{a,b} = G_a$, since $G_a \subseteq G_{a,b}$, $G_{a,b} \neq F^\times$ and G_a has index 2 in $F^\times$. Since $f_{a,b} \sim f_{b,a}$, we have also $G_{a,b} = G_b$. Thus $(a,c)_F = (b,c)_F$ for all $c \in F^\times$, and hence $(ab,c)_F = 1$ for all $c \in F^\times$. Thus ab is a square and $(a,a)_F = (a,b)_F = -1$. Since $(a,-a)_F = 1$, it follows that $(a,-1)_F = -1$. Hence $f_{a,b} \sim f_{a,a} \sim f_{a,-1}$. Replacing a,b by $-1,a$ we now obtain $(-1,-1)_F = -1$ and $f_{a,-1} \sim f_{-1,-1}$.

Thus the form

$$f = \xi_1^2 + \xi_2^2 + \xi_3^2 + \xi_4^2$$

is not universal and the subgroup P of all elements of $F^\times$ represented by f coincides with the set of all elements of $F^\times$ represented by $\xi^2 + \eta^2$. Hence $P + P \subseteq P$ and P is the set of all $c \in F^\times$ such that $(-1,c)_F = 1$. Consequently $-1 \notin P$ and F is the disjoint union of the sets $\{O\}$, P and $-P$. Thus F is an ordered field with P as the set of positive elements.

For any $c \in F^\times$, $c^2 \in P$. It follows that if $a,b \in P$ then $(-a,-b)_F = -1$, since $a\xi^2 + b\eta^2$ does not represent -1. Consequently, if $a,b \in P$, then $(-a,-b)_F = -1 = (-1,-b)_F$ and $(-a,b)_F = 1 = (-1,b)_F$. Thus, for all $c \in F^\times$, $(-a,c)_F = (-1,c)_F$ and hence $(a,c)_F = 1$. Therefore a is a square and the Hilbert field F is of type (A). $\square$

PROPOSITION 31 *If F is a Hilbert field of type* (B), *then any quadratic form f in more than 4 variables is isotropic.*

For any prime p, the field $\mathbb{Q}_p$ of p-adic numbers is a Hilbert field of type (B).

Proof The quadratic form f is equivalent to a diagonal form $a_1\xi_1^2 + \ldots + a_n\xi_n^2$, where $n > 4$. If $g = a_1\xi_1^2 + \ldots + a_4\xi_4^2$ is isotropic, then so also is f. If g is anisotropic then, since F is of type (B), it is universal and represents $-a_5$. This proves the first part of the proposition.

We already know that $\mathbb{Q}_p$ is a Hilbert field and we have already shown, after the proof of Corollary VI.17, that $\mathbb{Q}_p$ is not an ordered field. Hence $\mathbb{Q}_p$ is a Hilbert field of type (B). $\square$

Proposition 10 shows that two non-singular quadratic forms in n variables, with coefficients from a Hilbert field of type (A), are equivalent over F if and only if they have the same positive index. We consider next the equivalence of quadratic forms with coefficients from a Hilbert field of type (B). We will show that they are classified by their determinant and their Hasse invariant.

If a non-singular quadratic form f, with coefficients from a Hilbert field F, is equivalent to a diagonal form $a_1\xi_1^2 + \ldots + a_n\xi_n^2$, then its *Hasse invariant* is defined to be the product of Hilbert symbols

$$s_F(f) = \prod_{1 \le j < k \le n}(a_j,a_k)_F.$$

We write $s_p(f)$ for $s_F(f)$ when $F = \mathbb{Q}_p$. (It should be noted that some authors define the Hasse invariant with $\prod_{j \le k}$ in place of $\prod_{j < k}$). It must first be shown that this is indeed an invariant of f, and for this we make use of *Witt's chain equivalence theorem*:

LEMMA 32 *Let V be a non-singular quadratic space over an arbitrary field F. If $\mathcal{B} = \{u_1,\ldots,u_n\}$ and $\mathcal{B}' = \{u_1',\ldots,u_n'\}$ are both orthogonal bases of V, then there exists a chain of orthogonal bases $\mathcal{B}_0,\mathcal{B}_1,\ldots,\mathcal{B}_m$, with $\mathcal{B}_0 = \mathcal{B}$ and $\mathcal{B}_m = \mathcal{B}'$, such that $\mathcal{B}_{j-1}$ and $\mathcal{B}_j$ differ by at most 2 vectors for each $j \in \{1,\ldots,m\}$.*

Proof Since there is nothing to prove if dim $V = n \leq 2$, we assume that $n \geq 3$ and the result holds for all smaller values of n. Let $p = p(\mathscr{B})$ be the number of nonzero coefficients in the representation of u_1' as a linear combination of $u_1,...,u_n$. Without loss of generality we may suppose

$$u_1' = \sum_{j=1}^{p} a_j u_j,$$

where $a_j \neq 0$ $(1 \leq j \leq p)$. If $p = 1$, we may replace u_1 by u_1' and the result now follows by applying the induction hypothesis to the subspace of all vectors orthogonal to u_1'. Thus we now assume $p \geq 2$. We have

$$a_1^2(u_1,u_1) + ... + a_p^2(u_p,u_p) = (u_1',u_1') \neq 0,$$

and each summand on the left is nonzero. If the sum of the first two terms is zero, then $p > 2$ and either the sum of the first and third terms is nonzero or the sum of the second and third terms is nonzero. Hence we may suppose without loss of generality that

$$a_1^2(u_1,u_1) + a_2^2(u_2,u_2) \neq 0.$$

If we put

$$v_1 = a_1 u_1 + a_2 u_2, \ v_2 = u_1 + b u_2, \ v_j = u_j \text{ for } 3 \leq j \leq n,$$

where $b = - a_1(u_1,u_1)/a_2(u_2,u_2)$, then $\mathscr{B}_1 = \{v_1,...,v_n\}$ is an orthogonal basis and $u_1' = v_1 + a_3 v_3 + ... + a_p v_p$. Thus $p(\mathscr{B}_1) < p(\mathscr{B})$. By replacing $\mathscr{B}$ by $\mathscr{B}_1$ and repeating the procedure, we must arrive after $s < n$ steps at an orthogonal basis $\mathscr{B}_s$ for which $p(\mathscr{B}_s) = 1$. The induction hypothesis can now be applied to $\mathscr{B}_s$ in the same way as for $\mathscr{B}$. $\square$

PROPOSITION 33 *Let F be a Hilbert field. If the non-singular diagonal forms $a_1\xi_1^2 + ... + a_n\xi_n^2$ and $b_1\xi_1^2 + ... + b_n\xi_n^2$ are equivalent over F, then*

$$\prod_{1 \leq j < k \leq n}(a_j,a_k)_F = \prod_{1 \leq j < k \leq n}(b_j,b_k)_F.$$

Proof Suppose first that $n = 2$. Since $a_1\xi_1^2 + a_2\xi_2^2$ represents b_1, $\xi_1^2 + a_1^{-1}a_2\xi_2^2$ represents $a_1^{-1}b_1$ and hence $(- a_1^{-1}a_2, a_1^{-1}b_1)_F = 1$. Thus $(a_1 b_1, - a_1 a_2 b_1^2)_F = 1$ and hence $(a_1 b_1, a_2 b_1)_F = 1$. But (Proposition 28 (ii)) the Hilbert symbol is multiplicative, since F is a Hilbert field. It follows that $(a_1,a_2)_F (b_1,a_1 a_2 b_1)_F = 1$. Since the determinants $a_1 a_2$ and $b_1 b_2$ are in the same square class, this implies $(a_1,a_2)_F = (b_1,b_2)_F$, as we wished to prove.

Suppose now that $n > 2$. Since the Hilbert symbol is symmetric, the product $\prod_{1 \leq j < k \leq n}(a_j,a_k)_F$ is independent of the ordering of $a_1,...,a_n$. It follows from Lemma 32 that we may restrict attention to the case where $a_1\xi_1^2 + a_2\xi_2^2$ is equivalent to $b_1\xi_1^2 + b_2\xi_2^2$ and

$a_j = b_j$ for all $j > 2$. Then $(a_1,a_2)_F = (b_1,b_2)_F$, by what we have already proved, and it is enough to show that

$$(a_1,c)_F \, (a_2,c)_F \; = \; (b_1,c)_F \, (b_2,c)_F \; \text{ for any } c \in F^\times.$$

But this follows from the multiplicativity of the Hilbert symbol and the fact that a_1a_2 and b_1b_2 are in the same square class. $\square$

Proposition 33 shows that the Hasse invariant is well-defined.

PROPOSITION 34 *Two non-singular quadratic forms in n variables, with coefficients from a Hilbert field F of type* (B), *are equivalent over F if and only if their determinants are in the same square class and they have the same Hasse invariant.*

Proof Only the sufficiency of the conditions needs to be proved. Since this is trivial for $n = 1$, we suppose first that $n = 2$. It is enough to show that if

$$f = a(\xi_1^2 + d\xi_2^2), \; g = b(\eta_1^2 + d\eta_2^2),$$

where $(a,ad)_F = (b,bd)_F$, then f is equivalent to g. The hypothesis implies $(-d,a)_F = (-d,b)_F$ and hence $(-d,ab)_F = 1$. Thus $\xi_1^2 + d\xi_2^2$ represents ab and f represents b. Since $\det f$ and $\det g$ are in the same square class, it follows that f is equivalent to g.

Suppose next that $n \geq 3$ and the result holds for all smaller values of n. Let $f(\xi_1,...,\xi_n)$ and $g(\eta_1,...,\eta_n)$ be non-singular quadratic forms with $\det f = \det g = d$ and $s_F(f) = s_F(g)$. By Proposition 31, the quadratic form

$$h(\xi_1,...,\xi_n,\eta_1,...,\eta_n) \; = \; f(\xi_1,...,\xi_n) - g(\eta_1,...,\eta_n)$$

is isotropic and hence, by Proposition 7, there exists some $a_1 \in F^\times$ which is represented by both f and g. Thus

$$f \sim a_1\xi_1^2 + f^*, \; g \sim a_1\eta_1^2 + g^*,$$

where

$$f^* = a_2\xi_2^2 + ... + a_n\xi_n^2, \; g^* = b_2\eta_2^2 + ... + b_n\eta_n^2.$$

Evidently $\det f^*$ and $\det g^*$ are in the same square class and $s_F(f) = c s_F(f^*)$, $s_F(g) = c' s_F(g^*)$, where

$$c = (a_1,a_2\cdots a_n)_F = (a_1,a_1)_F \, (a_1,d)_F = (a_1,b_2\cdots b_n)_F = c'.$$

Hence $s_F(f^*) = s_F(g^*)$. It follows from the induction hypothesis that $f^* \sim g^*$, and so $f \sim g$. $\square$

3 The Hasse–Minkowski theorem

Let a,b,c be nonzero squarefree integers which are relatively prime in pairs. It was proved by Legendre (1785) that the equation

$$ax^2 + by^2 + cz^2 = 0$$

has a nontrivial solution in integers x,y,z if and only if a,b,c are not all of the same sign and the congruences

$$u^2 \equiv -bc \bmod a, \quad v^2 \equiv -ca \bmod b, \quad w^2 \equiv -ab \bmod c$$

are all soluble.

It was first completely proved by Gauss (1801) that every positive integer which is not of the form $4^n(8k+7)$ can be represented as a sum of three squares. Legendre had given a proof, based on the assumption that if a and m are relatively prime positive integers, then the arithmetic progression

$$a,\ a+m,\ a+2m,\ \dots$$

contains infinitely many primes. Although his proof of this assumption was faulty, his intuition that it had a role to play in the arithmetic theory of quadratic forms was inspired. The assumption was first proved by Dirichlet (1837) and will be referred to here as 'Dirichlet's theorem on primes in an arithmetic progression'. In the present chapter Dirichlet's theorem will simply be assumed, but it will be proved (in a quantitative form) in Chapter X.

It was shown by Meyer (1884), although the published proof was incomplete, that a quadratic form in five or more variables with integer coefficients is isotropic if it is neither positive definite nor negative definite.

The preceding results are all special cases of the *Hasse–Minkowski theorem*, which is the subject of this section. Let $\mathbb{Q}$ denote the field of rational numbers. By Ostrowski's theorem (Proposition VI.4), the completions $\mathbb{Q}_v$ of $\mathbb{Q}$ with respect to an arbitrary absolute value $|\ |_v$ are the field $\mathbb{Q}_\infty = \mathbb{R}$ of real numbers and the fields $\mathbb{Q}_p$ of p-adic numbers, where p is an arbitrary prime. The Hasse–Minkowski theorem has the following statement:

A non-singular quadratic form $f(\xi_1,\dots,\xi_n)$ with coefficients from $\mathbb{Q}$ is isotropic in $\mathbb{Q}$ if and only if it is isotropic in every completion of $\mathbb{Q}$.

This concise statement contains, and to some extent conceals, a remarkable amount of information. (Its equivalence to Legendre's theorem when $n = 3$ may be established by elementary arguments.) The theorem was first stated and proved by Hasse (1923). Minkowski (1890) had derived necessary and sufficient conditions for the equivalence over $\mathbb{Q}$ of two non-singular quadratic forms with rational coefficients by using known results on quadratic forms with integer coefficients. The role of p-adic numbers was taken by congruences modulo prime powers. Hasse drew attention to the simplifications obtained by studying from the outset quadratic forms over the field $\mathbb{Q}$, rather than the ring $\mathbb{Z}$, and soon afterwards (1924) he showed that the theorem continues to hold if the rational field $\mathbb{Q}$ is replaced by an arbitrary algebraic number field (with its corresponding completions).

The condition in the statement of the theorem is obviously necessary and it is only its sufficiency which requires proof. Before embarking on this we establish one more property of the Hilbert symbol for the field $\mathbb{Q}$ of rational numbers.

PROPOSITION 35 *For any* $a,b \in \mathbb{Q}^\times$, *the number of completions* $\mathbb{Q}_v$ *for which* $(a,b)_v = -1$ *(where* v *denotes either* ∞ *or an arbitrary prime* p*) is finite and even.*

Proof By Proposition 23, it is sufficient to establish the result when a and b are square-free integers such that ab is also square-free. Then $(a,b)_r = 1$ for any odd prime r which does not divide ab, by Proposition 25. We wish to show that $\prod_v (a,b)_v = 1$. Since the Hilbert symbol is multiplicative, it is sufficient to establish this in the following special cases: for $a = -1$ and $b = -1,2,p$; for $a = 2$ and $b = p$; for $a = p$ and $b = q$, where p and q are distinct odd primes. But it follows from Propositions 24, 25 and 27 that

$$\prod_v (-1,-1)_v = (-1,-1)_\infty (-1,-1)_2 = (-1)(-1) = 1;$$
$$\prod_v (-1,2)_v = (-1,2)_\infty (-1,2)_2 = 1 \cdot 1 = 1;$$
$$\prod_v (-1,p)_v = (-1,p)_p (-1,p)_2 = (-1/p) (-1)^{(p-1)/2};$$
$$\prod_v (2,p)_v = (2,p)_p (2,p)_2 = (2/p) (-1)^{(p^2-1)/8};$$
$$\prod_v (p,q)_v = (p,q)_p (p,q)_q (p,q)_2 = (q/p) (p/q) (-1)^{(p-1)(q-1)/4}.$$

Hence the proposition holds if and only if

$$(-1/p) = (-1)^{(p-1)/2}, \quad (2/p) = (-1)^{(p^2-1)/8}, \quad (q/p) (p/q) = (-1)^{(p-1)(q-1)/4}.$$

Thus it is actually equivalent to the law of quadratic reciprocity and its two 'supplements'. $\square$

We are now ready to prove the Hasse–Minkowski theorem:

THEOREM 36 *A non-singular quadratic form* $f(\xi_1,...,\xi_n)$ *with rational coefficients is isotropic in* $\mathbb{Q}$ *if and only if it is isotropic in every completion* $\mathbb{Q}_v$.

Proof We may assume that the quadratic form is diagonal:

$$f = a_1\xi_1^2 + \ldots + a_n\xi_n^2,$$

where $a_k \in \mathbb{Q}^\times$ ($k = 1,\ldots,n$). Moreover, by replacing ξ_k by $r_k\xi_k$, we may assume that each coefficient a_k is a square-free integer.

The proof will be broken into three parts, according as $n = 2$, $n = 3$ or $n \geq 4$. The proofs for $n = 2$ and $n = 3$ are quite independent. The more difficult proof for $n \geq 4$ uses induction on n and Dirichlet's theorem on primes in an arithmetic progression.

(i) $n = 2$: We show first that if $a \in \mathbb{Q}^\times$ is a square in $\mathbb{Q}_v^\times$ for all v, then a is already a square in $\mathbb{Q}^\times$. Since a is a square in $\mathbb{Q}_\infty^\times$, we have $a > 0$. Let $a = \prod_p p^{\alpha_p}$ be the factorization of a into powers of distinct primes, where $\alpha_p \in \mathbb{Z}$ and $\alpha_p \neq 0$ for at most finitely many primes p. Since $|a|_p = p^{-\alpha_p}$ and a is a square in $\mathbb{Q}_p$, α_p must be even. But if $\alpha_p = 2\beta_p$ then $a = b^2$, where $b = \prod_p p^{\beta_p}$.

Suppose now that $f = a_1\xi_1^2 + a_2\xi_2^2$ is isotropic in $\mathbb{Q}_v$ for all v. Then $a: = -a_1a_2$ is a square in $\mathbb{Q}_v$ for all v and hence, by what we have just proved, a is a square in $\mathbb{Q}$. But if $a = b^2$, then $a_1a_2^2 + a_2b^2 = 0$ and thus f is isotropic in $\mathbb{Q}$.

(ii) $n = 3$: By replacing f by $-a_3f$ and ξ_3 by $a_3\xi_3$, we see that it is sufficient to prove the theorem for

$$f = a\xi^2 + b\eta^2 - \zeta^2,$$

where a and b are nonzero square-free integers. The quadratic form f is isotropic in $\mathbb{Q}_v$ if and only if $(a,b)_v = 1$. If $a = 1$ or $b = 1$, then f is certainly isotropic in $\mathbb{Q}$. Since f is not isotropic in $\mathbb{Q}_\infty$ if $a = b = -1$, this proves the result if $|ab| = 1$. We will assume that the result does not hold for some pair a,b and derive a contradiction. Choose a pair a,b for which the result does not hold and for which $|ab|$ has its minimum value. Then $a \neq 1$, $b \neq 1$ and $|ab| \geq 2$. Without loss of generality we may assume $|a| \leq |b|$, and then $|b| \geq 2$.

We are going to show that there exists an integer c such that $c^2 \equiv a \bmod b$. Since $\pm b$ is a product of distinct primes, it is enough to show that the congruence $x^2 \equiv a \bmod p$ is soluble for each prime p which divides b (by Corollary II.38). Since this is obvious if $a \equiv 0$ or $1 \bmod p$, we may assume that p is odd and a not divisible by p. Then, since f is isotropic in $\mathbb{Q}_p$, $(a,b)_p = 1$. Hence a is a square mod p by Proposition 25.

Consequently there exist integers c,d such that $a = c^2 - bd$. Moreover, by adding to c a suitable multiple of b we may assume that $|c| \leq |b|/2$. Then

$$|d| = |c^2 - a|/|b| \le |b|/4 + 1 < |b|$$

and $d \ne 0$, since a is square-free and $a \ne 1$. We have

$$bd(a\xi^2 + b\eta^2 - \zeta^2) = aX^2 + dY^2 - Z^2,$$

where

$$X = c\xi + \zeta, \ Y = b\eta, \ Z = a\xi + c\zeta.$$

Moreover the linear transformation $\xi,\eta,\zeta \to X,Y,Z$ is invertible in any field of zero characteristic, since $c^2 - a \ne 0$. Hence, since f is isotropic in $\mathbb{Q}_v$ for all v, so also is $g = a\xi^2 + d\eta^2 - \zeta^2$. Since f is not isotropic in $\mathbb{Q}$, by hypothesis, neither is g. But this contradicts the original choice of f, since $|ad| < |ab|$.

It may be noted that for $n = 3$ it need only be assumed that f is isotropic in $\mathbb{Q}_p$ for all primes p. For the preceding proof used the fact that f is isotropic in $\mathbb{Q}_\infty$ only to exclude from consideration the quadratic form $-\xi^2 - \eta^2 - \zeta^2$ and this quadratic form is anisotropic also in $\mathbb{Q}_2$, by Proposition 27. In fact for $n = 3$ it need only be assumed that f is isotropic in $\mathbb{Q}_v$ for all v with at most one exception since, by Proposition 35, the number of exceptions must be even.

(iii) $n \ge 4$: We have

$$f = a_1\xi_1^2 + \dots + a_n\xi_n^2,$$

where $a_1,\dots,a_n$ are square-free integers. We write $f = g - h$, where

$$g = a_1\xi_1^2 + a_2\xi_2^2, \quad h = -a_3\xi_3^2 - \dots - a_n\xi_n^2.$$

Let S be the finite set consisting of ∞ and all primes p which divide $2a_1 \cdots a_n$. By Proposition 7, for each $v \in S$ there exists $c_v \in \mathbb{Q}_v^\times$ which is represented in $\mathbb{Q}_v$ by both g and h. We will show that we can take c_v to be the same nonzero integer c for every $v \in S$.

Let $v = p$ be a prime in S. By multiplying by a square in $\mathbb{Q}_p^\times$ we may assume that $c_p = p^{\varepsilon_p} c_p'$, where $\varepsilon_p = 0$ or 1 and $|c_p'|_p = 1$. If p is odd and if b_p is an integer such that $|c_p - b_p|_p \le p^{-\varepsilon_p - 1}$, then $|b_p|_p = |c_p|_p$ and $|b_p c_p^{-1} - 1|_p \le p^{-1}$. Hence $b_p c_p^{-1}$ is a square in $\mathbb{Q}_p^\times$, by Proposition VI.16, and we can replace c_p by b_p. Similarly if $p = 2$ and if b_2 is an integer such that $|c_2 - b_2|_2 \le 2^{-\varepsilon_2 - 3}$, then $|b_2|_2 = |c_2|_2$ and $|b_2 c_2^{-1} - 1|_2 \le 2^{-3}$. Hence $b_2 c_2^{-1}$ is a square in $\mathbb{Q}_2^\times$ and we can replace c_2 by b_2.

By the Chinese remainder theorem (Corollary II.38), the simultaneous congruences

$$c \equiv b_2 \bmod 2^{\varepsilon_2 + 3}, \quad c \equiv b_p \bmod p^{\varepsilon_p + 1} \text{ for every odd } p \in S,$$

have a solution $c \in \mathbb{Z}$, which is uniquely determined mod m, where $m = 4\prod_{p \in S} p^{\varepsilon_p + 1}$. In exactly the same way as before we can replace b_p by c for all primes $p \in S$. By choosing c to have the same sign as c_∞, we can take $c_v = c$ for all $v \in S$.

If $d = \prod_{p \in S} p^{\varepsilon_p}$ is the greatest common divisor of c and m then, by Dirichlet's theorem on primes in an arithmetic progression, there exists an integer k with the same sign as c such that

$$c/d + km/d = \pm q,$$

where q is a prime. If we put

$$a = c + km = \pm dq,$$

then q is the only prime divisor of a which is not in S and the quadratic forms

$$g^* = -a\xi_0^2 + a_1\xi_1^2 + a_2\xi_2^2, \quad h^* = a_3\xi_3^2 + \ldots + a_n\xi_n^2 + a\xi_{n+1}^2$$

are isotropic in $\mathbb{Q}_v$ for every $v \in S$, since $c^{-1}a$ is a square in $\mathbb{Q}_v^\times$.

For all primes p not in S, except $p = q$, a is not divisible by p. Hence, by the definition of S and Corollary 26, g^* is isotropic in $\mathbb{Q}_v$ for all v, except possibly $v = q$. Consequently, by the final remark of part (ii) of the proof, g^* is isotropic in $\mathbb{Q}$.

Suppose first that $n = 4$. In this case, by the same argument, $h^* = a_3\xi_3^2 + a_4\xi_4^2 + a\xi_5^2$ is also isotropic in $\mathbb{Q}$. Hence, by Proposition 6, there exist $y_1,\ldots,y_4 \in \mathbb{Q}$ such that

$$a_1y_1^2 + a_2y_2^2 = a = -a_3y_3^2 - a_4y_4^2.$$

Thus f is isotropic in $\mathbb{Q}$.

Suppose next that $n \geq 5$ and the result holds for all smaller values of n. Then the quadratic form h^* is isotropic in $\mathbb{Q}_v$, not only for $v \in S$, but for all v. For if p is a prime which is not in S, then a_3,a_4,a_5 are not divisible by p. It follows from Corollary 26 that the quadratic form $a_3\xi_3^2 + a_4\xi_4^2 + a_5\xi_5^2$ is isotropic in $\mathbb{Q}_p$, and hence h^* is also. Since h^* is a non-singular quadratic form in $n - 1$ variables, it follows from the induction hypothesis that h^* is isotropic in $\mathbb{Q}$. The proof can now be completed in the same way as for $n = 4$. $\blacksquare$

COROLLARY 37 *A non-singular rational quadratic form in $n \geq 5$ variables is isotropic in $\mathbb{Q}$ if and only if it is neither positive definite nor negative definite.*

Proof This follows at once from Theorem 36, on account of Propositions 10 and 31. $\blacksquare$

COROLLARY 38 *A non-singular quadratic form over the rational field $\mathbb{Q}$ represents a nonzero rational number c in $\mathbb{Q}$ if and only if it represents c in every completion $\mathbb{Q}_v$.*

Proof Only the sufficiency of the condition requires proof. But if the rational quadratic form $f(\xi_1,...,\xi_n)$ represents c in $\mathbb{Q}_v$ for all v then, by Theorem 36, the quadratic form

$$f^*(\xi_0,\xi_1,...,\xi_n) = -c\xi_0^2 + f(\xi_1,...,\xi_n)$$

is isotropic in $\mathbb{Q}$. Hence f represents c in $\mathbb{Q}$, by Proposition 6. $\square$

PROPOSITION 39 *Two non-singular quadratic forms with rational coefficients are equivalent over $\mathbb{Q}$ if and only if they are equivalent over all completions $\mathbb{Q}_v$.*

Proof Again only the sufficiency of the condition requires proof. Let f and g be non-singular rational quadratic forms in n variables which are equivalent over $\mathbb{Q}_v$ for all v.

Suppose first that $n = 1$ and that $f = a\xi^2$, $g = b\eta^2$. By hypothesis, for every v there exists $t_v \in \mathbb{Q}_v^\times$ such that $b = at_v^2$. Thus ba^{-1} is a square in $\mathbb{Q}_v^\times$ for every v, and hence ba^{-1} is a square in $\mathbb{Q}^\times$, by part (i) of the proof of Theorem 36. Therefore f is equivalent to g over $\mathbb{Q}$.

Suppose now that $n > 1$ and the result holds for all smaller values of n. Choose some $c \in \mathbb{Q}^\times$ which is represented by f in $\mathbb{Q}$. Then f certainly represents c in $\mathbb{Q}_v$ and hence g represents c in $\mathbb{Q}_v$, since g is equivalent to f over $\mathbb{Q}_v$. Since this holds for all v, it follows from Corollary 38 that g represents c in $\mathbb{Q}$.

Thus, by the remark after the proof of Proposition 2, f is equivalent over $\mathbb{Q}$ to a quadratic form $c\xi_1^2 + f^*(\xi_2,...,\xi_n)$ and g is equivalent over $\mathbb{Q}$ to a quadratic form $c\xi_1^2 + g^*(\xi_2,...,\xi_n)$. Since f is equivalent to g over $\mathbb{Q}_v$, it follows from Witt's cancellation theorem that $f^*(\xi_2,...,\xi_n)$ is equivalent to $g^*(\xi_2,...,\xi_n)$ over $\mathbb{Q}_v$. Since this holds for every v, it follows from the induction hypothesis that f^* is equivalent to g^* over $\mathbb{Q}$, and so f is equivalent to g over $\mathbb{Q}$. $\square$

COROLLARY 40 *Two non-singular quadratic forms f and g in n variables with rational coefficients are equivalent over the rational field $\mathbb{Q}$ if and only if*

(i) $(\det f)/(\det g)$ *is a square in* $\mathbb{Q}^\times$,

(ii) $\text{ind}^+ f = \text{ind}^+ g$,

(iii) $s_p(f) = s_p(g)$ *for every prime* p.

Proof This follows at once from Proposition 39, on account of Propositions 10 and 34. $\square$

The *strong Hasse principle* (Theorem 36) says that a quadratic form is *isotropic* over the global field $\mathbb{Q}$ if (and only if) it is isotropic over all its local completions $\mathbb{Q}_v$. The *weak Hasse principle* (Proposition 39) says that two quadratic forms are *equivalent* over $\mathbb{Q}$ if (and only if) they are equivalent over all $\mathbb{Q}_v$. These *local-global principles* have proved remarkably fruitful.

They organize the subject, they can be extended to other situations and, even when they fail, they are still a useful guide. We describe some results which illustrate these remarks.

As mentioned at the beginning of this section, the strong Hasse principle continues to hold when the rational field is replaced by any algebraic number field. Waterhouse (1976) has established the weak Hasse principle for pairs of quadratic forms: if over every completion $\mathbb{Q}_v$ there is a change of variables taking both f_1 to g_1 and f_2 to g_2, then there is also such a change of variables over $\mathbb{Q}$. For quadratic forms over the field $F = K(t)$ of rational functions in one variable with coefficients from a field K, the weak Hasse principle always holds, and the strong Hasse principle holds for $K = \mathbb{R}$, but not for all fields K.

The strong Hasse principle also fails for polynomial forms over $\mathbb{Q}$ of degree > 2. For example, Selmer (1951) has shown that the cubic equation $3x^3 + 4y^3 + 5z^3 = 0$ has no nontrivial solutions in $\mathbb{Q}$, although it has nontrivial solutions in every completion $\mathbb{Q}_v$. However, Gusic' (1995) has proved the weak Hasse principle for non-singular ternary cubic forms.

Finally, we draw attention to a remarkable local-global principle of Rumely (1986) for algebraic integer solutions of arbitrary systems of polynomial equations

$$f_1(\xi_1,...,\xi_n) = ... = f_r(\xi_1,...,\xi_n) = 0$$

with rational coefficients.

We now give some applications of the results which have been established.

PROPOSITION 41   *A positive integer can be represented as the sum of the squares of three integers if and only if it is not of the form* $4^n b$, *where* $n \geq 0$ *and* $b \equiv 7 \bmod 8$.

Proof   The necessity of the condition is easily established. Since the square of any integer is congruent to $0,1$ or $4 \bmod 8$, the sum of three squares cannot be congruent to 7. For the same reason, if there exist integers x,y,z such that $x^2 + y^2 + z^2 = 4^n b$, where $n \geq 1$ and b is odd, then x,y,z must all be even and thus $(x/2)^2 + (y/2)^2 + (z/2)^2 = 4^{n-1}b$. By repeating the argument n times, we see that there is no such representation if $b \equiv 7 \bmod 8$.

We show next that any positive integer which satisfies this necessary condition is the sum of three squares of *rational* numbers. We need only show that any positive integer $a \not\equiv 7 \bmod 8$, which is not divisible by 4, is represented in $\mathbb{Q}$ by the quadratic form

$$f = \xi_1^2 + \xi_2^2 + \xi_3^2.$$

For every odd prime p, f is isotropic in $\mathbb{Q}_p$, by Corollary 26, and hence any integer is represented in $\mathbb{Q}_p$ by f, by Proposition 5. By Corollary 38, it only remains to show that f represents a in $\mathbb{Q}_2$.

It is easily seen that if $a \equiv 1,3$ or $5 \bmod 8$, then there exist integers $x_1,x_2,x_3 \in \{0,1,2\}$ such that

$$x_1^2 + x_2^2 + x_3^2 \equiv a \ \bmod 8.$$

Hence $a^{-1}(x_1^2 + x_2^2 + x_3^2)$ is a square in $\mathbb{Q}_2^\times$ and f represents a in $\mathbb{Q}_2$.

Again, if $a \equiv 2$ or $6 \bmod 8$, then $a \equiv 2,6,10$ or $14 \bmod 2^4$ and it is easily seen that there exist integers $x_1,x_2,x_3 \in \{0,1,2,3\}$ such that

$$x_1^2 + x_2^2 + x_3^2 \equiv a \ \bmod 2^4.$$

Hence $a^{-1}(x_1^2 + x_2^2 + x_3^2)$ is a square in $\mathbb{Q}_2^\times$ and f represents a in $\mathbb{Q}_2$.

To complete the proof of the proposition we now show, by an elegant argument due to Aubry (1912), that if f represents c in $\mathbb{Q}$ then it also represents c in $\mathbb{Z}$.

Let

$$(x,y) \ = \ \{f(x+y) - f(x) - f(y)\}/2$$

be the symmetric bilinear form associated with f, so that $f(x) = (x,x)$, and assume there exists a point $x \in \mathbb{Q}^3$ such that $(x,x) = c \in \mathbb{Z}$. If $x \notin \mathbb{Z}^3$, we can choose $z \in \mathbb{Z}^3$ so that each coordinate of z differs in absolute value by at most $1/2$ from the corresponding coordinate of x. Hence if we put $y = x - z$, then $y \neq 0$ and $0 < (y,y) \le 3/4$.

If $x' = x - \lambda y$, where $\lambda = 2(x,y)/(y,y)$, then $x' \in \mathbb{Q}^3$ and $(x',x') = (x,x) = c$. Substituting $y = x - z$, we obtain

$$(y,y)x' \ = \ (y,y)x - 2(x,y)y \ = \ \{(z,z) - (x,x)\}x + 2\{(x,x) - (x,z)\}z.$$

If $m > 0$ is the least common denominator of the coordinates of x, so that $mx \in \mathbb{Z}^3$, it follows that

$$m(y,y)x' = \{(z,z) - c)\}mx + 2\{mc - (mx,z)\}z \in \mathbb{Z}^3.$$

But

$$m(y,y) \ = \ m\{(x,x) - 2(x,z) + (z,z)\} \ = \ mc - 2(mx,z) + m(z,z) \ \in \ \mathbb{Z}.$$

Thus if $m' > 0$ is the least common denominator of the coordinates of x', then m' divides $m(y,y)$. Hence $m' \le (3/4)m$. If $x' \notin \mathbb{Z}^3$, we can repeat the argument with x replaced by x'. After performing the process finitely many times we must obtain a point $x^* \in \mathbb{Z}^3$ such that $(x^*,x^*) = c$. $\square$

As another application of the preceding results we now prove

PROPOSITION 42 *Let n,a,b be integers with n and a positive. Then there exists an $n \times n$ nonsingular rational matrix A such that*

$$A^t A = aI_n + bJ_n,$$ (3)

where J_n is the $n \times n$ matrix with all entries 1, if and only if $a + bn > 0$ and

(i) *for n odd*: *$a + bn$ is a square and the quadratic form*

$$a\xi^2 + (-1)^{(n-1)/2} n\eta^2 - \zeta^2$$

is isotropic in $\mathbb{Q}$;

(ii) *for n even*: *$a(a + bn)$ is a square and either $n \equiv 0 \bmod 4$, or $n \equiv 2 \bmod 4$ and a is a sum of two squares.*

Proof Since $\det(A^t A) = (\det A)^2$ and $\det(aI_n + bJ_n) = a^{n-1}(a + bn)$, by Lemma V.15, it is necessary that $a + bn$ be a nonzero square if n is odd and that $a(a + bn)$ be a nonzero square if n is even. In either event, $a + bn > 0$.

If we put

$$B = \begin{bmatrix} 1 & 1 & \dots & 1 & 1 \\ -1 & 1 & \dots & 1 & 1 \\ 0 & -2 & \dots & 1 & 1 \\ & .. & \dots & .. & \\ 0 & 0 & \dots & 1-n & 1 \end{bmatrix},$$

then $D: = B^t B$ and $E: = B^t JB$ are diagonal matrices:

$$D = \text{diag } [d_1,\dots,d_{n-1},n], \quad E = \text{diag } [0,\dots,0,n^2],$$

where $d_j = j(j + 1)$ for $1 \leq j < n$. Hence, if $C = D^{-1}B^t AB$, then

$$C^t DC = B^t A^t AB.$$

Thus the rational matrix A satisfies (3) if and only if the rational matrix C satisfies

$$C^t DC = aD + bE,$$

and consequently if and only if the diagonal quadratic forms

$$f = d_1\xi_1^2 + \dots + d_{n-1}\xi_{n-1}^2 + n\xi_n^2, \quad g = ad_1\eta_1^2 + \dots + ad_{n-1}\eta_{n-1}^2 + n(a + bn)\eta_n^2$$

are equivalent over $\mathbb{Q}$.

We now apply Corollary 40. Since $(\det g)/(\det f) = a^{n-1}(a + bn)$, the condition that $\det g/\det f$ be a square in $\mathbb{Q}^\times$ reproduces the conditions already obtained that $a + bn$ or $a(a + bn)$ be a nonzero square. Evidently also $\text{ind}^+ f = \text{ind}^+ g = n$. The relation $s_p(g) = s_p(f)$ takes the form

$$\prod_{1 \le i < j < n}(ad_i, ad_j)_p \; \prod_{1 \le i < n}(ad_i, n(a + bn))_p \;=\; \prod_{1 \le i < j < n}(d_i, d_j)_p \; \prod_{1 \le i < n}(d_i, n)_p.$$

The multiplicativity and symmetry of the Hilbert symbol imply that

$$(ad_i, ad_j)_p \;=\; (a, a)_p \, (a, d_i d_j)_p \, (d_i, d_j)_p.$$

Since $(a, a)_p = (a, -1)_p$, it follows that $s_p(g) = s_p(f)$ if and only if

$$(a, -1)_p^{(n-1)(n-2)/2} \, (a, n)_p^{n-1} \, \prod_{1 \le i < n}(ad_i, a + bn)_p \, \prod_{1 \le i < j < n}(a, d_i d_j)_p \;=\; 1.$$

But

$$\prod_{1 \le i < j < n} d_i d_j \;=\; (d_1 \cdots d_{n-1})^{n-2}$$

and, by the definition of d_j, $d_1 \cdots d_{n-1}$ is in the same rational square class as n. Hence $s_p(g) = s_p(f)$ if and only if

$$(a, -1)_p^{(n-1)(n-2)/2} \, (a, n)_p \, (an, a + bn)_p \;=\; 1. \tag{4}$$

If n is odd, then $a + bn$ is a square and (4) reduces to $(a, (-1)^{(n-1)/2} n)_p = 1$. Since $a > 0$, this holds for all p if and only if (i) is satisfied.

If n is even, then $a(a + bn)$ is a square and (4) reduces to $(a, (-1)^{(n-2)/2} a)_p = 1$. Since $a > 0$, this holds for all p if and only if the ternary quadratic form

$$a\xi^2 + (-1)^{(n-2)/2} a\eta^2 - \zeta^2,$$

is isotropic in $\mathbb{Q}$. Thus it is certainly satisfied if $n \equiv 0 \bmod 4$. If $n \equiv 2 \bmod 4$ it is satisfied if and only if the quadratic form $\xi^2 + \eta^2 - a\zeta^2$ is isotropic. Thus it is satisfied if a is a sum of two squares. It is not satisfied if a is not a sum of two squares since then, by Proposition II.39, for some prime $p \equiv 3 \bmod 4$, the highest power of p which divides a is odd and

$$(a, a)_p \;=\; (a, -1)_p \;=\; (p, -1)_p \;=\; (-1)^{(p-1)/2} \;=\; -1. \quad \square$$

It follows at once from Proposition 42 that, for any positive integer n, there is an $n \times n$ *rational* matrix A such that $A^t A = nI_n$ if and only if either n is an odd square, or $n \equiv 2$ mod 4 and n is a sum of two squares, or $n \equiv 0$ mod 4 (the Hadamard matrix case).

In Chapter V we considered not only Hadamard matrices, but also designs. We now use Proposition 42 to derive the necessary conditions for the existence of square 2-designs which were obtained by Bruck, Ryser and Chowla (1949/50). Let v, k, λ be integers such that $0 < \lambda < k < v$ and $k(k-1) = \lambda(v-1)$. Since $k - \lambda + \lambda v = k^2$, it follows from Proposition 42 that there exists a $v \times v$ rational matrix A such that

$$A^t A = (k - \lambda)I_v + \lambda J_v$$

if and only if, *either v is even and $k - \lambda$ is a square, or v is odd and* the quadratic form

$$(k - \lambda)\xi^2 + (-1)^{(v-1)/2}v\eta^2 - \zeta^2$$

is isotropic in $\mathbb{Q}$. Instead of the latter condition we can just as well require that *the quadratic form*

$$(k - \lambda)\xi^2 + (-1)^{(v-1)/2}\lambda\eta^2 - \zeta^2$$

is isotropic in $\mathbb{Q}$. For $(k - \lambda, \lambda v)_p = 1$ for all p, since

$$(k - \lambda)\xi^2 + \lambda v\eta^2 - \zeta^2 = 0$$

has the nontrivial rational solution $\xi = k - 1 + \lambda$, $\eta = k - 1 - \lambda$, $\zeta = \lambda(k - 1 - v)$, and so $(k - \lambda, (-1)^{(v-1)/2}v)_p = 1$ for all p if and only if $(k - \lambda, (-1)^{(v-1)/2}\lambda)_p = 1$ for all p.

A projective plane of order d corresponds to a $(d^2+d+1, d+1, 1)$ (square) 2-design. In this case Proposition 42 tells us that there is no projective plane of order d if d is not a sum of two squares and $d \equiv 1$ or 2 mod 4. In particular, there is no projective plane of order 6.

The existence of projective planes of any prime power order follows from the existence of finite fields of any prime power order. (All known projective planes are of prime power order, but even for $d = 9$ there are projective planes of the same order d which are not isomorphic.) Since there is no projective plane of order 6, the least order in doubt is $d = 10$. The condition derived from Proposition 42 is obviously satisfied in this case, since

$$10\xi^2 - \eta^2 - \zeta^2 = 0$$

has the solution $\xi = \eta = 1$, $\zeta = 3$. However, Lam, Thiel and Swiercz (1989) have announced that, nevertheless, there is no projective plane of order 10. The result was obtained by a search involving thousands of hours time on a supercomputer and does not appear to have been independently verified.

4 Supplements

It was shown in the proof of Proposition 41 that if an integer can be represented as a sum of 3 squares of rational numbers, then it can be represented as a sum of 3 squares of integers. A similar argument was used by Cassels (1964) to show that if a polynomial can be represented as a sum of n squares of rational functions, then it can be represented as a sum of n squares of polynomials. This was immediately generalized by Pfister (1965) in the following way:

PROPOSITION 43 *For any field F, if there exist scalars $\alpha_1,...,\alpha_n \in F$ and rational functions $r_1(t),...,r_n(t) \in F(t)$ such that*

$$p(t) = \alpha_1 r_1(t)^2 + ... + \alpha_n r_n(t)^2$$

is a polynomial, then there exist polynomials $p_1(t),...,p_n(t) \in F[t]$ such that

$$p(t) = \alpha_1 p_1(t)^2 + ... + \alpha_n p_n(t)^2.$$

Proof Suppose first that $n = 1$. We can write $r_1(t) = p_1(t)/q_1(t)$, where $p_1(t)$ and $q_1(t)$ are relatively prime polynomials and $q_1(t)$ has leading coefficient 1. Since

$$p(t)q_1(t)^2 = \alpha_1 p_1(t)^2,$$

we must actually have $q_1(t) = 1$.

Suppose now that $n > 1$ and the result holds for all smaller values of n. We may assume that $\alpha_j \neq 0$ for all j, since otherwise the result follows from the induction hypothesis. Suppose first that the quadratic form

$$\phi = \alpha_1 \xi_1^2 + ... + \alpha_n \xi_n^2$$

is isotropic over F. Then there exists an invertible linear change of variables $\xi_j = \sum_{k=1}^n \tau_{jk} \eta_k$ with $\tau_{jk} \in F$ $(1 \leq j,k \leq n)$ such that

$$\phi = \eta_1^2 - \eta_2^2 + \beta_3 \eta_3^2 + ... + \beta_n \eta_n^2,$$

where $\beta_j \in F$ for all $j > 2$. If we substitute

$$\eta_1 = \{p(t) + 1\}/2, \ \eta_2 = \{p(t) - 1\}/2, \ \eta_j = 0 \text{ for all } j > 2,$$

we obtain a representation for $p(t)$ of the required form.

Thus we now suppose that ϕ is anisotropic over F. This implies that ϕ is also anisotropic over $F(t)$, since otherwise there would exist a nontrivial representation

$$\alpha_1 q_1(t)^2 + \ldots + \alpha_n q_n(t)^2 = 0,$$

where $q_j(t) \in F[t]$ $(1 \le j \le n)$, and by considering the terms of highest degree we would obtain a contradiction.

By hypothesis there exists a representation

$$p(t) = \alpha_1 \{f_1(t)/f_0(t)\}^2 + \ldots + \alpha_n \{f_n(t)/f_0(t)\}^2,$$

where $f_0(t), f_1(t), \ldots, f_n(t) \in F[t]$. Assume that f_0 does not divide f_j for some $j \in \{1, \ldots, n\}$. Then $d := \deg f_0 > 0$ and we can write

$$f_j(t) = g_j(t) f_0(t) + h_j(t),$$

where $g_j(t), h_j(t) \in F[t]$ and $\deg h_j < d$ $(1 \le j \le n)$.

Let

$$(x,y) = \{\phi(x+y) - \phi(x) - \phi(y)\}/2$$

be the symmetric bilinear form associated with the quadratic form ϕ and put

$$f = (f_1, \ldots, f_n), \quad g = (g_1, \ldots, g_n), \quad h = (h_1, \ldots, h_n).$$

If

$$f_0^* = \{(g,g) - p\} f_0 - 2\{(f,g) - pf_0\}, \quad f^* = \{(g,g) - p\} f - 2\{(f,g) - pf_0\} g,$$

and $f^* = (f_1^*, \ldots, f_n^*)$, then clearly $f_0^*, f_1^*, \ldots, f_n^* \in F[t]$. Since $(f,f) = pf_0^2$ and $g = (f - h)/f_0$, we can also write

$$f_0^* = (h,h)/f_0, \quad f^* = \{(h,h)f - 2(f,h)h\}/f_0^2.$$

It follows that $\deg f_0^* < d$ and $(f^*, f^*) = pf_0^{*2}$. Also $f_0^* \ne 0$, since $h \ne 0$ and ϕ is anisotropic. Thus

$$p(t) = \alpha_1 \{f_1^*(t)/f_0^*(t)\}^2 + \ldots + \alpha_n \{f_n^*(t)/f_0^*(t)\}^2.$$

If f_0^* does not divide f_j^* for some $j \in \{1, \ldots, n\}$, we can repeat the process. After at most d steps we must obtain a representation for $p(t)$ of the required form. $\square$

It was already known to Hilbert (1888) that there is no analogue of Proposition 43 for polynomials in more than one variable. Motzkin (1967) gave the simple example

$$p(x,y) = 1 - 3x^2y^2 + x^4y^2 + x^2y^4,$$

which is a sum of 4 squares in $\mathbb{R}(x,y)$, but is not a sum of any finite number of squares in $\mathbb{R}[x,y]$.

In the same paper in which he proved Proposition 43 Pfister introduced his *multiplicative forms*. The quadratic forms $f_a, f_{a,b}$ in §2 are examples of such forms. Pfister (1966) used his multiplicative forms to obtain several new results on the structure of the Witt ring and then (1967) to give a strong solution to Hilbert's 17th Paris problem. We restrict attention here to the latter application.

Let $g(x), h(x) \in \mathbb{R}[x]$ be polynomials in n variables $x = (\xi_1,...,\xi_n)$ with real coefficients. The rational function $f(x) = g(x)/h(x)$ is said to be *positive definite* if $f(a) \geq 0$ for every $a \in \mathbb{R}^n$ such that $h(a) \neq 0$. Hilbert's 17th problem asks if every positive definite rational function can be represented as a sum of squares:

$$f(x) = f_1(x)^2 + ... + f_s(x)^2,$$

where $f_1(x),...,f_s(x) \in \mathbb{R}(x)$. The question was answered affirmatively by Artin (1927). Artin's solution allowed the number s of squares to depend on the function f, and left open the possibility that there might be no uniform bound. Pfister showed that one can always take $s = 2^n$.

Finally we mention a conjecture of Oppenheim (1929-1953), that if $f(\xi_1,...,\xi_n)$ is a non-singular isotropic real quadratic form in $n \geq 3$ variables, which is not a scalar multiple of a rational quadratic form, then $f(\mathbb{Z}^n)$ is dense in $\mathbb{R}$, i.e. for each $\alpha \in \mathbb{R}$ and $\varepsilon > 0$ there exist $z_1,...,z_n \in \mathbb{Z}$ such that $|f(z_1,...,z_n) - \alpha| < \varepsilon$. (It is not difficult to show that this is not always true for $n = 2$.) Raghunathan (1980) made a general conjecture about Lie groups, which he observed would imply Oppenheim's conjecture. Oppenheim's conjecture was then proved in this way by Margulis (1987), using deep results from the theory of Lie groups and ergodic theory. The full conjecture of Raghunathan has now also been proved by Ratner (1991).

5 Further remarks

Lam [18] gives a good introduction to the arithmetic theory of quadratic spaces. The Hasse–Minkowski theorem is also proved in Serre [29]. Additional information is contained in the books of Cassels [4], Kitaoka [16], Milnor and Husemoller [20], O'Meara [22] and Scharlau [28].

Quadratic spaces were introduced (under the name 'metric spaces') by Witt [32]. This noteworthy paper also made several other contributions: Witt's cancellation theorem, the Witt ring, Witt's chain equivalence theorem and the Hasse invariant in its most general form (as described below). Quadratic spaces are treated not only in books on the arithmetic of quadratic forms, but also in works of a purely algebraic nature, such as Artin [1], Dieudonné [8] and Jacobson [15].

An important property of the Witt ring was established by Merkur'ev (1981). In one formulation it says that every element of order 2 in the *Brauer group* of a field F is represented by the Clifford algebra of some quadratic form over F. For a clear account, see Lewis [19].

Our discussion of Hilbert fields is based on Fröhlich [9]. It may be shown that any locally compact non-archimedean valued field is a Hilbert field. Fröhlich gives other examples, but rightly remarks that the notion of Hilbert field clarifies the structure of the theory, even if one is interested only in the p-adic case. (The name 'Hilbert field' is also given to fields for which Hilbert's irreducibility theorem is valid.)

In the study of quadratic forms over an arbitrary field F, the Hilbert symbol $(a,b/F)$ is a generalized quaternion algebra (more strictly, an equivalence class of such algebras) and the Hasse invariant is a tensor product of Hilbert symbols. See, for example, Lam [18].

Hasse's original proof of the Hasse–Minkowski theorem is reproduced in Hasse [13]. In principle it is the same as that given here, using a reduction argument due to Lagrange for $n = 3$ and Dirichlet's theorem on primes in an arithmetic progression for $n \geq 4$.

The book of Cassels contains a proof of Theorem 36 which does not use Dirichlet's theorem, but it uses intricate results on genera of quadratic forms and is not so 'clean'. However, Conway [6] has given an elementary approach to the *equivalence* of quadratic forms over $\mathbb{Q}$ (Proposition 39 and Corollary 40).

The book of O'Meara gives a proof of the Hasse–Minkowski theorem over any algebraic number field which avoids Dirichlet's theorem and is 'cleaner' than ours, but it uses deep results from *class field theory*. For the latter, see Cassels and Fröhlich [5], Garbanati [10] and Neukirch [21].

To determine if a rational quadratic form $f(\xi_1,...,\xi_n) = \sum_{j,k=1}^{n} a_{jk}\xi_j\xi_k$ is isotropic by means of Theorem 36 one has to show that it is isotropic in infinitely many completions. Nevertheless, the problem is a finite one. Clearly one may assume that the coefficients a_{jk} are integers and, if the equation $f(x_1,...,x_n) = 0$ has a nontrivial solution in rational numbers, then it also has a nontrivial solution in integers. But Cassels has shown by elementary arguments that if $f(x_1,...,x_n) = 0$ for some $x_j \in \mathbb{Z}$, not all zero, then the x_j may be chosen so that

$$\max_{1 \le j \le n} |x_j| \le (3H)^{(n-1)/2},$$

where $H = \sum_{j,k=1}^{n} |a_{jk}|$. See Lemma 8.1 of [4].

Williams [31] gives a sharper result for the ternary quadratic form

$$g(\xi,\eta,\zeta) = a\xi^2 + b\eta^2 + c\zeta^2,$$

where a,b,c are integers with greatest common divisor $d > 0$. If $g(x,y,z) = 0$ for some integers x,y,z, not all zero, then these integers may be chosen so that

$$|x| \le |bc|^{1/2}/d, \quad |y| \le |ca|^{1/2}/d, \quad |z| \le |ab|^{1/2}/d.$$

The necessity of the Bruck–Ryser–Chowla conditions for the existence of symmetric block designs may also be established in a more elementary way, without also proving their sufficiency for rational equivalence. See, for example, Beth *et al.* [2]. For the non-existence of a projective plane of order 10, see C. Lam [17].

For various manifestations of the local-global principle, see Waterhouse [30], Hsia [14], Gusic' [12] and Green *et al.* [11].

The work of Pfister instigated a flood of papers on the algebraic theory of quadratic forms. The books of Lam and Scharlau give an account of these developments. For Hilbert's 17th problem, see also Pfister [23],[24] and Rajwade [25].

Although a positive integer which is a sum of n rational squares is also a sum of n squares of integers, the same does not hold for higher powers. For example,

$$5906 = (149/17)^4 + (25/17)^4,$$

but there do not exist positive integers m,n such that $5906 = m^4 + n^4$, since $9^4 > 5906$, $2 \cdot 7^4 < 5906$ and $5906 - 8^4 = 1810$ is not a fourth power. For the representation of a polynomial as a sum of squares of polynomials, see Rudin [27].

For Oppenheim's conjecture, see Dani and Margulis [7], Borel [3] and Ratner [26].

6 Selected References

[1] E. Artin, *Geometric algebra*, reprinted, Wiley, New York, 1988. [Original edition, 1957]

[2] T. Beth, D. Jungnickel and H. Lenz, *Design theory*, 2nd ed., 2 vols., Cambridge University Press, 1999.

[3]   A. Borel, Values of indefinite quadratic forms at integral points and flows on spaces of lattices, *Bull. Amer. Math. Soc. (N.S.)* **32** (1995), 184-204.

[4]   J.W.S. Cassels, *Rational quadratic forms*, Academic Press, London, 1978.

[5]   J.W.S. Cassels and A. Fröhlich (ed.), *Algebraic number theory*, Academic Press, London, 1967.

[6]   J.H. Conway, Invariants for quadratic forms, *J. Number Theory* **5** (1973), 390-404.

[7]   S.G. Dani and G.A. Margulis, Values of quadratic forms at integral points: an elementary approach, *Enseign. Math.* **36** (1990), 143-174.

[8]   J. Dieudonné, *La géométrie des groupes classiques*, 2nd ed., Springer-Verlag, Berlin, 1963.

[9]   A. Fröhlich, Quadratic forms 'à la' local theory, *Proc. Camb. Phil. Soc.* **63** (1967), 579-586.

[10] D. Garbanati, Class field theory summarized, *Rocky Mountain J. Math.* **11** (1981), 195-225.

[11] B. Green, F. Pop and P. Roquette, On Rumely's local-global principle, *Jahresber. Deutsch. Math.–Verein.* **97** (1995), 43-74.

[12] I. Gusic', Weak Hasse principle for cubic forms, *Glas. Mat. Ser. III* **30** (1995), 17-24.

[13] H. Hasse, *Mathematische Abhandlungen* (ed. H.W. Leopoldt and P. Roquette), Band I, de Gruyter, Berlin, 1975.

[14] J.S. Hsia, On the Hasse principle for quadratic forms, *Proc. Amer. Math. Soc.* **39** (1973), 468-470.

[15] N. Jacobson, *Basic Algebra I*, 2nd ed., Freeman, New York, 1985.

[16] Y. Kitaoka, *Arithmetic of quadratic forms*, Cambridge University Press, 1993.

[17] C.W.H. Lam, The search for a finite projective plane of order 10, *Amer. Math. Monthly* **98** (1991), 305-318.

[18] T.Y. Lam, *The algebraic theory of quadratic forms*, revised 2nd printing, Benjamin, Reading, Mass., 1980.

[19] D.W. Lewis, The Merkuryev–Suslin theorem, *Irish Math. Soc. Newsletter* **11** (1984), 29-37.

[20] J. Milnor and D. Husemoller, *Symmetric bilinear forms*, Springer-Verlag, Berlin, 1973.

[21] J. Neukirch, *Class field theory*, Springer-Verlag, Berlin, 1986.

[22] O.T. O'Meara, *Introduction to quadratic forms*, corrected reprint, Springer-Verlag, New York, 1999. [Original edition, 1963]

[23] A. Pfister, Hilbert's seventeenth problem and related problems on definite forms, *Mathematical developments arising from Hilbert problems* (ed. F.E. Browder), pp. 483-489, Proc. Symp. Pure Math. **28**, Part 2, Amer. Math. Soc., Providence, Rhode Island, 1976.

[24] A. Pfister, *Quadratic forms with applications to algebraic geometry and topology*, Cambridge University Press, 1995.

[25] A.R. Rajwade, *Squares*, Cambridge University Press, 1993.

[26] M. Ratner, Interactions between ergodic theory, Lie groups, and number theory, *Proceedings of the International Congress of Mathematicians: Zürich 1994*, pp. 157-182, Birkhäuser, Basel, 1995.

[27] W. Rudin, Sums of squares of polynomials, *Amer. Math. Monthly* **107** (2000), 813-821.

[28] W. Scharlau, *Quadratic and Hermitian forms*, Springer-Verlag, Berlin, 1985.

[29] J.-P. Serre, *A course in arithmetic*, Springer-Verlag, New York, 1973.

[30] W.C. Waterhouse, Pairs of quadratic forms, *Invent. Math.* **37** (1976), 157-164.

[31] K.S. Williams, On the size of a solution of Legendre's equation, *Utilitas Math.* **34** (1988), 65-72.

[32] E. Witt, Theorie der quadratischen Formen in beliebigen Körpern, *J. Reine Angew. Math.* **176** (1937), 31-44.

VIII
The geometry of numbers

It was shown by Hermite (1850) that if

$$f(x) = x^t A x$$

is a positive definite quadratic form in n real variables, then there exists a vector x with *integer* coordinates, not all zero, such that

$$f(x) \leq c_n (\det A)^{1/n},$$

where c_n is a positive constant depending only on n. Minkowski (1891) found a new and more geometric proof of Hermite's result, which gave a much smaller value for the constant c_n. Soon afterwards (1893) he noticed that his proof was valid not only for an n-dimensional ellipsoid $f(x) \leq$ const., but for any convex body which was symmetric about the origin. This led him to a large body of results, to which he gave the somewhat paradoxical name 'geometry of numbers'. It seems fair to say that Minkowski was the first to realize the importance of convexity for mathematics, and it was in his lattice point theorem that he first encountered it.

1 Minkowski's lattice point theorem

A set $C \subseteq \mathbb{R}^n$ is said to be *convex* if $x_1, x_2 \in C$ implies $\theta x_1 + (1 - \theta) x_2 \in C$ for $0 < \theta < 1$. Geometrically, this means that whenever two points belong to the set the whole line segment joining them is also contained in the set.

The *indicator function* or 'characteristic function' of a set $S \subseteq \mathbb{R}^n$ is defined by $\chi(x) = 1$ or 0 according as $x \in S$ or $x \notin S$. If the indicator function is Lebesgue integrable, then the set S is said to have *volume*

$$\lambda(S) = \int_{\mathbb{R}^n} \chi(x)\, dx.$$

The indicator function of a convex set C is actually Riemann integrable. It is easily seen that if a convex set C is not contained in a hyperplane of $\mathbb{R}^n$, then its *interior* int C (see §4 of

Chapter I) is not empty. It follows that $\lambda(C) = 0$ if and only if C is contained in a hyperplane, and $0 < \lambda(C) < \infty$ if and only if C is bounded and is not contained in a hyperplane.

A set $S \subseteq \mathbb{R}^n$ is said to be *symmetric* (with respect to the origin) if $x \in S$ implies $-x \in S$. Evidently any (nonempty) symmetric convex set contains the origin.

A point $x = (\xi_1,...,\xi_n) \in \mathbb{R}^n$ whose coordinates $\xi_1,...,\xi_n$ are all integers will be called a *lattice point*. Thus the set of all lattice points in $\mathbb{R}^n$ is $\mathbb{Z}^n$.

These definitions are the ingredients for Minkowski's *lattice point theorem*:

THEOREM 1 *Let C be a symmetric convex set in $\mathbb{R}^n$. If $\lambda(C) > 2^n$, or if C is compact and $\lambda(C) = 2^n$, then C contains a nonzero point of $\mathbb{Z}^n$.*

The proof of Theorem 1 will be deferred to §3. Here we illustrate the utility of the result by giving several applications, all of which go back to Minkowski himself.

PROPOSITION 2 *If A is an $n \times n$ positive definite real symmetric matrix, then there exists a nonzero point $x \in \mathbb{Z}^n$ such that*

$$x^t A x \le c_n (\det A)^{1/n},$$

where $c_n = (4/\pi)\{(n/2)!\}^{2/n}$.

Proof For any $\rho > 0$ the ellipsoid $x^t A x \le \rho$ is a compact symmetric convex set. By putting $A = T^t T$, for some nonsingular matrix T, it may be seen that the volume of this set is $\kappa_n \rho^{n/2} (\det A)^{-1/2}$, where κ_n is the volume of the n-dimensional unit ball. It follows from Theorem 1 that the ellipsoid contains a nonzero lattice point if $\kappa_n \rho^{n/2} (\det A)^{-1/2} = 2^n$. But, as we will see in §4 of Chapter IX, $\kappa_n = \pi^{n/2}/(n/2)!$, where $x! = \Gamma(x + 1)$. This gives the value c_n for ρ. $\blacksquare$

It follows from Stirling's formula (Chapter IX, §4) that $c_n \sim 2n/\pi e$ for $n \to \infty$. Hermite had proved Proposition 2 with $c_n = (4/3)^{(n-1)/2}$. Hermite's value is smaller than Minkowski's for $n \le 8$, but much larger for large n.

As a second application of Theorem 1 we prove Minkowski's *linear forms theorem*:

PROPOSITION 3 *Let A be an $n \times n$ real matrix with determinant ± 1. Then there exists a nonzero point $x \in \mathbb{Z}^n$ such that $Ax = y = (\eta_k)$ satisfies*

$$|\eta_1| \le 1, \quad |\eta_k| < 1 \text{ for } 1 < k \le n.$$

Proof For any positive integer m, let C_m be the set of all $x \in \mathbb{R}^n$ such that $Ax \in D_m$, where

$$D_m = \{y = (\eta_k) \in \mathbb{R}^n : |\eta_1| \le 1 + 1/m, |\eta_k| < 1 \text{ for } 2 \le k \le n\}.$$

Then C_m is a symmetric convex set, since A is linear and D_m is symmetric and convex. Moreover $\lambda(C_m) = 2^n(1 + 1/m)$, since $\lambda(D_m) = 2^n(1 + 1/m)$ and A is volume-preserving. Therefore, by Theorem 1, C_m contains a lattice point $x_m \neq O$. Since $C_m \subset C_1$ for all $m > 1$ and the number of lattice points in C_1 is finite, there exist only finitely many distinct points x_m. Thus there exists a lattice point $x \neq O$ which belongs to C_m for infinitely many m. Evidently x has the required properties. $\square$

The continued fraction algorithm enables one to find rational approximations to irrational numbers. The subject of *Diophantine approximation* is concerned with the more general problem of solving inequalities in integers. From Proposition 3 we can immediately obtain a result in this area due to Dirichlet (1842):

PROPOSITION 4 *Let $A = (\alpha_{jk})$ be an $n \times m$ real matrix and let $t > 1$ be real. Then there exist integers $q_1,\dots,q_m,p_1,\dots,p_n$, with $0 < \max (|q_1|,\dots,|q_m|) < t^{n/m}$, such that*

$$|\textstyle\sum_{k=1}^m \alpha_{jk}q_k - p_j| \leq 1/t \quad (1 \leq j \leq n).$$

Proof Since the matrix

$$\begin{pmatrix} t^{-n/m}I_m & 0 \\ tA & tI_n \end{pmatrix}$$

has determinant 1, it follows from Proposition 3 that there exists a nonzero vector

$$x = \begin{pmatrix} q \\ -p \end{pmatrix} \in \mathbb{Z}^{n+m}$$

such that

$$|q_k| < t^{n/m} \quad (k = 1,\dots,m),$$

$$|\textstyle\sum_{k=1}^m \alpha_{jk}q_k - p_j| \leq 1/t \quad (j = 1,\dots,n).$$

Since $q = O$ would imply $|p_j| < 1$ for all j and hence $p = O$, which contradicts $x \neq O$, we must have $\max_k |q_k| > 0$. $\square$

COROLLARY 5 *Let $A = (\alpha_{jk})$ be an $n \times m$ real matrix such that $Az \notin \mathbb{Z}^n$ for any nonzero vector $z \in \mathbb{Z}^m$. Then there exist infinitely many $(m+n)$-tuples $q_1,\dots,q_m,p_1,\dots,p_n$ of integers with greatest common divisor 1 and with arbitrarily large values of*

$$\|q\| = \max (|q_1|,\dots,|q_m|)$$

such that

$$|\textstyle\sum_{k=1}^m \alpha_{jk}q_k - p_j| < \|q\|^{-m/n} \quad (1 \leq j \leq n).$$

Proof Let $q_1,...,q_m,p_1,...,p_n$ be integers satisfying the conclusions of Proposition 4 for some $t > 1$. Evidently we may assume that $q_1,...,q_m,p_1,...,p_n$ have no common divisor greater than 1. For given $q_1,...,q_m$, let δ_j be the distance of $\sum_{k=1}^{m} \alpha_{jk}q_k$ from the nearest integer and put $\delta = \max \delta_j$ $(1 \le j \le n)$. By hypothesis $0 < \delta < 1$, and by construction

$$\delta \le 1/t < \|q\|^{-m/n}.$$

Choosing some $t' > 2/\delta$, we find a new set of integers $q_1',...,q_m',p_1',...,p_n'$ satisfying the same requirements with t replaced by t', and hence with $\delta' \le 1/t' < \delta/2$. Proceeding in this way, we obtain a sequence of $(m + n)$-tuples of integers $q_1^{(\nu)},...,q_m^{(\nu)},p_1^{(\nu)},...,p_n^{(\nu)}$ for which $\delta^{(\nu)} \to 0$ and hence $\|q^{(\nu)}\| \to \infty$, since we cannot have $q^{(\nu)} = q$ for infinitely many ν. $\square$

The hypothesis of the corollary is certainly satisfied if $1,\alpha_{j1},...,\alpha_{jm}$ are linearly independent over the field $\mathbb{Q}$ of rational numbers for some $j \in \{1,...,n\}$.

Minkowski also used his lattice point theorem to give the first proof that the discriminant of any algebraic number field, other than $\mathbb{Q}$, has absolute value greater than 1. The proof is given in most books on algebraic number theory.

2 Lattices

In the previous section we defined the set of lattice points to be $\mathbb{Z}^n$. However, this definition is tied to a particular coordinate system in $\mathbb{R}^n$. It is useful to consider lattices from a more intrinsic point of view. The key property is 'discreteness'.

With vector addition as the group operation, $\mathbb{R}^n$ is an abelian group. A subgroup Λ is said to be *discrete* if there exists a ball with centre O which contains no other point of Λ. (More generally, a subgroup H of a topological group G is said to be discrete if there exists an open set $U \subseteq G$ such that $H \cap U = \{e\}$, where e is the identity element of G.)

If Λ is a discrete subgroup of $\mathbb{R}^n$, then any bounded subset of $\mathbb{R}^n$ contains at most finitely many points of Λ since, if there were infinitely many, they would have an accumulation point and their differences would accumulate at O. In particular, Λ is a closed subset of $\mathbb{R}^n$.

PROPOSITION 6 *If $x_1,...,x_m$ are linearly independent vectors in $\mathbb{R}^n$, then the set*

$$\Lambda = \{\zeta_1 x_1 + ... + \zeta_m x_m : \zeta_1,...,\zeta_m \in \mathbb{Z}\}$$

is a discrete subgroup of $\mathbb{R}^n$.

Proof It is clear that Λ is a subgroup of $\mathbb{R}^n$, since $x,y \in \Lambda$ implies $x - y \in \Lambda$. If Λ is not discrete, then there exist $y^{(\nu)} \in \Lambda$ with $|y^{(1)}| > |y^{(2)}| > \dots$ and $|y^{(\nu)}| \to 0$ as $\nu \to \infty$. Let V be the vector subspace of $\mathbb{R}^n$ with basis $x_1,\dots,x_m$ and for any vector

$$x = \alpha_1 x_1 + \dots + \alpha_m x_m,$$

where $\alpha_k \in \mathbb{R}$ ($1 \le k \le m$), put

$$\|x\| = \max(|\alpha_1|,\dots,|\alpha_m|).$$

This defines a norm on V. We have

$$y^{(\nu)} = \zeta_1^{(\nu)} x_1 + \dots + \zeta_m^{(\nu)} x_m,$$

where $\zeta_k^{(\nu)} \in \mathbb{Z}$ ($1 \le k \le m$). Since any two norms on a finite-dimensional vector space are equivalent (Lemma VI.7), it follows that $\zeta_k^{(\nu)} \to 0$ as $\nu \to \infty$ ($1 \le k \le m$). Since $\zeta_k^{(\nu)}$ is an integer, this is only possible if $y^{(\nu)} = O$ for all large ν, which is a contradiction. $\square$

The converse of Proposition 6 is also valid. In fact we will prove a sharper result:

PROPOSITION 7 *If Λ is a discrete subgroup of $\mathbb{R}^n$, then there exist linearly independent vectors $x_1,\dots,x_m$ in $\mathbb{R}^n$ such that*

$$\Lambda = \{\zeta_1 x_1 + \dots + \zeta_m x_m : \zeta_1,\dots,\zeta_m \in \mathbb{Z}\}.$$

Furthermore, if $y_1,\dots,y_m$ is any maximal set of linearly independent vectors in Λ, we can choose $x_1,\dots,x_m$ so that

$$\Lambda \cap \langle y_1,\dots,y_k \rangle = \{\zeta_1 x_1 + \dots + \zeta_k x_k : \zeta_1,\dots,\zeta_k \in \mathbb{Z}\} \quad (1 \le k \le m),$$

where $\langle Y \rangle$ denotes the vector subspace generated by the set Y.

Proof Let S_1 denote the set of all $\alpha_1 > 0$ such that $\alpha_1 y_1 \in \Lambda$ and let μ_1 be the infimum of all $\alpha_1 \in S_1$. We are going to show that $\mu_1 \in S_1$. If this is not the case there exist $\alpha_1^{(\nu)} \in S_1$ with $\alpha_1^{(1)} > \alpha_1^{(2)} > \dots$ and $\alpha_1^{(\nu)} \to \mu_1$ as $\nu \to \infty$. Since the ball $|x| \le (1 + \mu_1)|y_1|$ contains only finitely many points of Λ, this is a contradiction.

Any $\alpha_1 \in S_1$ can be written in the form $\alpha_1 = p\mu_1 + \theta$, where p is a positive integer and $0 \le \theta < \mu_1$. Since $\theta > 0$ would imply $\theta \in S_1$, contrary to the definition of μ_1, we must have $\theta = 0$. Hence if we put $x_1 = \mu_1 y_1$, then

$$\Lambda \cap \langle y_1 \rangle = \{\zeta_1 x_1 : \zeta_1 \in \mathbb{Z}\}.$$

Assume that, for some positive integer k $(1 \leq k < m)$, we have found vectors $x_1,...,x_k \in \Lambda$ such that

$$\Lambda \cap \langle y_1,...,y_k \rangle = \{ \zeta_1 x_1 + ... + \zeta_k x_k : \zeta_1,...,\zeta_k \in \mathbb{Z} \}.$$

We will prove the proposition by showing that this assumption continues to hold when k is replaced by $k + 1$.

Any $x \in \Lambda \cap \langle y_1,...,y_{k+1} \rangle$ has the form

$$x = \alpha_1 x_1 + ... + \alpha_k x_k + \alpha_{k+1} y_{k+1},$$

where $\alpha_1,...,\alpha_{k+1} \in \mathbb{R}$. Let S_{k+1} denote the set of all $\alpha_{k+1} > 0$ which arise in such representations and let μ_{k+1} be the infimum of all $\alpha_{k+1} \in S_{k+1}$. We will show that $\mu_{k+1} \in S_{k+1}$. If $\mu_{k+1} \notin S_{k+1}$, there exist $\alpha_{k+1}^{(\nu)} \in S_{k+1}$ with $\alpha_{k+1}^{(1)} > \alpha_{k+1}^{(2)} > ...$ and $\alpha_{k+1}^{(\nu)} \to \mu_{k+1}$ as $\nu \to \infty$. Then Λ contains a point

$$x^{(\nu)} = \alpha_1^{(\nu)} x_1 + ... + \alpha_k^{(\nu)} x_k + \alpha_{k+1}^{(\nu)} y_{k+1},$$

where $\alpha_j^{(\nu)} \in \mathbb{R}$ $(1 \leq j \leq k)$. In fact, by subtracting an integral linear combination of $x_1,...,x_k$ we may assume that $0 \leq \alpha_j^{(\nu)} < 1$ $(1 \leq j \leq k)$. Since only finitely many points of Λ are contained in the ball $|x| \leq |x_1| + ... + |x_k| + (1+\mu_{k+1})|y_{k+1}|$, this is a contradiction.

Hence $\mu_{k+1} > 0$ and Λ contains a vector

$$x_{k+1} = \alpha_1 x_1 + ... + \alpha_k x_k + \mu_{k+1} y_{k+1}.$$

As for S_1, it may be seen that S_{k+1} consists of all positive integer multiples of μ_{k+1}. Hence any $x \in \Lambda \cap \langle y_1,...,y_{k+1} \rangle$ has the form

$$x = \zeta_1 x_1 + ... + \zeta_k x_k + \zeta_{k+1} x_{k+1},$$

where $\zeta_1,...,\zeta_k \in \mathbb{R}$ and $\zeta_{k+1} \in \mathbb{Z}$. Since

$$x - \zeta_{k+1} x_{k+1} \in \Lambda \cap \langle y_1,...,y_k \rangle,$$

we must actually have $\zeta_1,...,\zeta_k \in \mathbb{Z}$. $\square$

By being more specific in the proof of Proposition 7 it may be shown that there is a *unique* choice of $x_1,...,x_m$ such that

$$y_1 = p_{11} x_1$$
$$y_2 = p_{21} x_1 + p_{22} x_2$$
$$.....$$
$$y_m = p_{m1} x_1 + p_{m2} x_2 + ... + p_{mm} x_m,$$

where $p_{ij} \in \mathbb{Z}$, $p_{ii} > 0$, and $0 \le p_{ij} < p_{ii}$ if $j < i$ (*Hermite's normal form*).

It is easily seen that in Proposition 7 we can choose $x_i = y_i$ $(1 \le i \le m)$ if and only if, for any $x \in \Lambda$ and any positive integer h, x is an integral linear combination of $y_1,...,y_m$ whenever hx is.

By combining Propositions 6 and 7 we obtain

PROPOSITION 8 *For a set $\Lambda \subseteq \mathbb{R}^n$ the following two conditions are equivalent:*

(i) *Λ is a discrete subgroup of $\mathbb{R}^n$ and there exists $R > 0$ such that, for each $y \in \mathbb{R}^n$, there is some $x \in \Lambda$ with $|y - x| < R$;*

(ii) *there exist n linearly independent vectors $x_1,...,x_n$ in $\mathbb{R}^n$ such that*

$$\Lambda = \{\zeta_1 x_1 + ... + \zeta_n x_n : \zeta_1,...,\zeta_n \in \mathbb{Z}\}.$$

Proof If (i) holds, then in the statement of Proposition 7 we must have $m = n$, i.e. (ii) holds. On the other hand, if (ii) holds then Λ is a discrete subgroup of $\mathbb{R}^n$, by Proposition 6. Moreover, for any $y \in \mathbb{R}^n$ we can choose $x \in \Lambda$ so that

$$y - x = \theta_1 x_1 + ... + \theta_n x_n,$$

where $0 \le \theta_j < 1$ $(j = 1,...,n)$, and hence

$$|y - x| < |x_1| + ... + |x_n|. \quad \square$$

A set $\Lambda \subseteq \mathbb{R}^n$ satisfying either of the two equivalent conditions of Proposition 8 will be called a *lattice* and any element of Λ a *lattice point*. The vectors $x_1,...,x_n$ in (ii) will be said to be a *basis* for the lattice.

A lattice is sometimes defined to be any discrete subgroup of $\mathbb{R}^n$, and what we have called a lattice is then called a 'nondegenerate' lattice. Our definition is chosen simply to avoid repetition of the word 'nondegenerate'. We may occasionally use the more general definition and, with this warning, believe it will be clear from the context when this occurs.

The basis of a lattice is not uniquely determined. In fact $y_1,...,y_n$ is also a basis if

$$y_j = \sum_{k=1}^{n} \alpha_{jk} x_k \quad (j = 1,...,n),$$

where $A = (\alpha_{jk})$ is an $n \times n$ matrix of integers such that $\det A = \pm 1$, since A^{-1} is then also a matrix of integers. Moreover, every basis $y_1,...,y_n$ is obtained in this way. For if

$$y_j = \sum_{k=1}^{n} \alpha_{jk} x_k, \quad x_i = \sum_{j=1}^{n} \beta_{ij} y_j, \quad (i,j = 1,...,n),$$

where $A = (\alpha_{jk})$ and $B = (\beta_{ij})$ are $n \times n$ matrices of integers, then $BA = I$ and hence $(\det B)(\det A) = 1$. Since $\det A$ and $\det B$ are integers, it follows that $\det A = \pm 1$.

Let $x_1,...,x_n$ be a basis for a lattice $\Lambda \subseteq \mathbb{R}^n$. If

$$x_k = \sum_{j=1}^{n} \gamma_{jk} e_j \quad (k = 1,...,n),$$

where $e_1,...,e_n$ is the canonical basis for $\mathbb{R}^n$ then, in terms of the nonsingular matrix $T = (\gamma_{jk})$, the lattice Λ is just the set of all vectors Tz with $z \in \mathbb{Z}^n$. The absolute value of the determinant of the matrix T does not depend on the choice of basis. For if $x_1',...,x_n'$ is any other basis, then

$$x_i' = \sum_{j=1}^{n} \alpha_{ij} x_j \quad (i = 1,...,n),$$

where $A = (\alpha_{ij})$ is an $n \times n$ matrix of integers with $\det A = \pm 1$. Thus

$$x_k' = \sum_{j=1}^{n} \gamma_{jk}' e_j \quad (k = 1,...,n),$$

where $T' = (\gamma_{jk}')$ satisfies $T' = TA^t$ and hence

$$|\det T'| = |\det T|.$$

The uniquely determined quantity $|\det T|$ will be called the *determinant* of the lattice Λ and denoted by $d(\Lambda)$. (Some authors, e.g. Conway and Sloane [14], call $|\det T|^2$ the determinant of Λ, but others prefer to call this the *discriminant* of Λ.)

The determinant $d(\Lambda)$ has a simple geometrical interpretation. In fact it is the volume of the parallelotope Π, consisting of all points $y \in \mathbb{R}^n$ such that

$$y = \theta_1 x_1 + ... + \theta_n x_n,$$

where $0 \leq \theta_k \leq 1$ $(k = 1,...,n)$. The interior of Π is a *fundamental domain* for the subgroup Λ, since

$$\mathbb{R}^n = \bigcup_{x \in \Lambda} (\Pi + x),$$

$$\text{int } (\Pi + x) \cap \text{int } (\Pi + x') = \varnothing \quad \text{if } x,x' \in \Lambda \text{ and } x \neq x'.$$

For any lattice $\Lambda \subseteq \mathbb{R}^n$, the set Λ^* of all vectors $y \in \mathbb{R}^n$ such that $y^t x \in \mathbb{Z}$ for every $x \in \Lambda$ is again a lattice, the *dual* (or 'polar' or 'reciprocal') of Λ. In fact,

$$\text{if } \Lambda = \{Tz : z \in \mathbb{Z}^n\}, \text{ then } \Lambda^* = \{(T^t)^{-1}w : w \in \mathbb{Z}^n\}.$$

Hence Λ is the dual of Λ^* and $d(\Lambda)d(\Lambda^*) = 1$. A lattice Λ is *self-dual* if $\Lambda^* = \Lambda$.

3 Proof of the lattice point theorem, and some generalizations

In this section we take up the proof of Minkowski's lattice point theorem. The proof will be based on a very general result, due to Blichfeldt (1914), which is not restricted to convex sets.

PROPOSITION 9 *Let S be a Lebesgue measurable subset of* $\mathbb{R}^n$, Λ *a lattice in* $\mathbb{R}^n$ *with determinant* $d(\Lambda)$ *and m a positive integer.*

If $\lambda(S) > m \, d(\Lambda)$, *or if S is compact and* $\lambda(S) = m \, d(\Lambda)$, *then there exist* $m + 1$ *distinct points* $x_1,...,x_{m+1}$ *of S such that the differences* $x_j - x_k$ $(1 \le j,k \le m + 1)$ *all lie in* Λ.

Proof Let $b_1,...,b_n$ be a basis for Λ and let P be the half-open parallelotope consisting of all points $x = \theta_1 b_1 + ... + \theta_n b_n$, where $0 \le \theta_i < 1$ $(i = 1,...,n)$. Then $\lambda(P) = d(\Lambda)$ and

$$\mathbb{R}^n = \bigcup_{z \in \Lambda} (P + z), \quad (P + z) \cap (P + z') = \varnothing \quad \text{if } z \ne z'.$$

Suppose first that $\lambda(S) > m \, d(\Lambda)$. If we put

$$S_z = S \cap (P + z), \quad T_z = S_z - z,$$

then $T_z \subseteq P$, $\lambda(T_z) = \lambda(S_z)$ and

$$\lambda(S) = \sum_{z \in \Lambda} \lambda(S_z).$$

Hence

$$\sum_{z \in \Lambda} \lambda(T_z) = \lambda(S) > m \, d(\Lambda) = m \, \lambda(P).$$

Since $T_z \subseteq P$ for every z, it follows that some point $y \in P$ is contained in at least $m + 1$ sets T_z. (In fact this must hold for all y in a subset of P of positive measure.) Thus there exist $m + 1$ distinct points $z_1,...,z_{m+1}$ of Λ and points $x_1,...,x_{m+1}$ of S such that $y = x_j - z_j$ $(j = 1,..,m+1)$. Then $x_1,...,x_{m+1}$ are distinct and

$$x_j - x_k = z_j - z_k \in \Lambda \quad (1 \le j,k \le m + 1).$$

Suppose next that S is compact and $\lambda(S) = m \, d(\Lambda)$. Let $\{\varepsilon_\nu\}$ be a decreasing sequence of positive numbers such that $\varepsilon_\nu \to 0$ as $\nu \to \infty$, and let S_ν denote the set of all points of $\mathbb{R}^n$ distant at most ε_ν from S. Then S_ν is compact, $\lambda(S_\nu) > \lambda(S)$ and

$$S_1 \supset S_2 \supset ..., \quad S = \bigcap_\nu S_\nu.$$

By what we have already proved, there exist $m + 1$ distinct points $x_1^{(\nu)},...,x_{m+1}^{(\nu)}$ of S_ν such that $x_j^{(\nu)} - x_k^{(\nu)} \in \Lambda$ for all j,k. Since $S_\nu \subseteq S_1$ and S_1 is compact, by restricting attention to a subsequence we may assume that $x_j^{(\nu)} \to x_j$ as $\nu \to \infty$ ($j = 1,...,m+1$). Then $x_j \in S$ and $x_j^{(\nu)} - x_k^{(\nu)} \to x_j - x_k$. Since $x_j^{(\nu)} - x_k^{(\nu)} \in \Lambda$, this is only possible if $x_j - x_k = x_j^{(\nu)} - x_k^{(\nu)}$ for all large ν. Hence $x_1,...,x_{m+1}$ are distinct. $\square$

Siegel (1935) has given an analytic formula which underlies Proposition 9 and enables it to be generalized. Although we will make no use of it, this formula will now be established. For notational simplicity we restrict attention to the (self-dual) lattice $\Lambda = \mathbb{Z}^n$.

PROPOSITION 10 *If $\psi: \mathbb{R}^n \to \mathbb{C}$ is a bounded measurable function which vanishes outside some compact set, then*

$$\int_{\mathbb{R}^n} \psi(x)\overline{\phi(x)}\, dx = \sum_{w \in \mathbb{Z}^n} \left| \int_{\mathbb{R}^n} \psi(x)e^{-2\pi i w^t x}\, dx \right|^2,$$

where

$$\phi(x) = \sum_{z \in \mathbb{Z}^n} \psi(x + z).$$

Proof Since ψ vanishes outside a compact set, there exists a finite set $T \subseteq \mathbb{Z}^n$ such that $\psi(x + z) = 0$ for all $x \in \mathbb{R}^n$ if $z \in \mathbb{Z}^n \setminus T$. Thus the sum defining $\phi(x)$ has only finitely many nonzero terms and ϕ also is a bounded measurable function which vanishes outside some compact set.

If we write

$$x = (\xi_1,...,\xi_n), \quad z = (\zeta_1,...,\zeta_n),$$

then the sum defining $\phi(x)$ is unaltered by the substitution $\zeta_j \to \zeta_j + 1$ and hence ϕ has period 1 in each of the variables ξ_j ($j = 1,...,n$). Let Π denote the fundamental parallelotope

$$\Pi = \{x = (\xi_1,...,\xi_n) \in \mathbb{R}^n : 0 \le \xi_j \le 1 \text{ for } j = 1,...,n\}.$$

Since the functions $e^{2\pi i w^t x}$ ($w \in \mathbb{Z}^n$) are an orthogonal basis for $L^2(\Pi)$, Parseval's equality (Chapter I, §10) holds:

$$\int_\Pi |\phi(x)|^2\, dx = \sum_{w \in \mathbb{Z}^n} |c_w|^2,$$

where

$$c_w = \int_\Pi \phi(x)e^{-2\pi i w^t x}\, dx.$$

But

$$c_w = \int_\Pi \sum_{z \in \mathbb{Z}^n} \psi(x + z)e^{-2\pi i w^t x}\, dx$$

$$= \int_\Pi \sum_{z \in \mathbb{Z}^n} \psi(x + z)e^{-2\pi i w^t(x+z)}\, dx,$$

since $e^{2k\pi i} = 1$ for any integer k. Hence

$$c_w = \int_{\mathbb{R}^n} \psi(y)e^{-2\pi i w^t y} \, dy.$$

On the other hand,

$$\int_\Pi |\phi(x)|^2 \, dx = \int_\Pi \sum_{z',z'' \in \mathbb{Z}^n} \psi(x + z') \overline{\psi(x + z'')} \, dx$$

$$= \int_\Pi \sum_{z,z' \in \mathbb{Z}^n} \psi(x + z') \overline{\psi(x + z'+z)} \, dx$$

$$= \int_{\mathbb{R}^n} \sum_{z \in \mathbb{Z}^n} \psi(y) \overline{\psi(y+z)} \, dy = \int_{\mathbb{R}^n} \psi(y) \overline{\phi(y)} \, dy.$$

Substituting these expressions in Parseval's equality, we obtain the result. □

Suppose, in particular, that ψ takes only real nonnegative values. Then so also does ϕ and

$$\int_{\mathbb{R}^n} \psi(x)\phi(x) \, dx \le \sup_{x \in \mathbb{R}^n} \phi(x) \int_{\mathbb{R}^n} \psi(x) \, dx.$$

On the other hand, omitting all terms with $w \ne 0$ we obtain

$$\sum_{w \in \mathbb{Z}^n} \left| \int_{\mathbb{R}^n} \psi(x)e^{-2\pi i w^t x} \, dx \right|^2 \ge \left(\int_{\mathbb{R}^n} \psi(x) \, dx \right)^2.$$

Hence, by Proposition 10,

$$\sup_{x \in \mathbb{R}^n} \phi(x) \ge \int_{\mathbb{R}^n} \psi(x) \, dx.$$

For example, let $S \subseteq \mathbb{R}^n$ be a measurable set with $\lambda(S) > m$. Then there exists a *bounded* measurable set $S' \subseteq S$ with $\lambda(S') > m$. If we take ψ to be the indicator function of S', then

$$\int_{\mathbb{R}^n} \psi(x) \, dx = \lambda(S') > m$$

and we conclude that there exists $y \in \mathbb{R}^n$ such that

$$\sum_{z \in \mathbb{Z}^n} \psi(y + z) = \phi(y) > m.$$

Since the only possible values of the summands on the left are 0 and 1, it follows that there exist $m + 1$ distinct points $z_1,...,z_{m+1} \in \mathbb{Z}^n = \Lambda$ such that $y + z_1,...,y + z_{m+1} \in S$. The proof of Proposition 9 can now be completed in the same way as before.

Let $\{K_\alpha\}$ be a family of subsets of $\mathbb{R}^n$, where each K_α is the *closure* of a nonempty open set G_α, i.e. K_α is the intersection of all closed sets containing G_α. The family $\{K_\alpha\}$ is said to be a *packing* of $\mathbb{R}^n$ if $\alpha \ne \alpha'$ implies $G_\alpha \cap G_{\alpha'} = \varnothing$ and is said to be a *covering* of $\mathbb{R}^n$ if $\mathbb{R}^n = \bigcup_\alpha K_\alpha$. It is said to be a *tiling* of $\mathbb{R}^n$ if it is both a packing and a covering.

For example, if Π is a fundamental parallelotope of a lattice Λ, then the family $\{\Pi + a: a \in \Lambda\}$ is a tiling of $\mathbb{R}^n$. More generally, if G is a nonempty open subset of $\mathbb{R}^n$ with closure K, we may ask whether the family $\{K + a: a \in \Lambda\}$ of all Λ-translates of K is either a packing or a covering of $\mathbb{R}^n$. Some necessary conditions may be derived with the aid of Proposition 9:

PROPOSITION 11 *Let K be the closure of a bounded nonempty open set $G \subseteq \mathbb{R}^n$ and let Λ be a lattice in $\mathbb{R}^n$.*

If the Λ-translates of K are a covering of $\mathbb{R}^n$ then $\lambda(K) \geq d(\Lambda)$, with strict inequality if they are not also a packing.

If the Λ-translates of K are a packing of $\mathbb{R}^n$ then $\lambda(K) \leq d(\Lambda)$, with strict inequality if they are not also a covering.

Proof Suppose first that the Λ-translates of K cover $\mathbb{R}^n$. Then every point of a fundamental parallelotope Π of Λ has the form $x - a$, where $x \in K$ and $a \in \Lambda$. Hence

$$\lambda(K) = \sum_{a \in \Lambda} \lambda(K \cap (\Pi + a))$$

$$= \sum_{a \in \Lambda} \lambda((K - a) \cap \Pi) \geq \lambda(\Pi) = d(\Lambda).$$

Suppose, in addition, that the Λ-translates of K are not a packing of $\mathbb{R}^n$. Then there exist distinct points x_1, x_2 in the interior G of K such that $a = x_1 - x_2 \in \Lambda$. Let

$$B_\varepsilon = \{x \in \mathbb{R}^n : |x| \leq \varepsilon\}.$$

We can choose $\varepsilon > 0$ so small that the balls $B_\varepsilon + x_1$ and $B_\varepsilon + x_2$ are disjoint and contained in G. Then $G' = G \setminus (B_\varepsilon + x_1)$ is a bounded nonempty open set with closure $K' = K \setminus (\text{int } B_\varepsilon + x_1)$. Since

$$B_\varepsilon + x_1 = B_\varepsilon + x_2 + a \subseteq K' + a,$$

the Λ-translates of K' contain K and therefore also cover $\mathbb{R}^n$. Hence, by what we have already proved, $\lambda(K') \geq d(\Lambda)$. Since $\lambda(K) > \lambda(K')$, it follows that $\lambda(K) > d(\Lambda)$.

Suppose now that the Λ-translates of K are a packing of $\mathbb{R}^n$. Then Λ does not contain the difference of two distinct points in the interior G of K, since $G + a$ and $G + b$ are disjoint if a,b are distinct points of Λ. It follows from Proposition 9 that

$$\lambda(K) = \lambda(G) \leq d(\Lambda).$$

Suppose, in addition, that the Λ-translates of K do not cover $\mathbb{R}^n$. Thus there exists a point $y \in \mathbb{R}^n$ which is not in any Λ-translate of K. We will show that we can choose $\varepsilon > 0$ so small that y is not in any Λ-translate of $K + B_\varepsilon$.

If this is not the case then, for any positive integer ν, there exists $a_\nu \in \Lambda$ such that

$$y \in K + B_{1/\nu} + a_\nu.$$

Evidently the sequence a_ν is bounded and hence there exists $a \in \Lambda$ such that $a_\nu = a$ for infinitely many ν. But then $y \in K + a$, which is contrary to hypothesis.

We may in addition assume ε chosen so small that $|x| > 2\varepsilon$ for every nonzero $x \in \Lambda$. Then the set $S = G \cup (B_\varepsilon + y)$ has the property that Λ does not contain the difference of any two distinct points of S. Hence, by Proposition 9, $\lambda(S) \leq d(\Lambda)$. Since

$$\lambda(K) = \lambda(G) < \lambda(S),$$

it follows that $\lambda(K) < d(\Lambda)$. $\square$

We next apply Proposition 9 to convex sets. Minkowski's lattice point theorem (Theorem 1) is the special case $m = 1$ (and $\Lambda = \mathbb{Z}^n$) of the following generalization, due to van der Corput (1936):

PROPOSITION 12 *Let C be a symmetric convex subset of $\mathbb{R}^n$, Λ a lattice in $\mathbb{R}^n$ with determinant $d(\Lambda)$, and m a positive integer.*

If $\lambda(C) > 2^n m\, d(\Lambda)$, or if C is compact and $\lambda(C) = 2^n m\, d(\Lambda)$, then there exist $2m$ distinct nonzero points $\pm y_1,\dots,\pm y_m$ of Λ such that

$$y_j \in C \quad (1 \leq j \leq m),$$
$$y_j - y_k \in C \quad (1 \leq j,k \leq m).$$

Proof The set $S = \{x/2 : x \in C\}$ has measure $\lambda(S) = \lambda(C)/2^n$. Hence, by Proposition 9, there exist $m + 1$ distinct points $x_1,\dots,x_{m+1} \in C$ such that $(x_j - x_k)/2 \in \Lambda$ for all j,k.

The vectors of $\mathbb{R}^n$ may be totally ordered by writing $x > x'$ if the first nonzero coordinate of $x - x'$ is positive. We assume the points $x_1,\dots,x_{m+1} \in C$ numbered so that

$$x_1 > x_2 > \dots > x_{m+1}.$$

Put

$$y_j = (x_j - x_{m+1})/2 \quad (j = 1,\dots, m).$$

Then, by construction, $y_j \in \Lambda$ $(j = 1,\dots, m)$. Moreover $y_j \in C$, since $x_1,\dots,x_{m+1} \in C$ and C is symmetric, and similarly $y_j - y_k = (x_j - x_k)/2 \in C$. Finally, since

$$y_1 > y_2 > \dots > y_m > O,$$

we have $y_j \neq O$ and $y_j \neq \pm y_k$ if $j \neq k$. $\square$

The conclusion of Proposition 12 need no longer hold if C is not compact and $\lambda(C) = 2^n m\, d(\Lambda)$. For example, take $\Lambda = \mathbb{Z}^n$ and let C be the symmetric convex set

$$C = \{x = (\xi_1,...,\xi_n) \in \mathbb{R}^n: |\xi_1| < m, |\xi_j| < 1 \text{ for } 2 \leq j \leq n\}.$$

Then $d(\Lambda) = 1$ and $\lambda(C) = 2^n m$. However, the only nonzero points of Λ in C are the $2(m-1)$ points $(\pm k, 0, ..., 0)$ $(1 \leq k \leq m-1)$.

To provide a broader view of the geometry of numbers we now mention without proof some further results. A different generalization of Minkowski's lattice point theorem was already proved by Minkowski himself. Let Λ be a lattice in $\mathbb{R}^n$ and let K be a compact symmetric convex subset of $\mathbb{R}^n$ with nonempty interior. Then ρK contains no nonzero point of Λ for small $\rho > 0$ and contains n linearly independent points of Λ for large $\rho > 0$. Let μ_i be the infimum of all $\rho > 0$ such that ρK contains at least i linearly independent points of Λ $(i = 1,...,n)$. The *successive minima* $\mu_i = \mu_i(K,\Lambda)$ evidently satisfy the inequalities

$$0 < \mu_1 \leq \mu_2 \leq ... \leq \mu_n < \infty.$$

Minkowski's lattice point theorem says that

$$\mu_1^n \lambda(K) \leq 2^n d(\Lambda).$$

Minkowski's *theorem on successive minima* strengthens this to

$$2^n d(\Lambda)/n! \leq \mu_1 \mu_2 \cdots \mu_n \lambda(K) \leq 2^n d(\Lambda).$$

The lower bound is quite easy to prove, but the upper bound is deeper-lying – notwithstanding simplifications of Minkowski's original proof. If $\Lambda = \mathbb{Z}^n$, then equality holds in the lower bound for the *cross-polytope* $K = \{(\xi_1,...,\xi_n) \in \mathbb{R}^n: \sum_{i=1}^n |\xi_i| \leq 1\}$ and in the upper bound for the *cube* $K = \{(\xi_1,...,\xi_n) \in \mathbb{R}^n: |\xi_i| \leq 1 \text{ for all } i\}$.

If K is a compact symmetric convex subset of $\mathbb{R}^n$ with nonempty interior, its *critical determinant* $\Delta(K)$ is defined to be the infimum of the determinant $d(\Lambda)$ for all lattices Λ with no nonzero point in the interior of K. A lattice Λ for which $d(\Lambda) = \Delta(K)$ is called a *critical lattice* for K. It will be shown in §6 that a critical lattice always exists.

It follows from Proposition 12 that $\Delta(K) \geq 2^{-n}\lambda(K)$. A conjectured sharpening of Minkowski's theorem on successive minima, which has been proved by Minkowski (1896) himself for $n = 2$ and for n-dimensional ellipsoids, and by Woods (1956) for $n = 3$, claims that

$$\mu_1 \mu_2 \cdots \mu_n \Delta(K) \leq d(\Lambda).$$

The successive minima of a convex body are connected with those of its dual body. If K is a compact symmetric convex subset of $\mathbb{R}^n$ with nonempty interior, then its *dual*

$$K^* = \{y \in \mathbb{R}^n: y^t x \leq 1 \text{ for all } x \in K\}$$

has the same properties, and K is the dual of K^*. Mahler (1939) showed that the successive minima of the dual body K^* with respect to the dual lattice Λ^* are related to the successive minima of K with respect to Λ by the inequalities

$$1 \leq \mu_i(K,\Lambda)\mu_{n-i+1}(K^*,\Lambda^*) \quad (i = 1,...,n),$$

and hence, by applying Minkowski's theorem on successive minima also to K^* and Λ^*, he obtained inequalities in the opposite direction:

$$\mu_i(K,\Lambda)\mu_{n-i+1}(K^*,\Lambda^*) \leq 4^n/\lambda(K)\lambda(K^*) \quad (i = 1,...,n).$$

By further proving that $\lambda(K)\lambda(K^*) \geq 4^n(n!)^{-2}$, he deduced that

$$\mu_i(K,\Lambda)\mu_{n-i+1}(K^*,\Lambda^*) \leq (n!)^2 \quad (i = 1,...,n).$$

Dramatic improvements of these bounds have recently been obtained. Banaszczyk (1996), using techniques from harmonic analysis, has shown that there is a numerical constant $C > 0$ such that, for all $n \geq 1$ and all $i \in \{1,...,n\}$,

$$\mu_i(K,\Lambda)\mu_{n-i+1}(K^*,\Lambda^*) \leq Cn(1 + \log n).$$

He had shown already (1993) that if $K = B_1$ is the n-dimensional closed unit ball, which is self-dual, then for all $n \geq 1$ and all $i \in \{1,...,n\}$,

$$\mu_i(B_1,\Lambda)\mu_{n-i+1}(B_1,\Lambda^*) \leq n.$$

This result is close to being best possible, since there exists a numerical constant $C' > 0$ and self-dual lattices $\Lambda_n \subseteq \mathbb{R}^n$ such that

$$\mu_1(B_1,\Lambda_n)\mu_n(B_1,\Lambda_n) \geq \mu_1(B_1,\Lambda_n)^2 \geq C'n.$$

Two other applications of Minkowski's theorem on successive minima will be mentioned here. The first is a sharp form, due to Bombieri and Vaaler (1983), of 'Siegel's lemma'. In his investigations on transcendental numbers Siegel (1929) used Dirichlet's pigeonhole principle to prove that if $A = (\alpha_{jk})$ is an $m \times n$ matrix of integers, where $m < n$, such that $|\alpha_{jk}| \leq \beta$ for all j,k, then the system of homogeneous linear equations

$$Ax = 0$$

has a solution $x = (\xi_k)$ in integers, not all 0, such that $|\xi_k| \leq 1 + (n\beta)^{m/(n-m)}$ for all k. Bombieri and Vaaler show that, if A has rank m and if $g > 0$ is the greatest common divisor of all $m \times m$

subdeterminants of A, then there are $n - m$ linearly independent integral solutions $x_j = (\xi_{jk})$ ($j = 1,...,n - m$) such that

$$\prod_{j=1}^{n-m} \|x_j\| \leq [\det (AA^t)]^{1/2} /g,$$

where $\|x_j\| = \max_k |\xi_{jk}|$.

The second application, due to Gillet and Soulé (1991), may be regarded as an arithmetic analogue of the Riemann–Roch theorem for function fields. Again let K be a compact symmetric convex subset of $\mathbb{R}^n$ with nonempty interior and let μ_i denote the infimum of all $\rho > 0$ such that ρK contains at least i linearly independent points of $\mathbb{Z}^n$ ($i = 1,...,n$). If $M(K)$ is the number of points of $\mathbb{Z}^n$ in K, and if h is the maximum number of linearly independent points of $\mathbb{Z}^n$ in the interior of K, then Gillet and Soulé show that $\mu_1 \cdots \mu_h/M(K)$ is bounded above and below by positive constants, which depend on n but not on K.

A number of results in this section have dealt with compact symmetric convex sets with nonempty interior. Since such sets may appear rather special, it should be pointed out that they arise very naturally in connection with normed vector spaces.

The vector space $\mathbb{R}^n$ is said to be *normed* if with each $x \in \mathbb{R}^n$ there is associated a real number $|x|$ with the properties

(i) $|x| \geq 0$, with equality if and only if $x = O$,
(ii) $|x + y| \leq |x| + |y|$ for all $x,y \in \mathbb{R}^n$,
(iii) $|\alpha x| = |\alpha| \, |x|$ for all $x \in \mathbb{R}^n$ and all $\alpha \in \mathbb{R}$.

Let K denote the set of all $x \in \mathbb{R}^n$ such that $|x| \leq 1$. Then K is bounded, since all norms on a finite-dimensional vector space are equivalent. In fact K is compact, since it follows from (ii) that K is closed. Moreover K is convex and symmetric, by (ii) and (iii). Furthermore, by (i) and (iii), $x/|x| \in K$ for each nonzero $x \in \mathbb{R}^n$. Hence the interior of K is nonempty and is actually the set of all $x \in \mathbb{R}^n$ such that $|x| < 1$.

Conversely, let K be a compact symmetric convex subset of $\mathbb{R}^n$ with nonempty interior. Then the origin is an interior point of K and for each nonzero $x \in \mathbb{R}^n$ there is a unique $\rho > 0$ such that ρx is on the boundary of K. If we put $|x| = \rho^{-1}$, and $|O| = 0$, then (i) obviously holds. Furthermore, since $|-x| = |x|$, it is easily seen that (iii) holds. Finally, if $y \in \mathbb{R}^n$ and $|y| = \sigma^{-1}$, then $\rho x, \sigma y \in K$ and hence, since K is convex,

$$\rho\sigma(\rho + \sigma)^{-1}(x + y) = \sigma(\rho + \sigma)^{-1}\rho x + \rho(\rho + \sigma)^{-1}\sigma y \in K.$$

Hence

$$|x + y| \leq (\rho + \sigma)/\rho\sigma = |x| + |y|.$$

Thus $\mathbb{R}^n$ is a normed vector space and K the set of all $x \in \mathbb{R}^n$ such that $|x| \le 1$.

4 Voronoi cells

Throughout this section we suppose $\mathbb{R}^n$ equipped with the *Euclidean metric*:

$$d(y,z) = \|y - z\|,$$

where $\|x\| = (x^t x)^{1/2}$. We call $\|x\|^2 = x^t x$ the *square-norm* of x and we denote the scalar product $y^t z$ by (y,z).

Fix some point $x_0 \in \mathbb{R}^n$. For any point $x \ne x_0$, the set of all points which are equidistant from x_0 and x is the hyperplane H_x which passes through the midpoint of the segment joining x_0 and x and is orthogonal to this segment. Analytically, H_x is the set of all $y \in \mathbb{R}^n$ such that

$$(x - x_0, y) = (x - x_0, x + x_0)/2,$$

which simplifies to

$$2(x - x_0, y) = \|x\|^2 - \|x_0\|^2.$$

The set of all points which are closer to x_0 than to x is the open half-space G_x consisting of all points $y \in \mathbb{R}^n$ such that

$$2(x - x_0, y) < \|x\|^2 - \|x_0\|^2.$$

The closed half-space $\overline{G}_x = H_x \cup G_x$ is the set of all points at least as close to x_0 as to x.

Let X be a subset of $\mathbb{R}^n$ containing more than one point which is *discrete*, i.e. for each $y \in \mathbb{R}^n$ there exists an open set containing y which contains at most one point of X. It follows that each bounded subset of $\mathbb{R}^n$ contains only finitely many points of X since, if there were infinitely many, they would have an accumulation point. Hence for each $y \in \mathbb{R}^n$ there exists an $x_0 \in X$ whose distance from y is minimal:

$$d(x_0, y) \le d(x, y) \quad \text{for every } x \in X. \tag{1}$$

For each $x_0 \in X$ we define its *Voronoi cell* $V(x_0)$ to be the set of all $y \in \mathbb{R}^n$ for which (1) holds. Voronoi cells are also called 'Dirichlet domains', since they were used by Dirichlet (1850) in $\mathbb{R}^2$ before Voronoi (1908) used them in $\mathbb{R}^n$.

If we choose $r > 0$ so that the open ball

$$\beta_r(x_0) := \{y \in \mathbb{R}^n : d(x_0, y) < r\}$$

contains no point of X except x_0, then $\beta_{r/2}(x_0) \subseteq V(x_0)$. Thus x_0 is an interior point of $V(x_0)$.

Since

$$\overline{G}_x = \{y \in \mathbb{R}^n : d(x_0, y) \le d(x, y)\},$$

we have $V(x_0) \subseteq \overline{G}_x$ and actually

$$V(x_0) = \bigcap_{x \in X \backslash x_0} \overline{G}_x. \tag{2}$$

It follows at once from (2) that $V(x_0)$ is closed and convex. Hence $V(x_0)$ is the closure of its nonempty interior.

According to the definitions of §3, the Voronoi cells form a tiling of $\mathbb{R}^n$, since

$$\mathbb{R}^n = \bigcup_{x \in X} V(x),$$

$$\text{int } V(x) \cap \text{int } V(x') = \varnothing \quad \text{if } x, x' \in X \text{ and } x \ne x'.$$

A subset A of a convex set C is said to be a *face* of C if A is convex and, for any $c, c' \in C$, $(c, c') \cap A \ne \varnothing$ implies $c, c' \in A$. The tiling by Voronoi cells has the additional property that $V(x) \cap V(x')$ is a face of both $V(x)$ and $V(x')$ if $x, x' \in X$ and $x \ne x'$. We will prove this by showing that if y_1, y_2 are distinct points of $V(x)$ and if $z \in (y_1, y_2) \cap V(x')$, then $y_1 \in V(x')$.

Since $z \in V(x) \cap V(x')$, we have $d(x, z) = d(x', z)$. Thus z lies on the hyperplane H which passes through the midpoint of the segment joining x and x' and is orthogonal to this segment. If $y_1 \notin V(x')$, then $d(x, y_1) < d(x', y_1)$. Thus y_1 lies in the open half-space G associated with the hyperplane H which contains x. But then y_2 lies in the open half-space G' which contains x', i.e. $d(x', y_2) < d(x, y_2)$, which contradicts $y_2 \in V(x)$.

We now assume that the set X is not only discrete, but also *relatively dense*, i.e.

(†)   there exists $R > 0$ such that, for each $y \in \mathbb{R}^n$, there is some $x \in X$ with $d(x, y) < R$.

It follows at once that $V(x_0) \subseteq \beta_R(x_0)$. Thus $V(x_0)$ is bounded and, since it is closed, even compact. The ball $\beta_{2R}(x_0)$ contains only finitely many points $x_1, \ldots, x_m$ of X apart from x_0. We are going to show that

$$V(x_0) = \bigcap_{i=1}^m \overline{G}_{x_i}. \tag{3}$$

By (2) we need only show that if $y \in \bigcap_{i=1}^m \overline{G}_{x_i}$, then $y \in \overline{G}_x$ for every $x \in X$.

Assume that $d(x_0, y) \ge R$ and choose z on the segment joining x_0 and y so that $d(x_0, z) = R$. For some $x \in X$ we have $d(x, z) < R$ and hence $0 < d(x_0, x) < 2R$. Consequently $x = x_i$ for some $i \in \{1, \ldots, m\}$. Since $d(x_i, z) < R = d(x_0, z)$, we have $z \notin \overline{G}_{x_i}$. But this is a contradiction, since $x_0, y \in \overline{G}_{x_i}$ and z is on the segment joining them.

We conclude that $d(x_0, y) < R$. If $x \in X$ and $x \ne x_0, x_1, \ldots, x_m$, then

$$d(x,y) \geq d(x_0,x) - d(x_0,y)$$
$$\geq 2R - R = R > d(x_0,y).$$

Consequently $y \in \overline{G}_x$ for every $x \in X$.

It follows from (3) that $V(x_0)$ is a polyhedron. Since $V(x_0)$ is bounded and has a nonempty interior, it is actually an *n-dimensional polytope*.

The faces of a polytope are an important part of its structure. An $(n-1)$-dimensional face of an *n*-dimensional polytope is said to be a *facet* and a 0-dimensional face is said to be a *vertex*. We now apply to $V(x_0)$ some properties common to all polytopes.

In the representation (3) it may be possible to omit some closed half-spaces $\overline{G}_{x_i}$ without affecting the validity of the representation. By omitting as many half-spaces as possible we obtain an *irredundant representation*, which by suitable choice of notation we may take to be

$$V(x_0) = \cap_{i=1}^{l} \overline{G}_{x_i}$$

for some $l \leq m$. The intersections $V(x_0) \cap H_{x_i}$ $(1 \leq i \leq l)$ are then the distinct facets of $V(x_0)$. Any nonempty proper face of $V(x_0)$ is contained in a facet and is the intersection of those facets which contain it. Furthermore, any nonempty face of $V(x_0)$ is the convex hull of those vertices of $V(x_0)$ which it contains.

It follows that for each x_i $(1 \leq i \leq l)$ there is a vertex v_i of $V(x_0)$ such that

$$d(x_0,v_i) = d(x_i,v_i).$$

For $d(x_0,v) \leq d(x_i,v)$ for every vertex v of $V(x_0)$. Assume that $d(x_0,v) < d(x_i,v)$ for every vertex v of $V(x_0)$. Then the open half-space G_{x_i} contains all vertices v and hence also their convex hull $V(x_0)$. But this is a contradiction, since $V(x_0) \cap H_{x_i}$ is a facet of $V(x_0)$.

To illustrate these results take $X = \mathbb{Z}^n$ and $x_0 = O$. Then the Voronoi cell $V(O)$ is the cube consisting of all points $y = (\eta_1,...,\eta_n) \in \mathbb{R}^n$ with $|\eta_i| \leq 1/2$ $(i = 1,...,n)$. It has the minimal number $2n$ of facets.

In fact any lattice Λ in $\mathbb{R}^n$ is discrete and has the property (†). *For a lattice Λ we can restrict attention to the Voronoi cell $V(\Lambda): = V(O)$*, since an arbitrary Voronoi cell is obtained from it by a translation: $V(x_0) = V(O) + x_0$. The Voronoi cell of a lattice has additional properties. Since $x \in \Lambda$ implies $-x \in \Lambda$, $y \in V(\Lambda)$ implies $-y \in V(\Lambda)$. Also, if x_i is a lattice vector determining a facet of $V(\Lambda)$ and if $y \in V(\Lambda) \cap H_{x_i}$, then $\|y\| = \|y - x_i\|$. Since $x \in \Lambda$ implies $x_i - x \in \Lambda$, it follows that $y \in V(\Lambda) \cap H_{x_i}$ implies $x_i - y \in V(\Lambda) \cap H_{x_i}$. Thus *the Voronoi cell $V(\Lambda)$ and all its facets are centrosymmetric.*

In addition, any orthogonal transformation of $\mathbb{R}^n$ which maps onto itself the lattice Λ also maps onto itself the Voronoi cell $V(\Lambda)$. Furthermore the Voronoi cell $V(\Lambda)$ has volume $d(\Lambda)$, by Proposition 11, since the lattice translates of $V(\Lambda)$ form a tiling of $\mathbb{R}^n$.

We define a *facet vector* or 'relevant vector' of a lattice Λ to be a vector $x_i \in \Lambda$ such that $V(\Lambda) \cap H_{x_i}$ is a facet of the Voronoi cell $V(\Lambda)$. If $V(\Lambda)$ is contained in the closed ball $B_R = \{x \in \mathbb{R}^n : \|x\| \leq R\}$, then every facet vector x_i satisfies $\|x_i\| \leq 2R$. For, if $y \in V(\Lambda) \cap H_{x_i}$ then, by Schwarz's inequality (Chapter I, §4),

$$\|x_i\|^2 = 2(x_i, y) \leq 2\|x_i\| \|y\|.$$

The facet vectors were characterized by Voronoi (1908) in the following way:

PROPOSITION 13 *A nonzero vector $x \in \Lambda$ is a facet vector of the lattice $\Lambda \subseteq \mathbb{R}^n$ if and only if every vector $x' \in x + 2\Lambda$, except $\pm x$, satisfies $\|x'\| > \|x\|$.*

Proof Suppose first that $\|x\| < \|x'\|$ for all $x' \neq \pm x$ such that $(x' - x)/2 \in \Lambda$. If $z \in \Lambda$ and $x' = 2z - x$, then $(x' - x)/2 \in \Lambda$. Hence if $z \neq O, x$ then

$$\|x/2\| < \|z - x/2\|,$$

i.e. $x/2 \in G_z$. Since $\|x/2\| = \|x - x/2\|$, it follows that $x/2 \in V(\Lambda)$ and x is a facet vector.

Suppose next that there exists $x' \neq \pm x$ such that $w = (x' - x)/2 \in \Lambda$ and $\|x'\| \leq \|x\|$. Then also $z = (x' + x)/2 \in \Lambda$ and $z, w \neq O$. If $y \in \overline{G}_z \cap \overline{G}_{-w}$, then

$$2(z, y) \leq \|z\|^2, \quad -2(w, y) \leq \|w\|^2.$$

Hence, by the parallelogram law (Chapter I, §10),

$$2(x, y) = 2(z, y) - 2(w, y) \leq \|z\|^2 + \|w\|^2$$

$$= \|x\|^2/2 + \|x'\|^2/2 \leq \|x\|^2.$$

That is, $y \in \overline{G}_x$. Thus $\overline{G}_x$ is not needed to define $V(\Lambda)$ and x is not a facet vector. $\square$

Any lattice Λ contains a nonzero vector with minimal square-norm. Such a vector will be called a *minimal vector*. Its square-norm will be called the *minimum* of Λ and will be denoted by $m(\Lambda)$.

PROPOSITION 14 *If $\Lambda \subseteq \mathbb{R}^n$ is a lattice with minimum $m(\Lambda)$, then any nonzero vector in Λ with square-norm $< 2m(\Lambda)$ is a facet vector. In particular, any minimal vector is a facet vector.*

Proof Put $r = m(\Lambda)$ and let x be a nonzero vector in Λ with $\|x\|^2 < 2r$. If x is not a facet vector, there exists $y \neq \pm x$ with $(y - x)/2 \in \Lambda$ such that $\|y\| \leq \|x\|$. Since $(y \pm x)/2 \in \Lambda$, $\|x \pm y\|^2 \geq 4r$. Thus

$$4r \leq \|x\|^2 + \|y\|^2 \pm 2(x,y) < 4r \pm 2(x,y),$$

which is impossible. $\blacksquare$

PROPOSITION 15 *For any lattice $\Lambda \subseteq \mathbb{R}^n$, the number of facets of its Voronoi cell $V(\Lambda)$ is at most $2(2^n - 1)$.*

Proof Let $x_1,...,x_n$ be a basis for Λ. Then any vector $x \in \Lambda$ has a unique representation $x = x' + x''$, where $x' \in 2\Lambda$ and

$$x'' = \alpha_1 x_1 + ... + \alpha_n x_n,$$

with $\alpha_j \in \{0,1\}$ for $j = 1,...,n$. Thus the number of cosets of 2Λ in Λ is 2^n. But, by Proposition 13, each coset contains at most one pair $\pm y$ of facet vectors. Since 2Λ itself does not contain any facet vectors, the total number of facet vectors is at most $2(2^n - 1)$. $\blacksquare$

There exist lattices $\Lambda \subseteq \mathbb{R}^n$ for which the upper bound of Proposition 15 is attained, e.g. the lattice $\Lambda = \{Tz : z \in \mathbb{Z}^n\}$ with $T = I + \beta J$, where J is the $n \times n$ matrix every element of which is 1 and $\beta = \{(1 + n)^{1/2} - 1\}/n$.

PROPOSITION 16 *Every vector of a lattice $\Lambda \subseteq \mathbb{R}^n$ is an integral linear combination of facet vectors.*

Proof Let $b_1,...,b_m$ be the facet vectors of Λ and put

$$\Lambda' = \{x = \beta_1 b_1 + ... + \beta_m b_m : \beta_1,...,\beta_m \in \mathbb{Z}\}.$$

Evidently Λ' is a subgroup of $\mathbb{R}^n$ and actually a discrete subgroup, since $\Lambda' \subseteq \Lambda$. If Λ' were contained in a hyperplane of $\mathbb{R}^n$ any point on the line through the origin orthogonal to this hyperplane would belong to the Voronoi cell V of Λ, which is impossible because V is bounded. Hence Λ' contains n linearly independent vectors.

Thus Λ' is a sublattice of Λ. It follows that the Voronoi cell V of Λ is contained in the Voronoi cell V' of Λ'. But if $y \in V'$, then

$$\|y\| \leq \|b_i - y\|, \quad (i = 1,...,m)$$

and hence $y \in V$. Thus $V' = V$. Hence the Λ'-translates of V and the Λ-translates of V are both tilings of $\mathbb{R}^n$. Since $\Lambda' \subseteq \Lambda$, this is possible only if $\Lambda' = \Lambda$. $\blacksquare$

Since every integral linear combination of facet vectors is in the lattice, Proposition 16 implies

COROLLARY 17 *Distinct lattices in $\mathbb{R}^n$ have distinct Voronoi cells.* □

Proposition 16 does not say that the lattice has a basis of facet vectors. It is known that every lattice in $\mathbb{R}^n$ has a basis of facet vectors if $n \leq 6$, but if $n > 6$ this is still an open question. It is known also that every lattice in $\mathbb{R}^n$ has a basis of minimal vectors when $n \leq 4$ but, when $n > 4$, there are lattices with no such basis. In fact a lattice may have no basis of minimal vectors, even though every lattice vector is an integral linear combination of minimal vectors.

Lattices and their Voronoi cells have long been used in crystallography. An n-dimensional *crystal* may be defined mathematically to be a subset of $\mathbb{R}^n$ of the form

$$F + \Lambda = \{x + y \colon x \in F, y \in \Lambda\},$$

where F is a finite set and Λ a lattice. Crystals may be studied by means of their symmetry groups.

An *isometry* of $\mathbb{R}^n$ is an invertible affine transformation which leaves unaltered the Euclidean distance between any two points. For example, any orthogonal transformation is an isometry and so is a translation by an arbitrary vector v. Any isometry is the composite of a translation and an orthogonal transformation. The *symmetry group* of a set $X \subseteq \mathbb{R}^n$ is the group of all isometries of $\mathbb{R}^n$ which map X to itself.

We define an n-dimensional *crystallographic group* to be a group G of isometries of $\mathbb{R}^n$ such that the vectors corresponding to translations in G form an n-dimensional lattice. It is not difficult to show that a subset of $\mathbb{R}^n$ is an n-dimensional crystal if and only if it is discrete and its symmetry group is an n-dimensional crystallographic group.

It was shown by Bieberbach (1911) that a group G of isometries of $\mathbb{R}^n$ is a crystallographic group if and only if it is discrete and has a compact fundamental domain D, i.e. the sets $\{g(D) \colon g \in G\}$ form a tiling of $\mathbb{R}^n$. He could then show that the translations in a crystallographic group form a torsion-free abelian normal subgroup of finite index. He showed later (1912) that two crystallographic groups G_1, G_2 are isomorphic if and only if there exists an invertible affine transformation A such that $G_2 = A^{-1}G_1A$. With the aid of results of Minkowski and Jordan it follows that, for a given dimension n, there are only finitely many non-isomorphic crystallographic groups. These results provided a positive answer to the first part of the 18th Paris problem of Hilbert (1900).

The structure of physical crystals is analysed by means of the corresponding 3-dimensional crystallographic groups. A stronger concept than isomorphism is useful for such applications.

Two crystallographic groups G_1, G_2 may be said to be *properly isomorphic* if there exists an orientation-preserving invertible affine transformation A such that $G_2 = A^{-1}G_1A$. An isomorphism class of crystallographic groups either coincides with a proper isomorphism class or splits into two distinct proper isomorphism classes.

Fedorov (1891) showed that there are 17 isomorphism classes of 2-dimensional crystallographic groups, none of which splits. Collating earlier work of Sohncke (1879), Schoenflies (1889) and himself, Fedorov (1892) also showed that there are 219 isomorphism classes of 3-dimensional crystallographic groups, 11 of which split. More recently, Brown *et al.* (1978) have shown that there are 4783 isomorphism classes of 4-dimensional crystallographic groups, 112 of which split.

5 Densest packings

The result of Hermite, mentioned at the beginning of the chapter, can be formulated in terms of lattices instead of quadratic forms. For any real non-singular matrix T, the matrix

$$A = T^t T$$

is a real positive definite symmetric matrix. Conversely, by a principal axes transformation or otherwise, it may be seen that any real positive definite symmetric matrix A may be represented in this way.

Let Λ be the lattice

$$\Lambda = \{y = Tx \in \mathbb{R}^n : x \in \mathbb{Z}^n\}$$

and put

$$\gamma(\Lambda) = m(\Lambda)/d(\Lambda)^{2/n},$$

where $d(\Lambda)$ is the determinant and $m(\Lambda)$ the minimum of Λ. Evidently $\gamma(\rho\Lambda) = \gamma(\Lambda)$ for any $\rho > 0$. Hermite's result that there exists a positive constant c_n, depending only on n, such that $0 < x^t A x \le c_n (\det A)^{1/n}$ for some $x \in \mathbb{Z}^n$ may be restated in the form

$$\gamma(\Lambda) \le c_n.$$

Hermite's constant γ_n is defined to be the least positive constant c_n such that this inequality holds for all $\Lambda \subseteq \mathbb{R}^n$.

It may be shown that $\gamma_n{}^n$ is a rational number for each n. It follows from Proposition 2 that $\overline{\lim}_{n\to\infty} \gamma_n/n \le 2/\pi e$. Minkowski (1905) showed also that

$$\underline{\lim}_{n\to\infty}\gamma_n/n \geq 1/2\pi e = 0.0585... ,$$

and it is possible that actually $\lim_{n\to\infty} \gamma_n/n = 1/2\pi e$. The significance of Hermite's constant derives from its connection with lattice packings of balls, as we now explain.

Let Λ be a lattice in $\mathbb{R}^n$ and K a subset of $\mathbb{R}^n$ which is the closure of a nonempty open set G. We say that Λ gives a *lattice packing* for K if the family of translates $K + x$ ($x \in \Lambda$) is a packing of $\mathbb{R}^n$, i.e. if for any two distinct points $x,y \in \Lambda$ the interiors $G + x$ and $G + y$ are disjoint. This is the same as saying that Λ does not contain the difference of any two distinct points of the interior of K, since $g + x = g' + y$ if and only if $g' - g = x - y$. If K is a compact symmetric convex set with nonempty interior G, it is the same as saying that the interior of the set $2K$ contains no nonzero point of Λ, since in this case $g,g' \in G$ implies $(g' - g)/2 \in G$ and $2g = g - (-g)$.

The *density* of the lattice packing, i.e. the fraction of the total space which is occupied by translates of K, is readily shown to be $\lambda(K)/d(\Lambda)$. Hence the maximum density of any lattice packing for K is

$$\delta(K) = \lambda(K)/\Delta(2K) = 2^{-n}\lambda(K)/\Delta(K),$$

where $\Delta(K)$ is the critical determinant of K, as defined in §3. The use of the word 'maximum' is justified, since it will be shown in §6 that the infimum involved in the definition of critical determinant is attained.

We are interested in the special case of a closed ball: $K = B_\rho = \{x \in \mathbb{R}^n : \|x\| \leq \rho\}$. By what we have said, Λ gives a lattice packing for B_ρ if and only if the interior of $B_{2\rho}$ contains no nonzero point of Λ, i.e. if and only if $m(\Lambda)^{1/2} \geq 2\rho$. Hence

$$\delta(B_\rho) = \sup \{\lambda(B_\rho)/d(\Lambda): m(\Lambda)^{1/2} = 2\rho\}$$

$$= \kappa_n\rho^n \sup \{d(\Lambda)^{-1}: m(\Lambda)^{1/2} = 2\rho\},$$

where $\kappa_n = \pi^{n/2}/(n/2)!$ again denotes the volume of the unit ball in $\mathbb{R}^n$. By homogeneity it follows that

$$\delta_n := \delta(B_\rho) = 2^{-n}\kappa_n \sup_\Lambda \gamma(\Lambda)^{n/2},$$

where the supremum is now over all lattices $\Lambda \subseteq \mathbb{R}^n$, i.e. in terms of Hermite's constant γ_n,

$$\delta_n = 2^{-n}\kappa_n\gamma_n^{n/2}.$$

Thus γ_n, like δ_n, measures the densest lattice packing of balls. A lattice $\Lambda \subseteq \mathbb{R}^n$ for which $\gamma(\Lambda) = \gamma_n$, i.e. a critical lattice for a ball, will be called simply a *densest lattice*.

The densest lattice in $\mathbb{R}^n$ is known for each $n \le 8$, and is uniquely determined apart from isometries and scalar multiples. In fact these densest lattices are all examples of indecomposable root lattices. These terms will now be defined.

A lattice Λ is said to be *decomposable* if there exist additive subgroups Λ_1, Λ_2 of Λ, each containing a nonzero vector, such that $(x_1, x_2) = 0$ for all $x_1 \in \Lambda_1$ and $x_2 \in \Lambda_2$, and every vector in Λ is the sum of a vector in Λ_1 and a vector in Λ_2. Since Λ_1 and Λ_2 are necessarily discrete, they are lattices in the wide sense (i.e. they are not full-dimensional). We say also that Λ is the *orthogonal sum* of the lattices Λ_1 and Λ_2. The orthogonal sum of any finite number of lattices is defined similarly. A lattice is *indecomposable* if it is not decomposable.

The following result was first proved by Eichler (1952).

PROPOSITION 18 *Any lattice Λ is an orthogonal sum of finitely many indecomposable lattices, which are uniquely determined apart from order.*

Proof (i) Define a vector $x \in \Lambda$ to be 'decomposable' if there exist nonzero $x_1, x_2 \in \Lambda$ such that $x = x_1 + x_2$ and $(x_1, x_2) = 0$. We show first that every nonzero $x \in \Lambda$ is a sum of finitely many indecomposable vectors.

By definition, x is either indecomposable or is the sum of two nonzero orthogonal vectors in Λ. Both these vectors have square-norm less than the square-norm of x, and for each of them the same alternative presents itself. Continuing in this way, we must eventually arrive at indecomposable vectors, since there are only finitely many vectors in Λ with square-norm less than that of x.

(ii) If Λ is the orthogonal sum of finitely many lattices L_ν then, by the definition of an orthogonal sum, every indecomposable vector of Λ lies in one of the sublattices L_ν. Hence if two indecomposable vectors are not orthogonal, they lie in the same sublattice L_ν.

(iii) Call two indecomposable vectors x, x' 'equivalent' if there exist indecomposable vectors $x = x_0, x_1, \ldots, x_{k-1}, x_k = x'$ such that $(x_j, x_{j+1}) \ne 0$ for $0 \le j < k$. Clearly 'equivalence' is indeed an equivalence relation and thus the set of all indecomposable vectors is partitioned into equivalence classes $\mathcal{C}_\mu$. Two vectors from different equivalence classes are orthogonal and, if Λ is an orthogonal sum of lattices L_ν as in (ii), then two vectors from the same equivalence class lie in the same sublattice L_ν.

(iv) Let Λ_μ be the subgroup of Λ generated by the vectors in the equivalence class $\mathcal{C}_\mu$. Then, by (i), Λ is generated by the sublattices Λ_μ. Since, by (iii), Λ_μ is orthogonal to $\Lambda_{\mu'}$ if $\mu \ne \mu'$, Λ is actually the orthogonal sum of the sublattices Λ_μ. If Λ is an orthogonal sum of lattices L_ν as in (ii), then each Λ_μ is contained in some L_ν. It follows that each Λ_μ is indecomposable and that these indecomposable sublattices are uniquely determined apart from order. $\square$

Let Λ be a lattice in $\mathbb{R}^n$. If $\Lambda \subseteq \Lambda^*$, i.e. if $(x,y) \in \mathbb{Z}$ for all $x,y \in \Lambda$, then Λ is said to be *integral*. If (x,x) is an even integer for every $x \in \Lambda$, then Λ is said to be *even*. (It follows that an even lattice is also integral.) If Λ is even and every vector in Λ is an integral linear combination of vectors in Λ with square-norm 2, then Λ is said to be a *root lattice*.

Thus in a root lattice the minimal vectors have square-norm 2. It may be shown by a long, but elementary, argument that any root lattice has a basis of minimal vectors such that every minimal vector is an integral linear combination of the basis vectors with coefficients which are all nonnegative or all nonpositive. Such a basis will be called a *simple* basis. The facet vectors of a root lattice are precisely the minimal vectors, and hence its Voronoi cell is the set of all $y \in \mathbb{R}^n$ such that $(y,x) \leq 1$ for every minimal vector x.

Any root lattice is an orthogonal sum of indecomposable root lattices. It was shown by Witt (1941) that the indecomposable root lattices can be completely enumerated; they are all listed in Table 1. We give also their minimal vectors in terms of the canonical basis $e_1,...,e_n$ of $\mathbb{R}^n$.

$$
\begin{aligned}
A_n &= \{x = (\xi_0,\xi_1,...,\xi_n) \in \mathbb{Z}^{n+1} : \xi_0 + \xi_1 + ... + \xi_n = 0\} \quad (n \geq 1); \\
D_n &= \{x = (\xi_1,...,\xi_n) \in \mathbb{Z}^n : \xi_1 + ... + \xi_n \text{ even}\} \quad (n \geq 3); \\
E_8 &= D_8 \cup D_8{}^\dagger, \text{ where } D_8{}^\dagger = (1/2,1/2,...,1/2) + D_8; \\
E_7 &= \{x = (\xi_1,...,\xi_8) \in E_8 : \xi_7 = -\xi_8\}; \\
E_6 &= \{x = (\xi_1,...,\xi_8) \in E_8 : \xi_6 = \xi_7 = -\xi_8\}.
\end{aligned}
$$

Table 1: *Indecomposable root lattices*

The lattice A_n has $n(n+1)$ minimal vectors, namely the vectors $\pm(e_j - e_k)$ $(0 \leq j < k \leq n)$, and the vectors $e_0 - e_1, e_1 - e_2, ... , e_{n-1} - e_n$ form a simple basis. By calculating the determinant of $B^t B$, where B is the $(n+1) \times n$ matrix whose columns are the vectors of this simple basis, it may be seen that the determinant of the lattice A_n is $(n+1)^{1/2}$.

The lattice D_n has $2n(n-1)$ minimal vectors, namely the vectors $\pm e_j \pm e_k$ $(1 \leq j < k \leq n)$. The vectors $e_1 - e_2, e_2 - e_3, ... , e_{n-1} - e_n, e_{n-1} + e_n$ form a simple basis and hence the lattice D_n has determinant 2.

The lattice E_8 has 240 minimal vectors, namely the 112 vectors $\pm e_j \pm e_k$ $(1 \leq j < k \leq 8)$ and the 128 vectors $(\pm e_1 \pm ... \pm e_8)/2$ with an even number of minus signs. The vectors

$$
v_1 = (e_1 - e_2 - ... - e_7 + e_8)/2, \ v_2 = e_1 + e_2, \ v_3 = e_2 - e_1, \ v_4 = e_3 - e_2, \ ... \ , \ v_8 = e_7 - e_6,
$$

form a simple basis and hence the lattice has determinant 1.

The lattice E_7 has 126 minimal vectors, namely the 60 vectors $\pm e_j \pm e_k$ $(1 \le j < k \le 6)$, the vectors $\pm (e_7 - e_8)$ and the 64 vectors $\pm (\sum_{i=1}^{6} (\pm e_i) - e_7 + e_8)/2$ with an odd number of minus signs in the sum. The vectors $v_1,...,v_7$ form a simple basis and the lattice has determinant $\sqrt{2}$.

The lattice E_6 has 72 minimal vectors, namely the 40 vectors $\pm e_j \pm e_k$ $(1 \le j < k \le 5)$ and the 32 vectors $\pm (\sum_{i=1}^{5} (\pm e_i) - e_6 - e_7 + e_8)/2$ with an even number of minus signs in the sum. The vectors $v_1,...,v_6$ form a simple basis and the lattice has determinant $\sqrt{3}$.

We now return to lattice packings of balls. The densest lattices for $n \le 8$ are given in Table 2. These lattices were shown to be densest by Lagrange (1773) for $n = 2$, Gauss (1831) for $n = 3$, Korkine and Zolotareff (1872,1877) for $n = 4,5$ and Blichfeldt (1925,1926,1934) for $n = 6,7,8$.

n	Λ	γ_n	δ_n
1	A_1	1	1
2	A_2	$(4/3)^{1/2} = 1.1547..$	$3^{1/2}\pi/6 = 0.9068..$
3	D_3	$2^{1/3} = 1.2599..$	$2^{1/2}\pi/6 = 0.7404..$
4	D_4	$2^{1/2} = 1.4142..$	$\pi^2/16 = 0.6168..$
5	D_5	$8^{1/5} = 1.5157..$	$2^{1/2}\pi^2/30 = 0.4652..$
6	E_6	$(64/3)^{1/6} = 1.6653..$	$3^{1/2}\pi^3/144 = 0.3729..$
7	E_7	$(64)^{1/7} = 1.8114..$	$\pi^3/105 = 0.2952..$
8	E_8	2	$\pi^4/384 = 0.2536..$

Table 2: Densest lattices in $\mathbb{R}^n$

Although the densest lattice in $\mathbb{R}^n$ is unknown for every $n > 8$, there are plausible candidates in some dimensions. In particular, a lattice discovered by Leech (1967) is believed to be densest in 24 dimensions. This lattice may be constructed in the following way. Let p be a prime such that $p \equiv 3 \bmod 4$ and let H_n be the Hadamard matrix of order $n = p + 1$ constructed by Paley's method (see Chapter V, §2). The columns of the matrix

$$T = (n/4 + 1)^{-1/2} \begin{pmatrix} (n/4+1)I_n & H_n - I_n \\ 0_n & I_n \end{pmatrix}$$

generate a lattice in $\mathbb{R}^{2n}$. For $p = 3$ we obtain the root lattice E_8 and for $p = 11$ the Leech lattice Λ_{24}.

Leech's lattice may be characterized as the unique even lattice Λ in $\mathbb{R}^{24}$ with $d(\Lambda) = 1$ and $m(\Lambda) > 2$. It was shown by Conway (1969) that, if G is the group of all orthogonal transformations of $\mathbb{R}^{24}$ which map the Leech lattice Λ_{24} onto itself, then the factor group $G/\{\pm I_{24}\}$ is a finite simple group, and two more finite simple groups are easily obtained as (stabilizer) subgroups. These are three of the 26 sporadic simple groups which were mentioned in §7 of Chapter V.

Leech's lattice has 196560 minimal vectors of square-norm 4. Thus the packing of unit balls associated with Λ_{24} is such that each ball touches 196560 other balls. It has been shown that 196560 is the maximal number of nonoverlapping unit balls in $\mathbb{R}^{24}$ which can touch another unit ball and that, up to isometry, there is only one possible arrangement.

Similarly, since E_8 has 240 minimal vectors of square-norm 2, the packing of balls of radius $2^{-1/2}$ associated with E_8 is such that each ball touches 240 other balls. It has been shown that 240 is the maximal number of nonoverlapping balls of fixed radius in $\mathbb{R}^8$ which can touch another ball of the same radius and that, up to isometry, there is only one possible arrangement.

In general, one may ask what is the *kissing number* of $\mathbb{R}^n$, i.e. the maximal number of nonoverlapping unit balls in $\mathbb{R}^n$ which can touch another unit ball? The question, for $n = 3$, first arose in 1694 in a discussion between Newton, who claimed that the answer was 12, and Gregory, who said 13. It was first shown by Hoppe (1874) that Newton was right, but the arrangement of the 12 balls in $\mathbb{R}^3$ is *not* unique up to isometry. One possibility is to take the centres of the 12 balls to be the vertices of a regular icosahedron, the centre of which is the centre of the unit ball they touch.

The kissing number of $\mathbb{R}^1$ is clearly 2. It is not difficult to show that the kissing number of $\mathbb{R}^2$ is 6 and that the centres of the six unit balls must be the vertices of a regular hexagon, the centre of which is the centre of the unit ball they touch. For $n > 3$ the kissing number of $\mathbb{R}^n$ is unknown, except for the two cases $n = 8$ and $n = 24$ already mentioned.

6 Mahler's compactness theorem

It is useful to study not only individual lattices, but also the family $\mathcal{L}_n$ of all lattices in $\mathbb{R}^n$. A sequence of lattices $\Lambda_k \in \mathcal{L}_n$ will be said to *converge* to a lattice $\Lambda \in \mathcal{L}_n$, in symbols $\Lambda_k \to \Lambda$, if there exist bases $b_{k1},\dots,b_{kn}$ of Λ_k ($k = 1,2,\dots$) and a basis $b_1,\dots,b_n$ of Λ such that

$$b_{kj} \to b_j \ \text{as } k \to \infty \ (j = 1,\dots,n).$$

Evidently this implies that $d(\Lambda_k) \to d(\Lambda)$ as $k \to \infty$. Also, for any $x \in \Lambda$ there exist $x_k \in \Lambda_k$

such that $x_k \to x$ as $k \to \infty$. In fact if $x = \alpha_1 b_1 + ... + \alpha_n b_n$, where $\alpha_i \in \mathbb{Z}$ $(i = 1,...,n)$, we can take $x_k = \alpha_1 b_{k1} + ... + \alpha_n b_{kn}$.

It is not obvious from the definition that the limit of a sequence of lattices is uniquely determined, but this follows at once from the next result.

PROPOSITION 19 *Let Λ be a lattice in $\mathbb{R}^n$ and let $\{\Lambda_k\}$ be a sequence of lattices in $\mathbb{R}^n$ such that $\Lambda_k \to \Lambda$ as $k \to \infty$. If $x_k \in \Lambda_k$ and $x_k \to x$ as $k \to \infty$, then $x \in \Lambda$.*

Proof With the above notation,

$$x = \alpha_1 b_1 + ... + \alpha_n b_n,$$

where $\alpha_i \in \mathbb{R}$ $(i = 1,...,n)$, and similarly

$$x_k = \alpha_{k1} b_1 + ... + \alpha_{kn} b_n,$$

where $\alpha_{ki} \in \mathbb{R}$ and $\alpha_{ki} \to \alpha_i$ as $k \to \infty$ $(i = 1,...,n)$.

The linear transformation T_k of $\mathbb{R}^n$ which maps b_i to b_{ki} $(i = 1,...,n)$ can be written in the form

$$T_k = I - A_k,$$

where $A_k \to O$ as $k \to \infty$. It follows that

$$T_k^{-1} = (I - A_k)^{-1} = I + A_k + A_k^2 + ... = I + C_k,$$

where also $C_k \to O$ as $k \to \infty$. Hence

$$x_k = T_k^{-1} (\alpha_{k1} b_{k1} + ... + \alpha_{kn} b_{kn})$$

$$= (\alpha_{k1} + \eta_{k1}) b_{k1} + ... + (\alpha_{kn} + \eta_{kn}) b_{kn},$$

where $\eta_{ki} \to 0$ as $k \to \infty$ $(i = 1,...,n)$. But $\alpha_{ki} + \eta_{ki} \in \mathbb{Z}$ for every k. Letting $k \to \infty$, we obtain $\alpha_i \in \mathbb{Z}$. That is, $x \in \Lambda$. $\square$

It is natural to ask if the Voronoi cells of a convergent sequence of lattices also converge in some sense. The required notion of convergence is in fact older than the notion of convergence of lattices and applies to arbitrary compact subsets of $\mathbb{R}^n$.

The *Hausdorff distance* h(K,K') between two compact subsets K,K' of $\mathbb{R}^n$ is defined to be the infimum of all $\rho > 0$ such that every point of K is distant at most ρ from some point of K' and every point of K' is distant at most ρ from some point of K. We will show that this defines a metric, the *Hausdorff metric*, on the space of all compact subsets of $\mathbb{R}^n$.

Evidently

$$0 \le h(K,K') = h(K',K) < \infty.$$

Moreover $h(K,K') = 0$ implies $K = K'$. For if $x' \in K'$, there exist $x_k \in K$ such that $x_k \to x'$ and hence $x' \in K$, since K is closed. Thus $K' \subseteq K$, and similarly $K \subseteq K'$.

Finally we prove the triangle inequality

$$h(K,K'') \le h(K,K') + h(K',K'').$$

To simplify writing, put $\rho = h(K,K')$ and $\rho' = h(K',K'')$. For any $\varepsilon > 0$, if $x \in K$ there exist $x' \in K'$ such that $\|x - x'\| < \rho + \varepsilon$ and then $x'' \in K''$ such that $\|x' - x''\| < \rho' + \varepsilon$. Hence

$$\|x - x''\| < \rho + \rho' + 2\varepsilon.$$

Similarly, if $x'' \in K''$ there exists $x \in K$ for which the same inequality holds. But ε can be arbitrarily small.

The definition of Hausdorff distance can also be expressed in the form

$$h(K,K') = \inf \{\rho \ge 0: K \subseteq K' + B_\rho, K' \subseteq K + B_\rho\},$$

where $B_\rho = \{x \in \mathbb{R}^n: \|x\| \le \rho\}$. A sequence K_j of compact subsets of $\mathbb{R}^n$ *converges* to a compact subset K of $\mathbb{R}^n$ if $h(K_j,K) \to 0$ as $j \to \infty$.

It was shown by Hausdorff (1927) that any uniformly bounded sequence of compact subsets of $\mathbb{R}^n$ has a convergent subsequence. In particular, any uniformly bounded sequence of compact convex subsets of $\mathbb{R}^n$ has a subsequence which converges to a compact convex set. This special case of Hausdorff's result, which is all that we will later require, had already been established by Blaschke (1916) and is known as *Blaschke's selection principle*.

PROPOSITION 20 *Let $\{\Lambda_k\}$ be a sequence of lattices in $\mathbb{R}^n$ and let V_k be the Voronoi cell of Λ_k. If there exists a compact convex set V with nonempty interior such that $V_k \to V$ in the Hausdorff metric as $k \to \infty$, then V is the Voronoi cell of a lattice Λ and $\Lambda_k \to \Lambda$ as $k \to \infty$.*

Proof Since every Voronoi cell V_k is symmetric, so also is the limit V. Since V has nonempty interior, it follows that the origin is itself an interior point of V. Thus there exists $\delta > 0$ such that the ball $B_\delta = \{x \in \mathbb{R}^n: \|x\| \le \delta\}$ is contained in V.

It follows that $B_{\delta/2} \subseteq V_k$ for all large k. The quickest way to see this is to use *Rådström's cancellation law*, which says that if A,B,C are nonempty compact convex subsets of $\mathbb{R}^n$ such that $A + C \subseteq B + C$, then $A \subseteq B$. In the present case we have

$$B_{\delta/2} + B_{\delta/2} \subseteq B_\delta \subseteq V \subseteq V_k + B_{\delta/2} \text{ for } k \geq k_0,$$

and hence $B_{\delta/2} \subseteq V_k$ for $k \geq k_0$. Since also $V_k \subseteq V + B_{\delta/2}$ for all large k, there exists $R > 0$ such that $V_k \subseteq B_R$ for all k.

The lattice Λ_k has at most $2(2^n - 1)$ facet vectors, by Proposition 15. Hence, by restriction to a subsequence, we may assume that all Λ_k have the same number m of facet vectors. Let $x_{k1},...,x_{km}$ be the facet vectors of Λ_k and choose the notation so that $x_{k1},...,x_{kn}$ are linearly independent. Since they all lie in the ball B_{2R}, by restriction to a further subsequence we may assume that

$$x_{kj} \to x_j \text{ as } k \to \infty \ (j = 1,...,m).$$

Evidently $\|x_j\| \geq \delta$ $(j = 1,...,m)$ since, for $k \geq k_0$, all nonzero $x \in \Lambda_k$ have $\|x\| \geq \delta$.

The set Λ of all integral linear combinations of $x_1,...,x_m$ is certainly an additive subgroup of $\mathbb{R}^n$. Moreover Λ is discrete. For suppose $y \in \Lambda$ and $\|y\| < \delta$. We have

$$y = \alpha_1 x_1 + ... + \alpha_m x_m,$$

where $\alpha_j \in \mathbb{Z}$ $(j = 1,...,m)$. If

$$y_k = \alpha_1 x_{k1} + ... + \alpha_m x_{km},$$

then $y_k \to y$ as $k \to \infty$ and hence $\|y_k\| < \delta$ for all large k. Since $y_k \in \Lambda_k$, it follows that $y_k = O$ for all large k and hence $y = O$.

Since the lattice $\Lambda_k{}'$ with basis $x_{k1},...,x_{kn}$ is a sublattice of Λ_k, we have

$$d(\Lambda_k{}') \geq d(\Lambda_k) = \lambda(V_k) \geq \lambda(B_{\delta/2}).$$

Since $d(\Lambda_k{}') = |\det (x_{k1},...,x_{kn})|$, it follows that also

$$|\det (x_1,...,x_n)| \geq \lambda(B_{\delta/2}) > 0.$$

Thus the vectors $x_1,...,x_n$ are linearly independent. Hence Λ is a lattice.

Let $b_1,...,b_n$ be a basis of Λ. Then, by the definition of Λ,

$$b_i = \alpha_{i1} x_1 + ... + \alpha_{im} x_m,$$

where $\alpha_{ij} \in \mathbb{Z}$ $(1 \leq i \leq n, 1 \leq j \leq m)$. Put

$$b_{ki} = \alpha_{i1} x_{k1} + ... + \alpha_{im} x_{km}.$$

Then $b_{ki} \in \Lambda_k$ and $b_{ki} \to b_i$ as $k \to \infty$ $(i = 1,...,n)$. Hence, for all large k, the vectors $b_{k1},...,b_{kn}$ are linearly independent. We are going to show that $b_{k1},...,b_{kn}$ is a basis of Λ_k for all large k.

Since $b_1,...,b_n$ is a basis of Λ, we have

$$x_j = \gamma_{j1}b_1 + ... + \gamma_{jn}b_n,$$

where $\gamma_{ji} \in \mathbb{Z}$ $(1 \le i \le n, 1 \le j \le m)$. Hence, if

$$y_{kj} = \gamma_{j1}b_{k1} + ... + \gamma_{jn}b_{kn},$$

then $y_{kj} \in \Lambda_k$ and $y_{kj} \to x_j$ as $k \to \infty$ $(j = 1,...,m)$. Thus, for all large k,

$$\|y_{kj} - x_{kj}\| < \delta \quad (j = 1,...,m).$$

Since $y_{kj} - x_{kj} \in \Lambda_k$, this implies that, for all large k, $y_{kj} = x_{kj}$ $(j = 1,...,m)$. Thus every facet vector of Λ_k is an integral linear combination of $b_{k1},...,b_{kn}$ and hence, by Proposition 16, every vector of Λ_k is an integral linear combination of $b_{k1},...,b_{kn}$. Since $b_{k1},...,b_{kn}$ are linearly independent, this shows that they are a basis of Λ_k.

Let W be the Voronoi cell of Λ. We wish to show that $V = W$. If $v \in V$, then there exist $v_k \in V_k$ such that $v_k \to v$. Assume $v \notin W$. Then $\|v\| > \|z - v\|$ for some $z \in \Lambda$, and so

$$\|v\| = \|z - v\| + \rho,$$

where $\rho > 0$. There exist $z_k \in \Lambda_k$ such that $z_k \to z$. Then, for all large k,

$$\|v\| > \|z_k - v\| + \rho/2$$

and hence, for all large k,

$$\|v_k\| > \|z_k - v_k\|.$$

But this contradicts $v_k \in V_k$.

This proves that $V \subseteq W$. On the other hand, V has volume

$$\lambda(V) = \lim_{k \to \infty} \lambda(V_k) = \lim_{k \to \infty} d(\Lambda_k)$$

$$= \lim_{k \to \infty} |\det (b_{k1},...,b_{kn})|$$

$$= |\det (b_1,...,b_n)| = d(\Lambda) = \lambda(W).$$

It follows that every interior point of W is in V, and hence $W = V$. Corollary 17 now shows that the same lattice Λ would have been obtained if we had restricted attention to some other subsequence of $\{\Lambda_k\}$.

Let $a_1,...,a_n$ be any basis of Λ. We are going to show that, for the sequence $\{\Lambda_k\}$ originally given, there exist $a_{ki} \in \Lambda_k$ such that

$$a_{ki} \to a_i \text{ as } k \to \infty \quad (i = 1,...,n).$$

If this is not the case then, for some $i \in \{1,...,n\}$ and some $\varepsilon > 0$, there exist infinitely many k such that

$$\|x - a_i\| > \varepsilon \quad \text{for all } x \in \Lambda_k.$$

From this subsequence we could as before pick a further subsequence $\Lambda_{k_\nu} \to \Lambda$. Then every $y \in \Lambda$ is the limit of a sequence $y_\nu \in \Lambda_{k_\nu}$. Taking $y = a_i$, we obtain a contradiction.

It only remains to show that $a_{k1},...,a_{kn}$ is a basis of Λ_k for all large k. Since

$$\lim_{k\to\infty} |\det (a_{k1},...,a_{kn})| = |\det (a_1,...,a_n)|$$

$$= d(\Lambda) = \lambda(V) = \lim_{k\to\infty} \lambda(V_k),$$

for all large k we must have

$$0 < |\det (a_{k1},...,a_{kn})| < 2\lambda(V_k).$$

But if $a_{k1},...,a_{kn}$ were not a basis of Λ_k for all large k, then for infinitely many k we would have

$$|\det (a_{k1},...,a_{kn})| \geq 2d(\Lambda_k) = 2\lambda(V_k). \quad \blacksquare$$

Proposition 20 has the following counterpart:

PROPOSITION 21 *Let $\{\Lambda_k\}$ be a sequence of lattices in $\mathbb{R}^n$ and let V_k be the Voronoi cell of Λ_k. If there exists a lattice Λ such that $\Lambda_k \to \Lambda$ as $k \to \infty$, and if V is the Voronoi cell of Λ, then $V_k \to V$ in the Hausdorff metric as $k \to \infty$.*

Proof By hypothesis, there exists a basis $b_1,...,b_n$ of Λ and a basis $b_{k1},...,b_{kn}$ of each Λ_k such that $b_{kj} \to b_j$ as $k \to \infty$ ($j = 1,...,n$). Choose $R > 0$ so that the fundamental parallelotope of Λ is contained in the ball $B_R = \{x \in \mathbb{R}^n : \|x\| \leq R\}$. Then, for all $k \geq k_0$, the fundamental parallelotope of Λ_k is contained in the ball B_{2R}. It follows that, for all $k \geq k_0$, every point of $\mathbb{R}^n$ is distant at most $2R$ from some point of Λ_k and hence $V_k \subseteq B_{2R}$.

Consequently, by Blaschke's selection principle, the sequence $\{V_k\}$ has a subsequence $\{V_{k_\nu}\}$ which converges in the Hausdorff metric to a compact convex set W. Moreover,

$$\lambda(W) = \lim_{\nu\to\infty} \lambda(V_{k_\nu}) = \lim_{\nu\to\infty} d(\Lambda_{k_\nu}) = d(\Lambda) > 0.$$

Consequently, since W is convex, it has nonempty interior. It now follows from Proposition 20 that $W = V$.

Thus any convergent subsequence of $\{V_k\}$ has the same limit V. If the whole sequence $\{V_k\}$ did not converge to V, there would exist $\rho > 0$ and a subsequence $\{V_{k_\nu}\}$ such that

$$h(V_{k_\nu},V) \geq \rho \quad \text{for all } \nu.$$

By the Blaschke selection principle again, this subsequence would itself have a convergent subsequence. Since its limit must be V, this yields a contradiction. $\square$

Suppose $\Lambda_k \in \mathscr{L}_n$ and $\Lambda_k \to \Lambda$ as $k \to \infty$. We will show that not only $d(\Lambda_k) \to d(\Lambda)$, but also $m(\Lambda_k) \to m(\Lambda)$ as $k \to \infty$. Since every $x \in \Lambda$ is the limit of a sequence $x_k \in \Lambda_k$, we must have $\overline{\lim}_{k \to \infty} m(\Lambda_k) \leq m(\Lambda)$. On the other hand, by Proposition 19, if $x_k \in \Lambda_k$ and $x_k \to x$, then $x \in \Lambda$. It follows that $\underline{\lim}_{k \to \infty} m(\Lambda_k) \geq m(\Lambda)$.

Suppose now that a subset $\mathscr{F}$ of $\mathscr{L}_n$ has the property that any infinite sequence Λ_k of lattices in $\mathscr{F}$ has a convergent subsequence. Then there exist positive constants ρ, σ such that

$$m(\Lambda) \geq \rho^2, \ d(\Lambda) \leq \sigma \quad \text{for all } \Lambda \in \mathscr{F}.$$

For otherwise there would exist a sequence Λ_k of lattices in $\mathscr{F}$ such that either $m(\Lambda_k) \to 0$ or $d(\Lambda_k) \to \infty$, and this sequence could have no convergent subsequence.

We now prove the fundamental *compactness theorem* of Mahler (1946), which says that this necessary condition on $\mathscr{F}$ is also sufficient.

PROPOSITION 22 *If $\{\Lambda_k\}$ is a sequence of lattices in $\mathbb{R}^n$ such that*

$$m(\Lambda_k) \geq \rho^2, \ d(\Lambda_k) \leq \sigma \ \text{ for all } k,$$

where ρ, σ are positive constants, then the sequence $\{\Lambda_k\}$ has a convergent subsequence.

Proof Let V_k denote the Voronoi cell of Λ_k. We show first that the ball $B_{\rho/2} = \{x \in \mathbb{R}^n : \|x\| \leq \rho/2\}$ is contained in every Voronoi cell V_k. In fact if $\|x\| \leq \rho/2$ then, for every nonzero $y \in \Lambda_k$,

$$\|x - y\| \geq \|y\| - \|x\| \geq \rho - \rho/2 = \rho/2 \geq \|x\|,$$

and hence $x \in V_k$.

Let v_k be a point of V_k which is furthest from the origin. Then V_k contains the convex hull C_k of the set $v_k \cup B_{\rho/2}$. Since the volume of V_k is bounded above by σ, so also is the volume of C_k. But this implies that the sequence v_k is bounded. Thus there exists $R > 0$ such that the ball B_R contains every Voronoi cell V_k.

By Blaschke's selection principle, the sequence $\{V_k\}$ has a subsequence $\{V_{k_v}\}$ which converges in the Hausdorff metric to a compact convex set V. Since $B_{\rho/2} \subseteq V$, it follows from Proposition 20 that $\Lambda_{k_v} \to \Lambda$, where Λ is a lattice with Voronoi cell V. $\square$

To illustrate the utility of Mahler's compactness theorem, we now show that, as stated in §3, any compact symmetric convex set K with nonempty interior has a critical lattice.

By the definition of the critical determinant $\Delta(K)$, there exists a sequence Λ_k of lattices with no nonzero points in the interior of K such that $d(\Lambda_k) \to \Delta(K)$ as $k \to \infty$. Since K contains a ball B_ρ with radius $\rho > 0$, we have $m(\Lambda_k) \geq \rho^2$ for all k. Hence, by Proposition 22, there is a subsequence Λ_{k_v} which converges to a lattice Λ as $v \to \infty$. Since every point of Λ is a limit of points of Λ_{k_v}, no nonzero point of Λ lies in the interior of K. Furthermore,

$$d(\Lambda) \,=\, \lim_{v \to \infty} d(\Lambda_{k_v}) \,=\, \Delta(K),$$

and hence Λ is a critical lattice for K.

7 Further remarks

The geometry of numbers is treated more extensively in Cassels [11], Erdős *et al.* [22] and Gruber and Lekkerkerker [27]. Minkowski's own account is available in [42]. Numerous references to the earlier literature are given in Keller [34]. Lagarias [36] gives an overview of lattice theory. For a simple proof that the indicator function of a convex set is Riemann integrable, see Szabo [57].

Diophantine approximation is studied in Cassels [12], Koksma [35] and Schmidt [50]. Minkowski's result that the discriminant of an algebraic number field other than $\mathbb{Q}$ has absolute value greater than 1 is proved in Narkiewicz [44], for example.

Minkowski's theorem on successive minima is proved in Bambah *et al.* [3]. For the results of Banaszczyk mentioned in §3, see [4] and [5]. Sharp forms of Siegel's lemma are proved not only in Bombieri and Vaaler [7], but also in Matveev [40]. The result of Gillet and Soulé appeared in [25]. Some interesting results and conjectures concerning the product $\lambda(K)\lambda(K^*)$ are described on pp. 425-427 of Schneider [51].

An algorithm of Lovász, which first appeared in Lenstra, Lenstra and Lovász [38], produces in finitely many steps a basis for a lattice Λ in $\mathbb{R}^n$ which is 'reduced'. Although the first vector of a reduced basis is in general not a minimal vector, it has square-norm at most $2^{n-1} m(\Lambda)$. This suffices for many applications and the algorithm has been used to solve a number of apparently unrelated computational problems, such as factoring polynomials in $\mathbb{Q}[t]$, integer linear programming and simultaneous Diophantine approximation. There is an account of the basis reduction algorithm in Schrijver [52]. The algorithmic geometry of numbers is surveyed in Kannan [33].

Mahler [39] has established an analogue of the geometry of numbers for formal Laurent series with coefficients from an arbitrary field F, the roles of $\mathbb{Z}, \mathbb{Q}$ and $\mathbb{R}$ being taken by $F[t]$,

$F(t)$ and $F((t))$. In particular, Eichler [19] has shown that the Riemann–Roch theorem for algebraic functions may be derived by geometry of numbers arguments.

There is also a generalization of Minkowski's lattice point theorem to locally compact groups, with Haar measure taking the place of volume; see Chapter 2 (Lemma 1) of Weil [60].

Voronoi *diagrams* and their uses are surveyed in Aurenhammer [1]. Proofs of the basic properties of polytopes referred to in §4 may be found in Brøndsted [9] and Coppel [15]. Planar tilings are studied in detail in Grünbaum and Shephard [28].

Mathematical crystallography is treated in Schwarzenberger [53] and Engel [21]. For the physicist's point of view, see Burckhardt [10], Janssen [32] and Birman [6]. There is much theoretical information, in addition to tables, in [31].

For Bieberbach's theorems, see Vince [59], Charlap [13] and Milnor [41]. Various equivalent forms for the definitions of crystal and crystallographic group are given in Dolbilin *et al.* [17]. It is shown in Charlap [13] that crystallographic groups may be abstractly characterized as groups containing a finitely generated maximal abelian torsion-free subgroup of finite index. (An abelian group is *torsion-free* if only the identity element has finite order.) The fundamental group of a compact flat Riemannian manifold is a torsion-free crystallographic group and all torsion-free crystallographic groups may be obtained in this way. For these connections with differential geometry, see Wolf [61] and Charlap [13].

In more than 4 dimensions the complete enumeration of all crystallographic groups is no longer practicable. However, algorithms for deciding if two crystallographic groups are equivalent in some sense have been developed by Opgenorth *et al.* [45]. An interesting subset of all crystallographic groups consists of those generated by reflections in hyperplanes, since Stiefel (1941/2) showed that they are in 1-1 correspondence with the compact simply-connected semi-simple Lie groups. See the 'Note historique' in Bourbaki [8].

There has recently been considerable interest in tilings of $\mathbb{R}^n$ which, although not lattice tilings, consist of translates of finitely many n-dimensional polytopes. The first example, in $\mathbb{R}^2$, due to Penrose (1974), was explained more algebraically by de Bruijn (1981). A substantial generalization of de Bruijn's construction was given by Katz and Duneau (1986), who showed that many such 'quasiperiodic' tilings may be obtained by a method of cut and projection from ordinary lattices in a higher-dimensional space. The subject gained practical significance with the discovery by Shechtman *et al.* (1984) that the diffraction pattern of an alloy of aluminium and magnesium has icosahedral symmetry, which is impossible for a crystal. Many other 'quasicrystals' have since been found. The papers referred to are reproduced, with others, in Steinhardt and Ostlund [56]. The mathematical theory of quasicrystals is surveyed in Le *et al.* [37].

Skubenko [54] has given an upper bound for Hermite's constant γ_n. Somewhat sharper bounds are known, but they have the same asymptotic behaviour and the proofs are much more complicated. A lower bound for γ_n was obtained with a new method by Ball [2].

For the densest lattices in $\mathbb{R}^n$ ($n \leq 8$), see Ryshkov and Baranovskii [49]. The enumeration of all root lattices is carried out in Ebeling [18]. (A more general problem is treated in Chap. 3 of Humphreys [30] and in Chap. 6 of Bourbaki [8].) For the Voronoi cells of root lattices, see Chap. 21 of Conway and Sloane [14] and Moody and Patera [43]. For the *Dynkin diagrams* associated with root lattices, see also Reiten [47].

Rajan and Shende [46] characterize root lattices as those lattices for which every facet vector is a minimal vector, but their definition of root lattice is not that adopted here. Their argument shows that if every facet vector of a lattice is a minimal vector then, after scaling to make the minimal vectors have square-norm 2, it is a root lattice in our sense.

There is a fund of information about lattice packings of balls in Conway and Sloane [14]. See also Thompson [58] for the Leech lattice and Coxeter [16] for the kissing number problem.

We have restricted attention to lattice packings and, in particular, to lattice packings of balls. Lattice packings of other convex bodies are discussed in the books on geometry of numbers cited above. Non-lattice packings have also received much attention. The notion of density is not so intuitive in this case and it should be realized that the density is unaltered if finitely many sets are removed from the packing.

Packings and coverings are discussed in the texts of Rogers [48] and Fejes Tóth [23],[24]. For packings of balls, see also Zong [62]. Sloane [55] and Elkies [20] provide introductions to the connections between lattice packings of balls and coding theory.

The third part of Hilbert's 18th problem, which is surveyed in Milnor [41], deals with the densest lattice or non-lattice packing of balls in $\mathbb{R}^n$. It is known that, for $n = 2$, the densest lattice packing is also a densest packing. The original proof by Thue (1882/1910) was incomplete, but a complete proof was given by L. Fejes Tóth (1940). The famous *Kepler conjecture* asserts that, also for $n = 3$, the densest lattice packing is a densest packing. A computer-aided proof has recently been announced by Hales [29]. It is unknown if the same holds for any $n > 3$.

Propositions 20 and 21 are due to Groemer [26], and are of interest quite apart from the application to Mahler's compactness theorem. Other proofs of the latter are given in Cassels [11] and Gruber and Lekkerkerker [27]. Blaschke's selection principle and Rådström's cancellation law are proved in [15] and [51], for example.

8 Selected references

[1] F. Aurenhammer, Voronoi diagrams – a survey of a fundamental geometric data structure, *ACM Computing Surveys* **23** (1991), 345-405.

[2] K. Ball, A lower bound for the optimal density of lattice packings, *Internat. Math. Res. Notices* 1992, no. 10, 217-221.

[3] R.P. Bambah, A.C. Woods and H. Zassenhaus, Three proofs of Minkowski's second inequality in the geometry of numbers, *J. Austral. Math. Soc.* **5** (1965), 453-462.

[4] W. Banaszczyk, New bounds in some transference theorems in the geometry of numbers, *Math. Ann.* **296** (1993), 625-635.

[5] W. Banaszczyk, Inequalities for convex bodies and polar reciprocal lattices in $\mathbb{R}^n$. II. Application of K-convexity, *Discrete Comput. Geom.* **16** (1996), 305-311.

[6] J.L. Birman, *Theory of crystal space groups and lattice dynamics,* Springer-Verlag, Berlin, 1984. [Original edition in *Handbuch der Physik*, 1974]

[7] E. Bombieri and J. Vaaler, On Siegel's lemma, *Invent. Math.* **73** (1983), 11-32.

[8] N. Bourbaki, *Groupes et algèbres de Lie, Chapitres 4,5 et 6,* Masson, Paris, 1981.

[9] A. Brøndsted, *An introduction to convex polytopes*, Springer-Verlag, New York, 1983.

[10] J.J. Burckhardt, *Die Bewegungsgruppen der Kristallographie*, 2nd ed., Birkhäuser, Basel, 1966.

[11] J.W.S. Cassels, *An introduction to the geometry of numbers*, corrected reprint, Springer-Verlag, Berlin, 1997. [Original edition, 1959]

[12] J.W.S. Cassels, *An introduction to Diophantine approximation*, Cambridge University Press, 1957.

[13] L.S. Charlap, *Bieberbach groups and flat manifolds*, Springer-Verlag, New York, 1986.

[14] J.H. Conway and N.J.A. Sloane, *Sphere packings, lattices and groups*, 3rd ed., Springer-Verlag, New York, 1999.

[15] W.A. Coppel, *Foundations of convex geometry*, Cambridge University Press, 1998.

[16] H.S.M. Coxeter, An upper bound for the number of equal nonoverlapping spheres that can touch another of the same size, *Convexity* (ed. V. Klee), pp. 53-71, Proc. Symp. Pure Math. **7**, Amer. Math. Soc., Providence, Rhode Island, 1963.

[17] N.P. Dolbilin, J.C. Lagarias and M. Senechal, Multiregular point systems, *Discrete Comput. Geom.* **20** (1998), 477-498.

[18] W. Ebeling, *Lattices and codes*, Vieweg, Braunschweig, 1994.

[19] M. Eichler, Ein Satz über Linearformen in Polynombereichen, *Arch. Math.* **10** (1959), 81-84.

[20] N.D. Elkies, Lattices, linear codes, and invariants, *Notices Amer. Math. Soc.* **47** (2000), 1238-1245 and 1382-1391.

[21] P. Engel, Geometric crystallography, *Handbook of convex geometry* (ed. P.M. Gruber and J.M. Wills), Volume B, pp. 989-1041, North-Holland, Amsterdam, 1993. (The same volume contains several other useful survey articles relevant to this chapter.)

[22] P. Erdös, P.M. Gruber and J. Hammer, *Lattice points*, Longman, Harlow, Essex, 1989.

[23] L. Fejes Tóth, *Regular Figures*, Pergamon, Oxford, 1964.

[24] L. Fejes Tóth, *Lagerungen in der Ebene auf der Kugel und im Raum*, 2nd ed., Springer-Verlag, Berlin, 1972.

[25] H. Gillet and C. Soulé, On the number of lattice points in convex symmetric bodies and their duals, *Israel J. Math.* **74** (1991), 347-357.

[26] H. Groemer, Continuity properties of Voronoi domains, *Monatsh. Math.* **75** (1971), 423-431.

[27] P.M. Gruber and C.G. Lekkerkerker, *Geometry of numbers*, 2nd ed., North-Holland, Amsterdam, 1987.

[28] B. Grünbaum and G.C. Shephard, *Tilings and patterns*, Freeman, New York, 1987.

[29] T.C. Hales, Cannonballs and honeycombs, *Notices Amer. Math. Soc.* **47** (2000), 440-449.

[30] J.E. Humphreys, *Introduction to Lie algebras and representation theory*, Springer-Verlag, New York, 1972.

[31] *International tables for crystallography, Vols. A-C*, Kluwer, Dordrecht, 1983-1993.

[32] T. Janssen, *Crystallographic groups*, North-Holland, Amsterdam, 1973.

[33] R. Kannan, Algorithmic geometry of numbers, *Annual review of computer science* **2** (1987), 231-267.

[34] O.-H. Keller, *Geometrie der Zahlen*, Enzyklopädie der mathematischen Wissenschaften I-2, 27, Teubner, Leipzig, 1954.

[35] J.F. Koksma, *Diophantische Approximationen*, Springer-Verlag, Berlin, 1936. [Reprinted Chelsea, New York, 1950]

[36] J.C. Lagarias, Point lattices, *Handbook of Combinatorics* (ed. R. Graham, M. Grötschel and L. Lovász), Vol. I, pp. 919-966, Elsevier, Amsterdam, 1995.

[37] T.Q.T. Le, S.A. Piunikhin and V.A. Sadov, The geometry of quasicrystals, *Russian Math. Surveys* **48** (1993), no. 1, 37-100.

[38] A.K. Lenstra, H.W. Lenstra and L. Lovász, Factoring polynomials with rational coefficients, *Math. Ann.* **261** (1982), 515-534.

[39] K. Mahler, An analogue to Minkowski's geometry of numbers in a field of series, *Ann. of Math.* **42** (1941), 488-522.

[40] E.M. Matveev, On linear and multiplicative relations, *Math. USSR-Sb.* **78** (1994), 411-425.

[41] J. Milnor, Hilbert's Problem 18: On crystallographic groups, fundamental domains, and on sphere packing, *Mathematical developments arising from Hilbert problems* (ed. F.E. Browder), pp. 491-506, Proc. Symp. Pure Math. **28**, Part 2, Amer. Math. Soc., Providence, Rhode Island, 1976.

[42] H. Minkowski, *Geometrie der Zahlen*, Teubner, Leipzig, 1896. [Reprinted Chelsea, New York, 1953]

[43] R.V. Moody and J. Patera, Voronoi and Delaunay cells of root lattices: classification of their faces and facets by Coxeter–Dynkin diagrams, *J. Phys. A* **25** (1992), 5089-5134.

[44] W. Narkiewicz, *Elementary and analytic theory of algebraic numbers*, 2nd ed., Springer-Verlag, Berlin, 1990.

[45] J. Opgenorth, W. Plesken and T. Schulz, Crystallographic algorithms and tables, *Acta Cryst. A* **54** (1998), 517-531.

[46] D.S. Rajan and A.M. Shende, A characterization of root lattices, *Discrete Math.* **161** (1996), 309-314.

[47] I. Reiten, Dynkin diagrams and the representation theory of Lie algebras, *Notices Amer. Math. Soc.* **44** (1997), 546-556.

[48] C.A. Rogers, *Packing and covering*, Cambridge University Press, 1964.

[49] S.S. Ryshkov and E.P. Baranovskii, Classical methods in the theory of lattice packings, *Russian Math. Surveys* **34** (1979), no. 4, 1-68.

[50] W.M. Schmidt, *Diophantine approximation*, Lecture Notes in Mathematics **785**, Springer-Verlag, Berlin, 1980.

[51] R. Schneider, *Convex bodies: the Brunn–Minkowski theory*, Cambridge University Press, 1993.

[52] A. Schrijver, *Theory of linear and integer programming*, corrected reprint, Wiley, Chichester, 1989.

[53] R.L.E. Schwarzenberger, *N-dimensional crystallography*, Pitman, London, 1980.

[54] B.F. Skubenko, A remark on an upper bound on the Hermite constant for the densest lattice packings of spheres, *J. Soviet Math.* **18** (1982), 960-961.

[55] N.J.A. Sloane, The packing of spheres, *Scientific American* **250** (1984), 92-101.

[56] P.J. Steinhardt and S. Ostlund (ed.), *The physics of quasicrystals*, World Scientific, Singapore, 1987.

[57] L. Szabo, A simple proof for the Jordan measurability of convex sets, *Elem. Math.* **52** (1997), 84-86.

[58] T.M. Thompson, *From error-correcting codes through sphere packings to simple groups*, Carus Mathematical Monograph No. 21, Mathematical Association of America, 1983.

[59] A. Vince, Periodicity, quasiperiodicity and Bieberbach's theorem, *Amer. Math. Monthly* **104** (1997), 27-35.

[60] A. Weil, *Basic number theory*, 2nd ed., Springer-Verlag, Berlin, 1973.

[61] J.A. Wolf, *Spaces of constant curvature*, 3rd ed., Publish or Perish, Boston, Mass., 1974.

[62] C. Zong, *Sphere packings*, Springer-Verlag, New York, 1999.

IX
The number of prime numbers

1 Finding the problem

It was already shown in Euclid's *Elements* (Book IX, Proposition 20) that there are infinitely many prime numbers. The proof is a model of simplicity: let $p_1,...,p_n$ be any finite set of primes and consider the integer $N = p_1 \cdots p_n + 1$. Then $N > 1$ and each prime divisor p of N is distinct from $p_1,...,p_n$, since $p = p_j$ would imply that p divides $N - p_1 \cdots p_n = 1$. It is worth noting that the same argument applies if we take $N = p_1^{\alpha_1} \cdots p_n^{\alpha_n} + 1$, with any positive integers $\alpha_1,...,\alpha_n$.

Euler (1737) gave an analytic proof of Euclid's result, which provides also quantitative information about the distribution of primes:

PROPOSITION 1 *The series* $\sum_p 1/p$, *where p runs through all primes, is divergent.*

Proof For any prime p we have

$$(1 - 1/p)^{-1} = 1 + p^{-1} + p^{-2} + ...$$

and hence

$$\prod_{p \leq x} (1 - 1/p)^{-1} = \prod_{p \leq x} (1 + p^{-1} + p^{-2} + ...) > \sum_{n \leq x} 1/n,$$

since any positive integer $n \leq x$ is a product of powers of primes $p \leq x$. Since

$$\sum_{n \leq x} 1/n > \sum_{n \leq x} \int_n^{n+1} dt/t > \log x,$$

it follows that

$$\prod_{p \leq x} (1 - 1/p)^{-1} > \log x.$$

On the other hand, since any positive integer has a unique representation as a product of prime powers,

$$\prod_{p \leq x} (1 - 1/p^2)^{-1} = \prod_{p \leq x} (1 + p^{-2} + p^{-4} + ...) \leq \sum_{n=1}^{\infty} 1/n^2 = :S,$$

and

$$S = 1 + \sum_{n=1}^{\infty} 1/(n+1)^2 < 1 + \sum_{n=1}^{\infty} \int_n^{n+1} dt/t^2 = 1 + \int_1^{\infty} dt/t^2 = 2.$$

(In fact $S = \pi^2/6$, as Euler (1735) also showed.) Since $1 - 1/p^2 = (1 - 1/p)(1 + 1/p)$, and since $1 + x \leq e^x$, it follows that

$$\prod_{p \leq x} (1 - 1/p)^{-1} \leq S \prod_{p \leq x} (1 + 1/p) < S \, e^{\sum_{p \leq x} 1/p}.$$

Combining this with the inequality of the previous paragraph, we obtain

$$\sum_{p \leq x} 1/p > \log \log x - \log S. \quad \square$$

Since the series $\sum_{n=1}^{\infty} 1/n^2$ is convergent, Proposition 1 says that 'there are more primes than squares'. Proposition 1 can be made more precise. It was shown by Mertens (1874) that

$$\sum_{p \leq x} 1/p = \log \log x + c + O(1/\log x),$$

where c is a constant ($c = 0.261497...$).

Let $\pi(x)$ denote the number of primes $\leq x$:

$$\pi(x) = \sum_{p \leq x} 1.$$

It may be asked whether $\pi(x)$ has some simple asymptotic behaviour as $x \to \infty$. It is not obvious that this is a sensible question. The behaviour of $\pi(x)$ for small values of x is quite irregular. Moreover the sequence of positive integers contains arbitrarily large blocks without primes; for example, none of the integers

$$n! + 2, \, n! + 3, \, ... \, , \, n! + n$$

is a prime. Indeed Euler (1751) expressed the view that "there reigns neither order nor rule" in the sequence of prime numbers.

From an analysis of tables of primes Legendre (1798) was led to conjecture that, for large values of x, $\pi(x)$ is given approximately by the formula

$$x/(A \log x - B),$$

where A,B are constants and $\log x$ again denotes the natural logarithm of x (i.e., to the base e). In 1808 he proposed the specific values $A = 1, B = 1.08366$.

The first significant results on the asymptotic behaviour of $\pi(x)$ were obtained by Chebyshev (1849). He proved that, for each positive integer n,

$$\underline{\lim}_{x \to \infty} \left(\pi(x) - \int_2^x dt/\log t \right) \log^n x / x \leq 0 \leq \overline{\lim}_{x \to \infty} \left(\pi(x) - \int_2^x dt/\log t \right) \log^n x / x,$$

where $\log^n x = (\log x)^n$. By repeatedly integrating by parts it may be seen that, for each positive integer n,

$\int_2^x dt/\log t = \{1 + 1!/\log x + 2!/\log^2 x + ... + (n-1)!/\log^{n-1} x\}x/\log x + n! \int_2^x dt/\log^{n+1} t + c_n,$

where c_n is a constant. Moreover, using the *Landau order symbol* defined under 'Notations',

since

$$\int_2^x dt/\log^{n+1} t = O(x/\log^{n+1} x),$$

$$\int_2^{x^{1/2}} dt/\log^{n+1} t < x^{1/2}/\log^{n+1} 2, \quad \int_{x^{1/2}}^x dt/\log^{n+1} t < 2^{n+1} x/\log^{n+1} x.$$

Thus Chebyshev's result shows that $A = B = 1$ are the best possible values for a formula of Legendre's type and suggests that

$$Li(x) = \int_2^x dt/\log t$$

is a better approximation to $\pi(x)$.

If we interpret this approximation as an asymptotic formula, then it implies that $\pi(x) \log x / x \to 1$ as $x \to \infty$, i.e., using another *Landau order symbol*,

$$\pi(x) \sim x/\log x . \tag{1}$$

The validity of the relation (1) is now known as the *prime number theorem*. If the n-th prime is denoted by p_n, then the prime number theorem can also be stated in the form $p_n \sim n \log n$:

PROPOSITION 2 $\pi(x) \sim x/\log x$ *if and only if* $p_n \sim n \log n$.

Proof If $\pi(x) \log x / x \to 1$, then

$$\log \pi(x) + \log \log x - \log x \to 0$$

and hence

$$\log \pi(x)/\log x \to 1.$$

Consequently

$$\pi(x) \log \pi(x) / x = \pi(x) \log x / x \cdot \log \pi(x) / \log x \to 1.$$

Since $\pi(p_n) = n$, this shows that $p_n \sim n \log n$.

Conversely, suppose $p_n/n \log n \to 1$. Since

$$(n+1)\log(n+1)/n \log n = (1 + 1/n)\{1 + \log(1 + 1/n)/\log n\} \to 1,$$

it follows that $p_{n+1}/p_n \to 1$. Furthermore

$$\log p_n - \log n - \log \log n \to 0,$$

and hence

$$\log p_n/\log n \to 1.$$

If $p_n \le x < p_{n+1}$, then $\pi(x) = n$ and

Since
$$n \log p_n / p_{n+1} \le \pi(x) \log x / x \le n \log p_{n+1} / p_n.$$

$$n \log p_n / p_{n+1} = p_n / p_{n+1} \cdot n \log n / p_n \cdot \log p_n / \log n \to 1$$

and similarly $n \log p_{n+1} / p_n \to 1$, it follows that also $\pi(x) \log x / x \to 1$. $\square$

Numerical evidence, both for the prime number theorem and for the fact that $Li(x)$ is a better approximation than $x/\log x$ to $\pi(x)$, is provided by Table 1.

x	$\pi(x)$	$x/\log x$	$Li(x)$	$\pi(x) \log x / x$	$\pi(x) / Li(x)$
10^3	168	144.	177.	1.16	0.94
10^4	1 229	1 085.	1 245.	1.132	0.987
10^5	9 592	8 685.	9 629.	1.1043	0.9961
10^6	78 498	72 382.	78 627.	1.08449	0.99835
10^7	664 579	620 420.	664 917.	1.07117	0.99949
10^8	5 761 455	5 428 681.	5 762 208.	1.06130	0.99987
10^9	50 847 534	48 254 942.	50 849 234.	1.05373	0.999966
10^{10}	455 052 511	434 294 481.	455 055 614.	1.04780	0.999993

TABLE 1

In a second paper Chebyshev (1852) made some progress towards proving the prime number theorem by showing that

$$a \le \underline{\lim}_{x \to \infty} \pi(x) \log x / x \le \overline{\lim}_{x \to \infty} \pi(x) \log x / x \le 6a/5,$$

where $a = 0.92129$. He used his results to give the first proof of *Bertrand's postulate*: for every real $x > 1$, there is a prime between x and $2x$.

New ideas were introduced by Riemann (1859), who linked the asymptotic behaviour of $\pi(x)$ with the behaviour of the function

$$\zeta(s) = \sum_{n=1}^{\infty} 1/n^s$$

for complex values of s. By developing these ideas, and by showing especially that $\zeta(s)$ has no zeros on the line $\Re s = 1$, Hadamard and de la Vallée Poussin independently proved the prime number theorem in 1896. Shortly afterwards de la Vallée Poussin (1899) confirmed that $Li(x)$ was a better approximation than $x/\log x$ to $\pi(x)$ by proving (in particular) that

$$\pi(x) = Li(x) + O(x/\log^\alpha x) \quad \text{for every } \alpha > 0. \tag{2}$$

Better error bounds than de la Vallée Poussin's have since been obtained, but they still fall far short of what is believed to be true.

Another approach to the prime number theorem was found by Wiener (1927-1933), as an application of his general theory of Tauberian theorems. A convenient form for this application was given by Ikehara (1931), and Bochner (1933) showed that in this case Wiener's general theory could be avoided.

It came as a great surprise to the mathematical community when in 1949 Selberg, assisted by Erdös, found a new proof of the prime number theorem which uses only the simplest facts of real analysis. Though elementary in a technical sense, this proof was still quite complicated. As a result of several subsequent simplifications it can now be given quite a clear and simple form. Nevertheless the Wiener–Ikehara proof will be presented here on account of its greater versatility. The error bound (2) can be obtained by both the Wiener and Selberg approaches, in the latter case at the cost of considerable complication.

2 Chebyshev's functions

In his second paper Chebyshev introduced two functions

$$\theta(x) = \sum_{p \le x} \log p, \quad \psi(x) = \sum_{p^\alpha \le x} \log p,$$

which have since played a major role. Although $\psi(x)$ has the most complicated definition, it is easier to treat analytically than either $\theta(x)$ or $\pi(x)$. As we will show, the asymptotic behaviour of $\theta(x)$ is essentially the same as that of $\psi(x)$, and the asymptotic behaviour of $\pi(x)$ may be deduced without difficulty from that of $\theta(x)$.

Evidently

$$\theta(x) = \psi(x) = 0 \text{ for } x < 2$$

and

$$0 < \theta(x) \le \psi(x) \text{ for } x \ge 2.$$

LEMMA 3 *The asymptotic behaviours of* $\psi(x)$ *and* $\theta(x)$ *are connected by*

(i) $\psi(x) - \theta(x) = O(x^{1/2} \log^2 x)$;
(ii) $\psi(x) = O(x)$ *if and only if* $\theta(x) = O(x)$, *and in this case* $\psi(x) - \theta(x) = O(x^{1/2} \log x)$.

Proof Since

$$\psi(x) = \sum_{p \le x} \log p + \sum_{p^2 \le x} \log p + \dots$$

and $k > \log x / \log 2$ implies $x^{1/k} < 2$, we have

$$\psi(x) = \theta(x) + \theta(x^{1/2}) + \dots + \theta(x^{1/m}),$$

where $m = \lfloor \log x / \log 2 \rfloor$. (As is now usual, we denote by $\lfloor y \rfloor$ the greatest integer $\leq y$.) But it is obvious from the definition of $\theta(x)$ that $\theta(x) = O(x \log x)$. Hence

$$\psi(x) - \theta(x) = O(\sum_{2 \leq k \leq m} x^{1/k} \log x) = O(x^{1/2} \log^2 x).$$

If $\theta(x) = O(x)$ the same argument yields $\psi(x) - \theta(x) = O(x^{1/2} \log x)$ and thus $\psi(x) = O(x)$. It is trivial that $\psi(x) = O(x)$ implies $\theta(x) = O(x)$. $\square$

The proof of Lemma 3 shows also that

$$\psi(x) = \theta(x) + \theta(x^{1/2}) + O(x^{1/3} \log^2 x).$$

LEMMA 4 $\psi(x) = O(x)$ *if and only if* $\pi(x) = O(x/\log x)$, *and then*

$$\pi(x) \log x / x = \psi(x)/x + O(1/\log x).$$

Proof Although their use can easily be avoided, it is more suggestive to use Stieltjes integrals. Suppose first that $\psi(x) = O(x)$. For any $x > 2$ we have

$$\pi(x) = \int_{2-}^{x+} 1/\log t \, d\theta(t)$$

and hence, on integrating by parts,

$$\pi(x) = \theta(x)/\log x + \int_2^x \theta(t)/t \log^2 t \, dt .$$

But

$$\int_2^x \theta(t)/t \log^2 t \, dt = O(x/\log^2 x),$$

since $\theta(t) = O(t)$ and, as we saw in §1,

$$\int_2^x dt/\log^2 t = O(x/\log^2 x).$$

Since

$$\theta(x)/\log x = \psi(x)/\log x + O(x^{1/2}),$$

by Lemma 3, it follows that

$$\pi(x) = \psi(x)/\log x + O(x/\log^2 x).$$

Suppose next that $\pi(x) = O(x/\log x)$. For any $x > 2$ we have

$$\theta(x) = \int_{2-}^{x+} \log t \, d\pi(t)$$

$$= \pi(x) \log x - \int_2^x \pi(t)/t \, dt = O(x),$$

and hence also $\psi(x) = O(x)$, by Lemma 3. $\quad\square$

It follows at once from Lemma 4 that *the prime number theorem, $\pi(x) \sim x/\log x$, is equivalent to $\psi(x) \sim x$.*

The method of argument used in Lemma 4 can be carried further. Put

$$\theta(x) = x + R(x), \quad \pi(x) = \int_2^x dt/\log t + Q(x).$$

Subtracting

$$\int_2^x dt/\log t = x/\log x - 2/\log 2 + \int_2^x dt/\log^2 t$$

from

we obtain

$$\pi(x) = \theta(x)/\log x + \int_2^x \theta(t)/t\log^2 t\, dt\ .$$

$$Q(x) = R(x)/\log x + \int_2^x R(t)/t\log^2 t\, dt + 2/\log 2\ . \qquad (3)_1$$

Also, adding

$$\int_2^x\!\left(\int_2^t du/\log u\right) dt/t = \int_2^x\!\left(\int_u^x dt/t\right) du/\log u$$

$$= \int_2^x (\log x - \log u)\, du/\log u$$

$$= \log x \int_2^x dt/\log t - x + 2$$

to

we obtain

$$\theta(x) = \pi(x)\log x - \int_2^x \pi(t)/t\, dt$$

$$R(x) = Q(x)\log x - \int_2^x Q(t)/t\, dt - 2. \qquad (3)_2$$

It follows from $(3)_1$-$(3)_2$ that $R(x) = O(x/\log^\alpha x)$ for some $\alpha > 0$ if and only if $Q(x) = O(x/\log^{\alpha+1} x)$. Consequently, by Lemma 3,

$$\psi(x) = x + O(x/\log^\alpha x) \ \textit{for every}\ \alpha > 0$$

if and only if

$$\pi(x) = \int_2^x dt/\log t + O(x/\log^\alpha x) \ \textit{for every}\ \alpha > 0,$$

and $\pi(x)$ then has the asymptotic expansion

$$\pi(x) \sim \{1 + 1!/\log x + 2!/\log^2 x + \dots\}x/\log x,$$

the error in breaking off the series after any finite number of terms having the order of magnitude of the first term omitted.

It follows from $(3)_1$-$(3)_2$ also that, *for a given α such that $1/2 \le \alpha < 1$,*

$$\psi(x) = x + O(x^\alpha \log^2 x),$$

if and only if

$$\pi(x) = \int_2^x dt/\log t + O(x^\alpha \log x).$$

The definition of $\psi(x)$ can be put in the form

$$\psi(x) = \sum_{n \leq x} \Lambda(n),$$

where the *von Mangoldt function* $\Lambda(n)$ is defined by

$$\Lambda(n) = \log p \text{ if } n = p^\alpha \text{ for some prime } p \text{ and some } \alpha > 0,$$

$$= 0 \text{ otherwise.}$$

For any positive integer n we have

$$\log n = \sum_{d|n} \Lambda(d), \tag{4}$$

since if $n = p_1^{\alpha_1} \cdots p_s^{\alpha_s}$ is the factorization of n into powers of distinct primes, then

$$\log n = \sum_{j=1}^s \alpha_j \log p_j.$$

3 Proof of the prime number theorem

The *Riemann zeta-function* is defined by

$$\zeta(s) = \sum_{n=1}^\infty 1/n^s. \tag{5}$$

This infinite series had already been considered by Euler, Dirichlet and Chebyshev, but Riemann was the first to study it for complex values of s. As customary, we write $s = \sigma + it$, where σ and t are real, and n^{-s} is defined for complex values of s by

$$n^{-s} = e^{-s \log n} = n^{-\sigma} (\cos(t \log n) - i \sin(t \log n)).$$

To show that the series (5) converges in the half-plane $\sigma > 1$ we compare the sum with an integral. If $\lfloor x \rfloor$ denotes again the greatest integer $\leq x$, then on integrating by parts we obtain

$$\int_1^N x^{-s} dx - \sum_{n=1}^N n^{-s} = \int_{1-}^{N+} x^{-s} d\{x - \lfloor x \rfloor\}$$

$$= -1 + s \int_1^N x^{-s-1} \{x - \lfloor x \rfloor\} dx.$$

Since

$$\int_1^N x^{-s} dx = (1 - N^{1-s})/(s - 1),$$

by letting $N \to \infty$ we see that $\zeta(s)$ is defined for $\sigma > 1$ and

$$\zeta(s) = 1/(s - 1) + 1 - s \int_1^\infty x^{-s-1} \{x - \lfloor x \rfloor\} dx.$$

But, since $x - \lfloor x \rfloor$ is bounded, the integral on the right is uniformly convergent in any half-plane $\sigma \geq \delta > 0$. It follows that the definition of $\zeta(s)$ can be extended to the half-plane $\sigma > 0$, so that it is holomorphic there except for a simple pole with residue 1 at $s = 1$.

The connection between the zeta-function and prime numbers is provided by *Euler's product formula*, which may be viewed as an analytic version of the fundamental theorem of arithmetic:

PROPOSITION 5 $\zeta(s) = \prod_p (1 - p^{-s})^{-1}$ *for* $\sigma > 1$, *where the product is taken over all primes p.*

Proof For $\sigma > 0$ we have

$$(1 - p^{-s})^{-1} = 1 + p^{-s} + p^{-2s} + \dots .$$

Since each positive integer can be uniquely expressed as a product of powers of distinct primes, it follows that

$$\prod_{p \leq x} (1 - p^{-s})^{-1} = \sum_{n \leq N_x} n^{-s},$$

where N_x is the set of all positive integers, including 1, whose prime factors are all $\leq x$. But N_x contains all positive integers $\leq x$. Hence

$$\left| \zeta(s) - \prod_{p \leq x} (1 - p^{-s})^{-1} \right| \leq \sum_{n > x} n^{-\sigma} \text{ for } \sigma > 1,$$

and the sum on the right tends to zero as $x \to \infty$. □

It follows at once from Proposition 5 that $\zeta(s) \neq 0$ for $\sigma > 1$, since the infinite product is convergent and each factor is nonzero.

PROPOSITION 6 $- \zeta'(s)/\zeta(s) = \sum_{n=1}^{\infty} \Lambda(n)/n^s$ *for* $\sigma > 1$, *where* $\Lambda(n)$ *is von Mangoldt's function.*

Proof The series $\omega(s) = \sum_{n=1}^{\infty} \Lambda(n) n^{-s}$ converges absolutely and uniformly in any half-plane $\sigma \geq 1 + \varepsilon$, where $\varepsilon > 0$, since

$$0 \leq \Lambda(n) \leq \log n < n^{\varepsilon/2} \text{ for all large } n.$$

Hence

$$\zeta(s)\omega(s) = \sum_{m=1}^{\infty} m^{-s} \sum_{k=1}^{\infty} \Lambda(k) k^{-s}$$

$$= \sum_{n=1}^{\infty} n^{-s} \sum_{d|n} \Lambda(d).$$

Since $\sum_{d|n} \Lambda(d) = \log n$, by (4), it follows that

$$\zeta(s)\omega(s) \ = \ \textstyle\sum_{n=1}^{\infty} n^{-s} \log n \ = \ - \ \zeta'(s).$$

Since $\zeta(s) \neq 0$ for $\sigma > 1$, the result follows. However, we can also prove directly that $\zeta(s) \neq 0$ for $\sigma > 1$, and thus make the proof of the prime number theorem independent of Proposition 5.

Obviously if $\zeta(s_0) = 0$ for some s_0 with $\Re s_0 > 1$ then $\zeta'(s_0) = 0$, and it follows by induction from Leibniz' formula for derivatives of a product that $\zeta^{(n)}(s_0) = 0$ for all $n \geq 0$. Since $\zeta(s)$ is holomorphic for $\sigma > 1$ and not identically zero, this is a contradiction. $\quad\square$

Proposition 6 may be restated in terms of Chebyshev's ψ-function:

$$-\zeta'(s)/\zeta(s) \ = \ \int_1^{\infty} u^{-s} \, d\psi(u) \ = \ \int_0^{\infty} e^{-sx} \, d\psi(e^x) \quad \text{for } \sigma > 1. \tag{6}$$

We are going to deduce from (6) that the function $\zeta(s)$ has no zeros on the line $\Re s = 1$. Actually we will prove a more general result:

PROPOSITION 7 *Let $f(s)$ be holomorphic in the closed half-plane $\Re s \geq 1$, except for a simple pole at $s = 1$. If, for $\Re s > 1$, $f(s) \neq 0$ and*

$$-f'(s)/f(s) \ = \ \int_0^{\infty} e^{-sx} \, d\phi(x),$$

where $\phi(x)$ is a nondecreasing function for $x \geq 0$, then

$$f(1 + it) \ \neq \ 0 \quad \text{for every real } t \neq 0.$$

Proof Put $s = \sigma + it$, where σ and t are real, and let

$$g(\sigma,t) \ = \ -\Re\{f'(s)/f(s)\}.$$

Thus

$$g(\sigma,t) \ = \ \int_0^{\infty} e^{-\sigma x} \cos(tx) \, d\phi(x) \text{ for } \sigma > 1.$$

Hence, by Schwarz's inequality (Chapter I, §10),

$$g(\sigma,t)^2 \ \leq \ \int_0^{\infty} e^{-\sigma x} \, d\phi(x) \int_0^{\infty} e^{-\sigma x} \cos^2(tx) \, d\phi(x)$$

$$= \ g(\sigma,0) \int_0^{\infty} e^{-\sigma x} \{1 + \cos(2tx)\} \, d\phi(x)/2$$

$$= \ g(\sigma,0) \, \{g(\sigma,0) + g(\sigma,2t)\}/2.$$

Since $f(s)$ has a simple pole at $s = 1$, by comparing the Laurent series of $f(s)$ and $f'(s)$ at $s = 1$ (see Chapter I, §5) we see that

$$(\sigma - 1)g(\sigma,0) \to 1 \quad \text{as } \sigma \to 1+.$$

Similarly if $f(s)$ has a zero of multiplicity $m(t) \geq 0$ at $1 + it$, where $t \neq 0$, then by comparing the Taylor series of $f(s)$ and $f'(s)$ at $s = 1 + it$ we see that

$$(\sigma - 1)g(\sigma,t) \to -m(t) \text{ as } \sigma \to 1+.$$

Thus if we multiply the inequality for $g(\sigma,t)^2$ by $(\sigma - 1)^2$ and let $\sigma \to 1+$, we obtain

$$m(t)^2 \leq \{1 - m(2t)\}/2 \leq 1/2.$$

Therefore, since $m(t)$ is an integer, $m(t) = 0$. $\quad\square$

For $f(s) = \zeta(s)$, Proposition 7 gives the result of Hadamard and de la Vallée Poussin:

COROLLARY 8 $\zeta(1 + it) \neq 0$ *for every real* $t \neq 0$. $\quad\square$

The use of Schwarz's inequality to prove Corollary 8 seems more natural than the usual proof by means of the inequality $3 + 4 \cos \theta + \cos 2\theta \geq 0$. It follows from Corollary 8 that $-\zeta'(s)/\zeta(s) - 1/(s - 1)$ is holomorphic in the closed half-plane $\sigma \geq 1$. Hence, by (6), the hypotheses of the following theorem, due to Ikehara (1931), are satisfied with

$$F(s) = -\zeta'(s)/\zeta(s), \quad \phi(x) = \psi(e^x), \quad h = A = 1.$$

THEOREM 9 *Let* $\phi(x)$ *be a nondecreasing function for* $x \geq 0$ *such that the Laplace transform*

$$F(s) = \int_0^\infty e^{-sx} \, d\phi(x)$$

is defined for $\Re s > h$, *where* $h > 0$. *If there exists a constant* A *and a function* $G(s)$, *which is continuous in the closed half-plane* $\Re s \geq h$, *such that*

$$G(s) = F(s) - Ah/(s - h) \text{ for } \Re s > h,$$

then

$$\phi(x) \sim Ae^{hx} \text{ for } x \to +\infty.$$

Proof For each $X > 0$ we have

$$\int_0^X e^{-sx} \, d\phi(x) = e^{-sX}\{\phi(X) - \phi(0)\} + s\int_0^X e^{-sx} \{\phi(x) - \phi(0)\} \, dx.$$

For real $s = \rho > h$ both terms on the right are nonnegative and the integral on the left has a finite limit as $X \to \infty$. Hence $e^{-\rho X}\phi(X)$ is a bounded function of X for each $\rho > h$. It follows that if $\Re s > h$ we can let $X \to \infty$ in the last displayed equation, obtaining

$$F(s) = s \int_0^\infty e^{-sx} \{\phi(x) - \phi(0)\} \, dx \text{ for } \Re s > h.$$

Hence

$$[G(s) - A]/s = F(s)/s - A/(s - h) = \int_0^\infty e^{-(s-h)x} \{\alpha(x) - A\} \, dx,$$

where $\alpha(x) = e^{-hx}\{\phi(x) - \phi(0)\}$. Thus we will prove the theorem if we prove the following statement:

Let $\alpha(x)$ be a nonnegative function for $x \geq 0$ such that

$$g(s) = \int_0^\infty e^{-sx} \{\alpha(x) - A\} \, dx,$$

where $s = \sigma + it$, is defined for every $\sigma > 0$ and the limit

$$\gamma(t) = \lim_{\sigma \to +0} g(s)$$

exists uniformly on any finite interval $-T \leq t \leq T$. If, for some $h > 0$, $e^{hx}\alpha(x)$ is a nondecreasing function, then

$$\lim_{x \to \infty} \alpha(x) = A.$$

In the proof of this statement we will use the fact that the Fourier transform

$$\hat{k}(u) = \int_{-\infty}^\infty e^{iut} k(t) \, dt$$

of the function

$$k(t) = 1 - |t| \text{ for } |t| \leq 1, \ = 0 \text{ for } |t| \geq 1,$$

has the properties

$$\hat{k}(u) \geq 0 \text{ for } -\infty < u < \infty, \quad C := \int_{-\infty}^\infty \hat{k}(u) \, du < \infty.$$

Indeed

$$\hat{k}(u) = \int_{-1}^1 e^{iut} (1 - |t|) \, dt$$

$$= 2\int_0^1 (1 - t) \cos ut \, dt$$

$$= 2(1 - \cos u)/u^2.$$

Let ε, λ, y be arbitrary positive numbers. If $s = \varepsilon + i\lambda t$, then

$$\lambda \int_{-1}^1 e^{i\lambda ty} k(t) g(s) \, dt = \lambda \int_{-1}^1 e^{i\lambda ty} k(t) \int_0^\infty e^{-\varepsilon x} e^{-i\lambda tx} \{\alpha(x) - A\} \, dx \, dt$$

$$= \lambda \int_0^\infty e^{-\varepsilon x} \{\alpha(x) - A\} \int_{-1}^1 e^{i\lambda t(y-x)} k(t) \, dt \, dx$$

$$= \lambda \int_0^\infty e^{-\varepsilon x} \alpha(x) \hat{k}(\lambda(y-x)) \, dx - \lambda A \int_0^\infty e^{-\varepsilon x} \hat{k}(\lambda(y-x)) \, dx.$$

When $\varepsilon \to +0$ the left side has the limit

$$\chi(y) := \lambda \int_{-1}^1 e^{i\lambda ty} k(t) \gamma(\lambda t) \, dt$$

and the second term on the right has the limit

$$\lambda A \int_0^\infty \hat{k}(\lambda(y-x))\, dx\,.$$

Consequently the first term on the right also has a finite limit. It follows that

$$\lambda \int_0^\infty \alpha(x)\, \hat{k}(\lambda(y-x))\, dx$$

is finite and is the limit of the first term on the right. Thus

$$\chi(y) = \lambda \int_0^\infty \{\alpha(x) - A\}\, \hat{k}(\lambda(y-x))\, dx$$

$$= \int_{-\infty}^{\lambda y} \{\alpha(y-v/\lambda) - A\}\, \hat{k}(v)\, dv\,.$$

By the 'Riemann–Lebesgue lemma', $\chi(y) \to 0$ as $y \to \infty$. In fact this may be proved in the following way. We have

$$\chi(y) = \int_{-\infty}^\infty e^{i\lambda ty}\, \omega(t)\, dt$$

where

$$\omega(t) = \lambda k(t)\gamma(\lambda t).$$

Changing the variable of integration to $t + \pi/\lambda y$, we obtain

$$\chi(y) = -\int_{-\infty}^\infty e^{i\lambda ty}\, \omega(t + \pi/\lambda y)\, dt\,.$$

Hence

$$2\chi(y) = \int_{-\infty}^\infty e^{i\lambda ty}\, \{\omega(t) - \omega(t + \pi/\lambda y)\}\, dt$$

and

$$2|\chi(y)| \le \int_{-\infty}^\infty |\omega(t) - \omega(t + \pi/\lambda y)|\, dt\,.$$

Since $\omega(t)$ is continuous and vanishes outside a finite interval, it follows that $\chi(y) \to 0$ as $y \to \infty$.

Since

$$\int_{-\infty}^{\lambda y} \hat{k}(v)\, dv \to C \text{ as } y \to \infty,$$

we deduce that

$$\lim_{y\to\infty} \int_{-\infty}^{\lambda y} \alpha(y - v/\lambda)\, \hat{k}(v)\, dv = AC \text{ for every } \lambda > 0.$$

We now use the fact that $e^{hx}\alpha(x)$ is a nondecreasing function. Choose any $\delta \in (0,1)$. If $y = x + \delta$, where $x \ge 0$, then for $|v| \le \lambda\delta$

$$\alpha(y - v/\lambda) \ge e^{-h(\delta - v/\lambda)}\alpha(x) \ge e^{-2h\delta}\alpha(x)$$

and hence

$$\int_{-\infty}^{\lambda y} \alpha(y - v/\lambda)\, \hat{k}(v)\, dv \ge e^{-2h\delta}\alpha(x) \int_{-\lambda\delta}^{\lambda\delta} \hat{k}(v)\, dv\,.$$

We can choose $\lambda = \lambda(\delta)$ so large that the integral on the right exceeds $(1 - \delta)C$. Then, letting $x \to \infty$ we obtain

$$AC \geq e^{-2h\delta}(1 - \delta)C \; \overline{\lim}_{x\to\infty} \; \alpha(x).$$

Since this holds for arbitrarily small $\delta > 0$, it follows that

$$\overline{\lim}_{x\to\infty} \alpha(x) \leq A.$$

Thus there exists a positive constant M such that

$$0 \leq \alpha(x) \leq M \text{ for all } x \geq 0.$$

On the other hand, if $y = x - \delta$, where $x \geq \delta$, then for $|v| \leq \lambda\delta$

$$\alpha(y - v/\lambda) \leq e^{h(\delta + v/\lambda)}\alpha(x) \leq e^{2h\delta}\alpha(x)$$

and hence

$$\int_{-\infty}^{\lambda y} \alpha(y - v/\lambda) \, \hat{k}(v) \, dv \leq e^{2h\delta}\alpha(x)\int_{-\lambda\delta}^{\lambda\delta} \hat{k}(v) \, dv + M\int_{|v|\geq\lambda\delta} \hat{k}(v) \, dv.$$

We can choose $\lambda = \lambda(\delta)$ so large that the second term on the right is less than δC. Then, letting $x \to \infty$ we obtain

$$AC \leq e^{2h\delta}C \; \underline{\lim}_{x\to\infty} \; \alpha(x) + \delta C.$$

Since this holds for arbitrarily small $\delta > 0$, it follows that

$$A \leq \underline{\lim}_{x\to\infty} \alpha(x).$$

Combining this with the inequality of the previous paragraph, we conclude that $\lim_{x\to\infty} \alpha(x) = A$. $\square$

Applying Theorem 9 to the special case mentioned before the statement of the theorem, we obtain $\psi(e^x) \sim e^x$. As we have already seen in §2, this is equivalent to the prime number theorem.

4 The Riemann hypothesis

In his celebrated paper on the distribution of prime numbers Riemann (1859) proved only two results. He showed that the definition of $\zeta(s)$ can be extended to the whole complex plane, so that $\zeta(s) - 1/(s - 1)$ is everywhere holomorphic, and he proved that the values of $\zeta(s)$ and $\zeta(1 - s)$ are connected by a certain functional equation. This functional equation will now be derived by one of the two methods which Riemann himself used. It is based on a remarkable identity which Jacobi (1829) used in his treatise on elliptic functions.

PROPOSITION 10 *For any t,y* $\in \mathbb{R}$ *with y > 0,*

$$\sum_{n=-\infty}^{\infty} e^{-(t+n)^2 \pi y} = y^{-1/2} \sum_{n=-\infty}^{\infty} e^{-n^2 \pi/y} e^{2\pi int}. \tag{7}$$

In particular,

$$\sum_{n=-\infty}^{\infty} e^{-n^2 \pi y} = y^{-1/2} \sum_{n=-\infty}^{\infty} e^{-n^2 \pi/y} . \tag{8}$$

Proof Put $f(v) = e^{-v^2 \pi y}$ and let

$$g(u) = \int_{-\infty}^{\infty} f(v) e^{-2\pi i u v} dv$$

be the Fourier transform of $f(v)$. We are going to show that

$$\sum_{n=-\infty}^{\infty} f(v+n) = \sum_{n=-\infty}^{\infty} g(n) e^{2\pi i n v}.$$

Let

$$F(v) = \sum_{n=-\infty}^{\infty} f(v+n).$$

This infinite series is uniformly convergent for $0 \le v \le 1$, and so also is the series obtained by term by term differentiation. Hence $F(v)$ is a continuously differentiable function. Consequently, since it is periodic with period 1, it is the sum of its own Fourier series:

$$F(v) = \sum_{m=-\infty}^{\infty} c_m e^{2\pi i m v},$$

where

$$c_m = \int_0^1 F(v) e^{-2\pi i m v} dv.$$

We can evaluate c_m by term by term integration:

$$c_m = \sum_{n=-\infty}^{\infty} \int_0^1 f(v+n) e^{-2\pi i m v} dv = \sum_{n=-\infty}^{\infty} \int_n^{n+1} f(v) e^{-2\pi i m v} dv$$

$$= \int_{-\infty}^{\infty} f(v) e^{-2\pi i m v} dv = g(m).$$

The argument up to this point is an instance of *Poisson's summation formula*. To evaluate $g(u)$ in the case $f(v) = e^{-v^2 \pi y}$ we differentiate with respect to u and integrate by parts, obtaining

$$g'(u) = -2\pi i \int_{-\infty}^{\infty} e^{-v^2 \pi y} v e^{-2\pi i u v} dv$$

$$= (i/y) \int_{-\infty}^{\infty} e^{-2\pi i u v} d e^{-v^2 \pi y}$$

$$= -(i/y) \int_{-\infty}^{\infty} e^{-v^2 \pi y} d e^{-2\pi i u v}$$

$$= -(2\pi u/y) g(u).$$

The solution of this first order linear differential equation is

$$g(u) = g(0)e^{-\pi u^2/y}.$$

Moreover

$$g(0) = \int_{-\infty}^{\infty} e^{-v^2\pi y} \, dv = (\pi y)^{-1/2}J,$$

where

$$J = \int_{-\infty}^{\infty} e^{-v^2} \, dv.$$

Thus we have proved that

$$\sum_{n=-\infty}^{\infty} e^{-(v+n)^2\pi y} = (\pi y)^{-1/2}J \sum_{n=-\infty}^{\infty} e^{-n^2\pi/y} \, e^{2\pi inv}.$$

Substituting $v = 0$, $y = 1$, we obtain $J = \pi^{1/2}$. $\blacksquare$

The *theta function*

$$\vartheta(x) = \sum_{n=-\infty}^{\infty} e^{-n^2\pi x} \quad (x > 0)$$

arises not only in the theory of elliptic functions, as we will see in Chapter XII, but also in problems of heat conduction and statistical mechanics. The transformation law

$$\vartheta(x) = x^{-1/2}\vartheta(1/x)$$

is very useful for computational purposes since, when x is small, the series for $\vartheta(x)$ converges extremely slowly but the series for $\vartheta(1/x)$ converges extremely rapidly.

Since the functional equation of Riemann's zeta function involves Euler's *gamma function*, we summarize here the main properties of the latter. Euler (1729) defined his function $\Gamma(z)$ by

$$1/\Gamma(z) = \lim_{n\to\infty} z(z + 1)\cdots(z + n)/n! \, n^z,$$

where $n^z = \exp(z \log n)$ and the limit exists for every $z \in \mathbb{C}$. It follows from the definition that $1/\Gamma(z)$ is everywhere holomorphic and that its only zeros are simple zeros at the points $z = 0$, $-1, -2, \ldots$. Moreover $\Gamma(1) = 1$ and

$$\Gamma(z + 1) = z \, \Gamma(z).$$

Hence $\Gamma(n + 1) = n!$ for any positive integer n. By putting $\Gamma(z + 1) = z!$ the definition of the factorial function may be extended to any $z \in \mathbb{C}$ which is not a negative integer. Wielandt (1939) has characterized $\Gamma(z)$ as the only solution of the functional equation

$$F(z + 1) = z \, F(z)$$

with $F(1) = 1$ which is holomorphic in the half-plane $\mathfrak{R}z > 0$ and bounded for $1 < \mathfrak{R}z < 2$.

It follows from the definition of $\Gamma(z)$ and the product formula for the sine function that

$$\Gamma(z)\,\Gamma(1-z) \;=\; \pi/\sin\,\pi z.$$

Many definite integrals may be evaluated in terms of the gamma function. By repeated integration by parts it may be seen that, if $\Re z > 0$ and $n \in \mathbb{N}$, then

$$n!\ n^z/z(z+1)\cdots(z+n) \;=\; \int_0^n (1-t/n)^n t^{z-1}\,dt,$$

where $t^{z-1} = \exp\{(z-1)\log t\}$. Letting $n \to \infty$, we obtain the integral representation

$$\Gamma(z) \;=\; \int_0^\infty e^{-t}\,t^{z-1}\,dt \quad \text{for } \Re z > 0. \tag{9}$$

It follows that $\Gamma(1/2) = \pi^{1/2}$, since

$$\int_0^\infty e^{-t}\,t^{-1/2}\,dt \;=\; \int_{-\infty}^\infty e^{-v^2}\,dv \;=\; \pi^{1/2},$$

by the proof of Proposition 10. It was already shown by Euler (1730) that

$$B(x,y)\!: \;=\; \int_0^1 t^{x-1}\,(1-t)^{y-1}\,dt \;=\; \Gamma(x)\Gamma(y)/\Gamma(x+y),$$

the relation holding for $\Re x > 0$ and $\Re y > 0$. The unit ball in $\mathbb{R}^n$ has volume $\kappa_n := \pi^{n/2}/(n/2)!$ and surface content $n\kappa_n$. *Stirling's formula*, $n! \approx (n/e)^n \sqrt{(2\pi n)}$, follows at once from the integral representation

$$\log\Gamma(z) \;=\; (z-1/2)\log z - z + (1/2)\log 2\pi - \int_0^\infty (t-\lfloor t \rfloor - 1/2)(z+t)^{-1}\,dt,$$

valid for any $z \in \mathbb{C}$ which is not zero or a negative integer. Euler's constant

$$\gamma \;=\; \lim_{n\to\infty}(1 + 1/2 + 1/3 + \ldots + 1/n - \log n) \approx 0.5772157$$

may also be defined by $\gamma = -\Gamma'(1)$.

We now return to the Riemann zeta function.

PROPOSITION 11 *The function* $Z(s) = \pi^{-s/2}\,\Gamma(s/2)\zeta(s)$ *satisfies the functional equation*

$$Z(s) \;=\; Z(1-s) \quad \text{for } 0 < \sigma < 1.$$

Proof From the representation (9) of the gamma function we obtain, for $\sigma > 0$ and $n \ge 1$,

$$\int_0^\infty x^{s/2-1}\,e^{-n^2\pi x}\,dx \;=\; \pi^{-s/2}\,\Gamma(s/2)n^{-s}.$$

Hence, if $\sigma > 1$,

$$Z(s) \;=\; \sum_{n=1}^\infty \int_0^\infty x^{s/2-1}\,e^{-n^2\pi x}\,dx$$

$$=\; \int_0^\infty x^{s/2-1}\,\phi(x)\,dx,$$

where

$$\phi(x) \;=\; \sum_{n=1}^\infty e^{-n^2\pi x}.$$

By Proposition 10,
$$2\phi(x) + 1 = x^{-1/2}[2\phi(1/x) + 1].$$
Hence
$$Z(s) = \int_1^\infty x^{s/2-1}\,\phi(x)\,dx + \int_0^1 x^{s/2-1}\,\{x^{-1/2}\phi(1/x) + (1/2)x^{-1/2} - 1/2\}dx$$
$$= \int_1^\infty x^{s/2-1}\,\phi(x)\,dx + \int_0^1 x^{s/2-3/2}\,\phi(1/x)\,dx + 1/(s-1) - 1/s$$
$$= \int_1^\infty (x^{s/2-1} + x^{-s/2-1/2})\,\phi(x)\,dx + 1/s(s-1).$$

The integral on the right is convergent for all s and thus provides the analytic continuation of $Z(s)$ to the whole plane. Moreover the right side is unchanged if s is replaced by $1 - s$. $\square$

The function $Z(s)$ in Proposition 11 is sometimes called the *completed* zeta function. In its product representation
$$Z(s) = \pi^{-s/2}\,\Gamma(s/2)\prod_p (1-p^{-s})^{-1}$$
it makes sense to regard $\pi^{-s/2}\Gamma(s/2)$ as an Euler factor at ∞, complementing the Euler factors $(1 - p^{-s})^{-1}$ at the primes p.

It follows from Proposition 11 and the previously stated properties of the gamma function that the definition of $\zeta(s)$ may be extended to the whole complex plane, so that $\zeta(s) - 1/(s-1)$ is everywhere holomorphic and $\zeta(s) = 0$ if $s = -2,-4,-6,...$. Since $\zeta(s) \neq 0$ for $\sigma \geq 1$ and $\zeta(0) = -1/2$, the functional equation shows that these 'trivial' zeros of $\zeta(s)$ are its only zeros in the half-plane $\sigma \leq 0$. Hence all 'nontrivial' zeros of $\zeta(s)$ lie in the strip $0 < \sigma < 1$ and are symmetrically situated with respect to the line $\sigma = 1/2$. The famous *Riemann hypothesis* asserts that all zeros in this strip actually lie on the line $\sigma = 1/2$.

The zeros of $\zeta(s)$ are also symmetric with respect to the real axis, since $\zeta(\bar{s}) = \overline{\zeta(s)}$. Furthermore $\zeta(s)$ has no real zeros in the strip $0 < \sigma < 1$, since
$$(1 - 2^{1-\sigma})\zeta(\sigma) = (1 - 2^{-\sigma}) + (3^{-\sigma} - 4^{-\sigma}) + ... \; > 0 \; \text{ for } 0 < \sigma < 1.$$

It has been verified by van de Lune *et al.* (1986), with the aid of a supercomputer, that the 1.5×10^9 zeros of $\zeta(s)$ in the rectangle $0 < \sigma < 1, 0 < t < T$, where $T = 545439823.215$, are all simple and lie on the line $\sigma = 1/2$.

The location of the zeros of $\zeta(s)$ is intimately connected with the asymptotic behaviour of $\pi(x)$. Let $\alpha*$ denote the least upper bound of the real parts of all zeros of $\zeta(s)$. Then $1/2 \leq \alpha* \leq 1$, since it is known that $\zeta(s)$ does have zeros in the strip $0 < \sigma < 1$, and the Riemann hypothesis is equivalent to $\alpha* = 1/2$. It was shown by von Koch (1901) that

$$\psi(x) \ = \ x + O(x^{\alpha*} \log^2 x)$$

and hence

$$\pi(x) \ = \ Li(x) + O(x^{\alpha*} \log x).$$

(Actually von Koch assumed $\alpha* = 1/2$, but his argument can be extended without difficulty.) It should be noted that these estimates are of interest only if $\alpha* < 1$.

On the other hand if, for some α such that $0 < \alpha < 1$,

$$\pi(x) \ = \ Li(x) + O(x^{\alpha} \log x),$$

then

$$\theta(x) \ = \ x + O(x^{\alpha} \log^2 x).$$

By the remark after the proof of Lemma 3, it follows that

$$\psi(x) \ = \ x + x^{1/2} + O(x^{\alpha} \log^2 x) + O(x^{1/3} \log^2 x).$$

But for $\sigma > 1$ we have

$$-\zeta'(s)/\zeta(s) \ = \ \int_1^{\infty} x^{-s} \, d\psi(x) \ = \ s \int_1^{\infty} \psi(x) x^{-s-1} \, dx$$

and hence

$$-\zeta'(s)/\zeta(s) - s/(s-1) - s/(s-1/2) \ = \ s \int_1^{\infty} \{\psi(x) - x - x^{1/2}\} x^{-s-1} \, dx.$$

The integral on the right is uniformly convergent in the half-plane $\sigma \geq \varepsilon + \max(\alpha, 1/3)$, for any $\varepsilon > 0$, and represents a holomorphic function. Consequently $1/2 \leq \alpha* \leq \max(\alpha, 1/3)$. It follows that $\alpha* \leq \alpha$ and $\psi(x) \ = \ x + O(x^{\alpha} \log^2 x)$.

Combining this with von Koch's result, we see that the Riemann hypothesis is equivalent to

$$\pi(x) \ = \ Li(x) + O(x^{1/2} \log x)$$

and to

$$\psi(x) \ = \ x + O(x^{1/2} \log^2 x).$$

Since it is still not known if $\alpha* < 1$, the error terms here are substantially smaller than any that have actually been established.

It has been shown by Cramér (1922) that

$$(\log x)^{-1} \int_2^x (\psi(t)/t - 1)^2 \, dt$$

has a finite limit as $x \to \infty$ if the Riemann hypothesis holds, and is unbounded otherwise. Similarly, for each $\alpha < 1$,

$$x^{-2(1-\alpha)} \int_2^x (\psi(t) - t)^2 \, t^{-2\alpha} dt$$

is bounded but does not have a finite limit as $x \to \infty$ if the Riemann hypothesis holds, and is unbounded otherwise.

For all values of x listed in Table 1 we have $\pi(x) < Li(x)$, and at one time it was conjectured that this inequality holds for all $x > 0$. However, Littlewood (1914) disproved the conjecture by showing that there exists a constant $c > 0$ such that

$$\pi(x_n) - Li(x_n) \; > \; c\, x_n^{1/2} \log \log \log x_n \,/\log x_n$$

for some sequence $x_n \to \infty$ and

$$\pi(\xi_n) - Li(\xi_n) \; < \; -c\, \xi_n^{1/2} \log \log \log \xi_n \,/\log \xi_n$$

for some sequence $\xi_n \to \infty$. This is a quite remarkable result, since no actual value of x is known for which $\pi(x) > Li(x)$. However, it is known that $\pi(x) > Li(x)$ for some x between 1.398201×10^{316} and 1.398244×10^{316}.

In this connection it may be noted that Rosser and Schoenfeld (1962) have shown that $\pi(x) > x/\log x$ for all $x \ge 17$. It had previously been shown by Rosser (1939) that $p_n > n \log n$ for all $n \ge 1$.

Not content with not being able to prove the Riemann hypothesis, Montgomery (1973) has assumed it and made a further conjecture. For given $\beta > 0$, let $N_T(\beta)$ be the number of zeros $1/2 + i\gamma$, $1/2 + i\gamma'$ of $\zeta(s)$ with $0 < \gamma' < \gamma \le T$ such that

$$\gamma - \gamma' \; \le \; 2\pi\beta/\log T.$$

Montgomery's conjecture is that, for each fixed $\beta > 0$,

$$N_T(\beta) \sim (T/2\pi) \log T \int_0^\beta \left\{ 1 - (\sin \pi u /\pi u)^2 \right\} du \quad \text{as } T \to \infty.$$

Goldston (1988) has shown that this is equivalent to

$$\int_1^{T^\beta} \left\{ \psi(x + x/T) - \psi(x) - x/T \right\}^2 x^{-2} dx \sim (\beta - 1/2) \log^2 T /T \quad \text{as } T \to \infty,$$

for each fixed $\beta \ge 1$, where $\psi(x)$ is Chebyshev's function.

In the language of physics Montgomery's conjecture says that $1 - (\sin \pi u /\pi u)^2$ is the *pair correlation function* of the zeros of $\zeta(s)$. Dyson pointed out that this is also the pair correlation function of the normalized eigenvalues of a random $N \times N$ Hermitian matrix in the limit $N \to \infty$. A great deal more is known about this so-called *Gaussian unitary ensemble*, which Wigner (1955) used to model the statistical properties of the spectra of complex nuclei. For example, if the eigenvalues are normalized so that the average difference between consecutive eigenvalues is

1, then the probability that the difference between an eigenvalue and the least eigenvalue greater than it does not exceed β converges as $N \to \infty$ to

$$\int_0^\beta p(u)\, du,$$

where the density function $p(u)$ can be explicitly specified.

It has been further conjectured that the spacings of the normalized zeros of the zeta-function have the same distribution. To make this precise, let the zeros $1/2 + i\gamma_n$ of $\zeta(s)$ with $\gamma_n > 0$ be numbered so that

$$\gamma_1 \leq \gamma_2 \leq \dots .$$

Since it is known that the number of γ's in an interval $[T, T+1]$ is asymptotic to $(\log T)/2\pi$ as $T \to \infty$, we put

$$\tilde{\gamma}_n = (\gamma_n \log \gamma_n)/2\pi ,$$

so that the average difference between consecutive $\tilde{\gamma}_n$ is 1. If $\delta_n = \tilde{\gamma}_{n+1} - \tilde{\gamma}_n$, and if $\nu_N(\beta)$ is the number of $\delta_n \leq \beta$ with $n \leq N$, then the conjecture is that for each $\beta > 0$

$$\nu_N(\beta)/N \to \int_0^\beta p(u)\, du \text{ as } N \to \infty.$$

This nearest neighbour conjecture and the Montgomery pair correlation conjecture have been extensively tested by Odlyzko (1987/9) with the aid of a supercomputer. There is good agreement between the conjectures and the numerical results.

5 Generalizations and analogues

The prime number theorem may be generalized to any algebraic number field in the following way. Let K be an algebraic number field, i.e. a finite extension of the field $\mathbb{Q}$ of rational numbers. Let R be the ring of all algebraic integers in K, $\mathcal{I}$ the set of all non-zero ideals of R, and $\mathcal{P}$ the subset of prime ideals. For any $A \in \mathcal{I}$, the quotient ring R/A is finite; its cardinality will be denoted by $|A|$ and called the *norm* of A.

It may be shown that the *Dedekind zeta-function*

$$\zeta_K(s) = \sum_{A \in \mathcal{I}} |A|^{-s}$$

is defined for $\mathfrak{R}s > 1$ and that the product formula

$$\zeta_K(s) = \prod_{P \in \mathcal{P}} (1 - |P|^{-s})^{-1}$$

holds in this open half-plane. Furthermore the definition of $\zeta_K(s)$ may be extended so that it is nonzero and holomorphic in the closed half-plane $\Re s \geq 1$, except for a simple pole at $s = 1$. By applying Ikehara's theorem we can then obtain the *prime ideal theorem*, which was first proved by Landau (1903):

$$\pi_K(x) \sim x/\log x,$$

where $\pi_K(x)$ denotes the number of prime ideals of R with norm $\leq x$.

It was shown by Hecke (1917) that the definition of the Dedekind zeta-function $\zeta_K(s)$ may also be extended so that it is holomorphic in the whole complex plane, except for the simple pole at $s = 1$, and so that, for some constant $A > 0$ and non-negative integers r_1, r_2 (which can all be explicitly described in terms of the structure of the algebraic number field K),

$$Z_K(s) = A\Gamma(s/2)^{r_1}\Gamma(s)^{r_2}\zeta_K(s)$$

satisfies the functional equation

$$Z_K(s) = Z_K(1-s).$$

The *extended Riemann hypothesis* asserts that, for every algebraic number field K,

$$\zeta_K(s) \neq 0 \quad \text{for } \Re s > 1/2.$$

The numerical evidence for the extended Riemann hypothesis is favourable, although in the nature of things it cannot be tested as extensively as the ordinary Riemann hypothesis. The extended Riemann hypothesis implies error bounds for the prime ideal theorem of the same order as those which the ordinary Riemann hypothesis implies for the prime number theorem. However, it also has many other consequences. We mention only two.

It has been shown by Bach (1990), making precise an earlier result of Ankeny (1952), that if the extended Riemann hypothesis holds then, for each prime p, there is a quadratic non-residue a mod p with $a < 2 \log^2 p$. Thus we do not have to search far in order to find a quadratic non-residue, or to disprove the extended Riemann hypothesis.

It will be recalled from Chapter II that if p is a prime and a an integer not divisible by p, then $a^{p-1} \equiv 1 \bmod p$. For each prime p there exists a *primitive root*, i.e. an integer a such that $a^k \not\equiv 1 \bmod p$ for $1 \leq k < p - 1$. It is easily seen that an even square is never a primitive root, that an odd square (including 1) is a primitive root only for the prime $p = 2$, and that -1 is a primitive root only for the primes $p = 2,3$.

Assuming the extended Riemann hypothesis, Hooley (1967) has proved a famous conjecture of Artin (1927): if the integer a is not a square or -1, then there exist infinitely many primes p for which a is a primitive root. Moreover, if $N_a(x)$ denotes the number of primes $p \le x$ for which a is a primitive root, then

$$N_a(x) \sim A_a x / \log x \ \text{ for } x \to \infty,$$

where A_a is a positive constant which can be explicitly described. (The expression for A_a which Artin conjectured requires modification in some cases.)

There are also analogues for function fields of these results for number fields. Let K be an arbitrary field. A *field of algebraic functions of one variable over K* is a field L which satisfies the following conditions:

(i) $K \subseteq L$,

(ii) L contains an element υ which is *transcendental* over K, i.e. υ satisfies no monic polynomial equation

$$u^n + a_1 u^{n-1} + \dots + a_n = 0$$

with coefficients $a_j \in K$,

(iii) L is a *finite extension* of the field $K(\upsilon)$ of rational functions of υ with coeffients from K, i.e. L is finite-dimensional as a vector space over $K(\upsilon)$.

Let R be a ring with $K \subseteq R \subset L$ such that $x \in L \setminus R$ implies $x^{-1} \in R$. Then the set P of all $a \in R$ such that $a = 0$ or $a^{-1} \notin R$ is an ideal of R, and actually the unique maximal ideal of R. Hence the quotient ring R/P is a field. Since R is the set of all $x \in L$ such that $xP \subseteq P$, it is uniquely determined by P. The ideal P will be called a *prime divisor* of the field L and R/P its *residue field*. It may be shown that the residue field R/P is a finite extension of (a field isomorphic to) K.

An arbitrary *divisor* of the field L is a formal product $A = \prod_P P^{\upsilon_P}$ over all prime divisors P of L, where the exponents υ_P are integers only finitely many of which are nonzero. The divisor is *integral* if $\upsilon_P \ge 0$ for all P.

The set K' of all elements of L which satisfy monic polynomial equations with coefficients from K is a subfield containing K, and L is also a field of algebraic functions of one variable over K'. It is easily shown that no element of $L \setminus R$ satisfies a monic polynomial equation with coefficients from R. Consequently $K' \subseteq R$ and the notion of prime divisor is the same whether we consider L to be over K or over K'. Since $(K')' = K'$, we may assume from the outset that $K' = K$. The elements of K will then be called *constants* and the elements of L *functions*.

Suppose now that the field of constants K is a finite field $\mathbb{F}_q$ containing q elements. We define the *norm* $N(P)$ of a prime divisor P to be the cardinality of the corresponding residue field R/P and the norm of an integral divisor $A = \prod_P P^{v_P}$ to be

$$N(A) = \prod_P N(P)^{v_P} .$$

It may be shown that, for each positive integer m, there exist only finitely many prime divisors of norm q^m. Moreover, for $\Re s > 1$ the zeta-function of L can be defined by

$$\zeta_L(s) = \sum_A N(A)^{-s},$$

where the sum is over all integral divisors of L, and then

$$\zeta_L(s) = \prod_P (1 - N(P)^{-s})^{-1},$$

where the product is over all prime divisors of L.

This seems quite similar to the number field case, but the function field case is actually simpler. F.K. Schmidt (1931) deduced from the Riemann–Roch theorem that there exists a polynomial $p(u)$ of even degree $2g$, with integer coefficients and constant term 1, such that

$$\zeta_L(s) = p(q^{-s})/(1 - q^{-s})(1 - q^{1-s}),$$

and that the zeta-function satisfies the functional equation

$$q^{(g-1)s}\zeta_L(s) = q^{(g-1)(1-s)}\zeta_L(1 - s).$$

The non-negative integer g is the *genus* of the field of algebraic functions.

The analogue of the Riemann hypothesis, that all zeros of $\zeta_L(s)$ lie on the line $\Re s = 1/2$, is equivalent to the statement that all zeros of the polynomial $p(u)$ have absolute value $q^{-1/2}$, or that the number N of prime divisors with norm q satisfies the inequality

$$|N - (q + 1)| \leq 2gq^{1/2}.$$

This analogue has been *proved* by Weil (1948). A simpler proof has been given by Bombieri (1974), using ideas of Stepanov (1969).

The theory of function fields can also be given a geometric formulation. The prime divisors of a function field L with field of constants K can be regarded as the points of a non-singular projective curve over K, and vice versa. Weil (1949) conjectured far-reaching generalizations of the preceding results for curves over a finite field to algebraic varieties of higher dimension.

Let V be a nonsingular projective variety of dimension d, defined by homogeneous polynomials with coefficients in $\mathbb{Z}$. For any prime p, let V_p be the (possibly singular) variety defined by reducing the coefficients mod p and consider the formal power series

$$Z_p(T): = \exp\left(\sum_{n \geq 1} N_n(p)T^n/n\right),$$

where $N_n(p)$ denotes the number of points of V_p defined over the finite field $\mathbb{F}_{p^n}$. Weil conjectured that, if V_p is a nonsingular projective variety of dimension d over $\mathbb{F}_p$, then

(i) $Z_p(T)$ is a rational function of T,

(ii) $Z_p(1/p^dT) = \pm p^{de/2}T^e Z_p(T)$ for some integer e,

(iii) $Z_p(T)$ has a factorization of the form

$$Z_p(T) = P_1(T)\cdots P_{2d-1}(T)/P_0(T)P_2(T)\cdots P_{2d}(T),$$

where $P_0(T) = 1 - T$, $P_{2d}(T) = 1 - p^dT$ and $P_j(T) \in \mathbb{Z}[T]$ $(0 < j < 2d)$,

(iv) $P_j(T) = \prod_{k=1}^{b_j}(1 - \alpha_{jk}T)$, where $|\alpha_{jk}| = p^{j/2}$ for $1 \leq k \leq b_j$, $(0 < j < 2d)$.

The Weil conjectures have a topological significance, since the integer e in (ii) is the Euler characteristic of the original variety V, regarded as a complex manifold, and b_j in (iv) is its j-th Betti number.

Conjecture (i) was proved by Dwork (1960). The remaining conjectures were proved by Deligne (1974), using ideas of Grothendieck. The most difficult part is the proof that $|\alpha_{jk}| = p^{j/2}$ (the Riemann hypothesis for varieties over finite fields). Deligne's proof is a major achievement of 20th century mathematics, but unfortunately of a different order of difficulty than anything which will be proved here.

An analogue for function fields of Artin's primitive root conjecture was already proved by Bilharz (1937), assuming the Riemann hypothesis for this case. Function fields have been used by Goppa (1981) to construct linear codes. Good codes are obtained when the number of prime divisors is large compared to the genus, and this can be guaranteed by means of the Riemann 'hypothesis'.

Carlitz and Uchiyama (1957) used the Riemann hypothesis for function fields to obtain useful estimates for exponential sums in one variable, and Deligne (1977) showed that these estimates could be extended to exponential sums in several variables. Let $\mathbb{F}_p$ be the field of p elements, where p is a prime, and let $f \in \mathbb{F}_p[u_1,...,u_n]$ be a polynomial in n variables of degree

$d \geq 1$ with coefficients from $\mathbb{F}_p$ which is not of the form $g^p - g + b$, where $b \in \mathbb{F}_p$ and $g \in \mathbb{F}_p[u_1,...,u_n]$. (This condition is certainly satisfied if $d < p$.) Then

$$\left| \sum_{x_1,...,x_n \in \mathbb{F}_p} e^{2\pi i f(x_1,...,x_n)/p} \right| \leq (d-1)p^{n-1/2}.$$

We mention one more application of the Weil conjectures. *Ramanujan's tau-function* is defined by

$$q \prod_{n=1}^{\infty} (1 - q^n)^{24} = \sum_{n=1}^{\infty} \tau(n)q^n.$$

It was conjectured by Ramanujan (1916), and proved by Mordell (1920), that

$$\sum_{n=1}^{\infty} \tau(n)/n^s = \prod_p (1 - \tau(p)p^{-s} + p^{11-2s})^{-1},$$

where the product is over all primes p. Ramanujan further conjectured that $|\tau(p)| \leq 2p^{11/2}$ for all p, and Deligne (1968/9) showed that this was a consequence of the (at that time unproven) Weil conjectures.

The prime number theorem also has an interesting analogue in the theory of dynamical systems. Let M be a compact Riemannian manifold with negative sectional curvatures, and let $N(T)$ denote the number of different (oriented) closed geodesics on M of length $\leq T$. It was first shown by Margulis (1970) that

$$N(T) \sim e^{hT}/hT \quad \text{as } T \to \infty,$$

where the positive constant h is the topological entropy of the associated geodesic flow.

Although much of the detail is specific to the problem, a proof may be given which has the same structure as the proof in §3 of the prime number theorem. If P is an arbitrary closed orbit of the geodesic flow and $\lambda(P)$ its least period, one shows that the zeta-function

$$\zeta_M(s) = \prod_P (1 - e^{-s\lambda(P)})^{-1}$$

is nonzero and holomorphic for $\Re s \geq h$, except for a simple pole at $s = h$, and then applies Ikehara's theorem. The study of geodesics on a surface of negative curvature was initiated by Hadamard (1898), but it is unlikely that he realized there was a connection with the prime number theorem which he had proved two years earlier!

6 Alternative formulations

There is an intimate connection between the *Dirichlet products* considered in §3 of Chapter III and *Dirichlet series*. It is easily seen that if the Dirichlet series

$$f(s) = \sum_{n=1}^{\infty} a(n)/n^s, \quad g(s) = \sum_{n=1}^{\infty} b(n)/n^s,$$

are absolutely convergent for $\Re s > \alpha$, then the product $h(s) = f(s)g(s)$ may also be represented by an absolutely convergent Dirichlet series for $\Re s > \alpha$:

$$h(s) = \sum_{n=1}^{\infty} c(n)/n^s,$$

where $c = a * b$, i.e.

$$c(n) = \sum_{d|n} a(d)b(n/d) = \sum_{d|n} a(n/d)b(d).$$

This implies, in particular, that for $\Re s > 1$

$$\zeta^2(s) = \sum_{n=1}^{\infty} \tau(n)/n^s, \quad \zeta(s-1)\zeta(s) = \sum_{n=1}^{\infty} \sigma(n)/n^s,$$

where as in Chapter III (not as in §5),

$$\tau(n) = \sum_{d|n} 1, \quad \sigma(n) = \sum_{d|n} d,$$

denote respectively the number of positive divisors of n and the sum of the positive divisors of n. The relation for Euler's phi-function,

$$\sigma(n) = \sum_{d|n} \tau(n/d)\varphi(d),$$

which was proved in Chapter III, now yields for $\Re s > 1$

$$\zeta(s-1)/\zeta(s) = \sum_{n=1}^{\infty} \varphi(n)/n^s.$$

From the property by which we defined the Möbius function we obtain also, for $\Re s > 1$,

$$1/\zeta(s) = \sum_{n=1}^{\infty} \mu(n)/n^s.$$

In view of this relation it is not surprising that the distribution of prime numbers is closely connected with the behaviour of the Möbius function. Put

$$M(x) = \sum_{n \le x} \mu(n).$$

Since $|\mu(n)| \le 1$, it is obvious that $|M(x)| \le \lfloor x \rfloor$ for $x > 0$. The next result is not so obvious:

PROPOSITION 12 $M(x)/x \to 0$ *as* $x \to \infty$.

Proof The function $f(s): = \zeta(s) + 1/\zeta(s)$ is holomorphic for $\sigma \ge 1$, except for a simple pole with residue 1 at $s = 1$. Moreover

$$f(s) = \sum_{n=1}^{\infty} \{1 + \mu(n)\}/n^s = \int_{1_-}^{\infty} x^{-s} \, d\phi(x) \quad \text{for } \sigma > 1,$$

where $\phi(x) = \lfloor x \rfloor + M(x)$ is a nondecreasing function. Since

$$f(s) = \int_{0_-}^{\infty} e^{-su} \, d\phi(e^u),$$

it follows from Ikehara's Theorem 9 that $\phi(x) \sim x$. $\square$

Proposition 12 is equivalent to the prime number theorem in the sense that either of the relations $M(x) = o(x)$, $\psi(x) \sim x$ may be deduced from the other by elementary (but not trivial) arguments.

The Riemann hypothesis also has an equivalent formulation in terms of the function $M(x)$. Suppose

$$M(x) = O(x^{\alpha}) \text{ as } x \to \infty,$$

for some α such that $0 < \alpha < 1$. For $\sigma > 1$ we have

$$1/\zeta(s) = \int_{1_-}^{\infty} x^{-s} \, dM(x) = s \int_{1}^{\infty} x^{-s-1} M(x) \, dx.$$

But for $\sigma > \alpha$ the integral on the right is convergent and defines a holomorphic function. Consequently it is the analytic continuation of $1/\zeta(s)$. Thus if $\alpha*$ again denotes the least upper bound of all zeros of $\zeta(s)$, then $\alpha \geq \alpha* \geq 1/2$. On the other hand, Littlewood (1912) showed that

$$M(x) = O(x^{\alpha*+\varepsilon}) \text{ for every } \varepsilon > 0.$$

It follows that *the Riemann hypothesis holds if and only if $M(x) = O(x^{\alpha})$ for every $\alpha > 1/2$.*

It has already been mentioned that the first 1.5×10^9 zeros of $\zeta(s)$ on the line $\sigma = 1/2$ are simple. It is likely that the Riemann hypothesis does not tell the whole story and that all zeros of $\zeta(s)$ on the line $\sigma = 1/2$ are simple. Thus it is of interest that this is guaranteed by a sufficiently sharp bound for $M(x)$. We will show that *if*

$$M(x) = O(x^{1/2} \log^{\alpha} x) \text{ as } x \to \infty,$$

for some $\alpha < 1$, then not only do all nontrivial zeros of $\zeta(s)$ lie on the line $\sigma = 1/2$ but they are all simple.

Let $\rho = 1/2 + i\gamma$ be a zero of $\zeta(s)$ of multiplicity $m \geq 1$ and take $s = \rho + h$, where $h > 0$. Then $\sigma = 1/2 + h$ and, since

$$1/\zeta(s) = s \int_{1}^{\infty} x^{-s-1} M(x) \, dx \text{ for } \sigma > 1/2,$$

we have

$$|1/\zeta(s)| \leq |s| \int_{1}^{\infty} x^{-\sigma-1} |M(x)| \, dx = O(|s|) \int_{1}^{\infty} x^{-h-1} \log^{\alpha} x \, dx$$

$$= O(|s|) \int_{0}^{\infty} e^{-hu} u^{\alpha} \, du = O(|s|) \, \Gamma(\alpha+1)/h^{\alpha+1}.$$

Thus $h^{\alpha+1}|1/\zeta(s)|$ is bounded for $h \to +0$ and hence $m \leq \alpha + 1$. Since m is an integer and $\alpha < 1$, this implies $m = 1$ and $\alpha \geq 0$.

The prime number theorem, in the equivalent form $M(x) = o(x)$, says that asymptotically $\mu(n)$ takes the values $+1$ and -1 with equal probability. By assuming that asymptotically the values $\mu(n)$ behave like independent random variables Good and Churchhouse (1968) have been led to two striking conjectures, analogous to the central limit theorem and the law of the iterated logarithm in the theory of probability:

CONJECTURE A   *If $N(n) \to \infty$ and $\log N/\log n \to 0$, then*

$$P_n \left\{ \frac{M(m+N) - M(m)}{\left(6N/\pi^2\right)^{1/2}} < t \right\} \to (2\pi)^{-1/2} \int_{-\infty}^{t} e^{-u^2/2} \, du,$$

where

$$P_n\{f(m) < t\} = \#\{m \leq n: f(m) < t\}/n.$$

CONJECTURE B $\overline{\lim}_{x\to\infty} M(x)(2x \log \log x)^{-1/2} = \sqrt{6}/\pi$

$$= -\underline{\lim}_{x\to\infty} M(x)(2x \log \log x)^{-1/2}.$$

From what has been said above, Conjecture B implies not only the Riemann hypothesis, but also that the zeros of $\zeta(s)$ are all simple. These probabilistic conjectures provide a more interesting reason than symmetry for believing in the validity of the Riemann hypothesis, but no progress has so far been made towards proving them.

7 Some further problems

A prime p is said to be a *twin prime* if $p + 2$ is also a prime. For example, 41 is a twin prime since both 41 and 43 are primes. It is still not known if there are infinitely many twin primes. However Brun (1919), using the sieve method which he devised for the purpose, showed that, if infinite, the sum of the reciprocals of all twin primes converges. Since the sum of the reciprocals of all primes diverges, this means that few primes are twin primes.

By a formal application of their circle method Hardy and Littlewood (1923) were led to conjecture that

$$\pi_2(x) \sim L_2(x) \text{ for } x \to \infty,$$

where $\pi_2(x)$ denotes the number of twin primes $\leq x$,

$$L_2(x) = 2C_2 \int_2^x dt/\log^2 t$$

and

$$C_2 = \prod_{p \geq 3} (1 - 1/(p-1)^2) = 0.66016181... .$$

This implies that $\pi_2(x)/\pi(x) \sim 2C_2/\log x$. Table 2, adapted from Brent (1975), shows that Hardy and Littlewood's formula agrees well with the facts. Brent also calculates

$$\sum_{\text{twin } p \leq 10^{10}} (1/p + 1/(p+2)) = 1.78748...$$

and, using the Hardy–Littlewood formula for the tail, obtains the estimate

$$\sum_{\text{all twin } p} (1/p + 1/(p+2)) = 1.90216... .$$

His calculations have been considerably extended by Nicely (1995).

x	$\pi_2(x)$	$L_2(x)$	$\pi_2(x)/L_2(x)$
10^3	35	46	0.76
10^4	205	214	0.96
10^5	1224	1249	0.980
10^6	8169	8248	0.9904
10^7	58980	58754	1.0038
10^8	440312	440368	0.99987
10^9	3424506	3425308	0.99977
10^{10}	27412679	27411417	1.000046

TABLE 2

Besides the twin prime formula many other asymptotic formulae were conjectured by Hardy and Littlewood. Most of them are contained in a general conjecture, which will now be described.

Let $f(t)$ be a polynomial in t of positive degree with integer coefficients. If $f(n)$ is prime for infinitely many positive integers n, then f has positive leading coefficient, f is irreducible over the field $\mathbb{Q}$ of rational numbers and, for each prime p, there is a positive integer n for which $f(n)$ is not divisible by p. It was conjectured by Bouniakowsky (1857) that conversely, if these three conditions are satisfied, then $f(n)$ is prime for infinitely many positive integers n. Schinzel (1958) extended the conjecture to several polynomials and Bateman and Horn (1962) gave Schinzel's conjecture the following quantitative form.

Let $f_j(t)$ be a polynomial in t of degree $d_j \geq 1$, with integer coefficients and positive leading coefficient, which is irreducible over the field $\mathbb{Q}$ of rational numbers ($j = 1,...,m$). Suppose also that the polynomials $f_1(t),...,f_m(t)$ are distinct and that, for each prime p, there is a positive integer n for which the product $f_1(n)\cdots f_m(n)$ is not divisible by p. Bateman and Horn's conjecture states that, if $N(x)$ is the number of positive integers $n \leq x$ for which $f_1(n),...,f_m(n)$ are all primes, then

$$N(x) \sim (d_1 \cdots d_m)^{-1} C(f_1,...,f_m) \int_2^x dt / \log^m t,$$

where

$$C(f_1,...,f_m) = \prod_p \{(1 - 1/p)^{-m}(1 - \omega(p)/p)\},$$

the product being taken over all primes p and $\omega(p)$ denoting the number of $u \in \mathbb{F}_p$ (the field of p elements) such that $f_1(u)\cdots f_m(u) = 0$. (The convergence of the infinite product when the primes are taken in their natural order follows from the prime ideal theorem.)

The twin prime formula is obtained by taking $m = 2$ and $f_1(t) = t, f_2(t) = t + 2$. By taking instead $f_1(t) = t, f_2(t) = 2t + 1$, the Bateman–Horn conjecture gives the same asymptotic formula $\pi_G(x) \sim L_2(x)$ for the number $\pi_G(x)$ of primes $p \leq x$ for which $2p + 1$ is also a prime ('Sophie Germain' primes). By taking $m = 1$ and $f_1(t) = t^2 + 1$ one obtains an asymptotic formula for the number of primes of the form $n^2 + 1$.

Bateman and Horn gave a heuristic derivation of their formula. However, the only case in which the formula has actually been proved is $m = 1, n_1 = 1$. This is the case of primes in an arithmetic progression which will be considered in the next chapter. When one considers the vast output of mathematical papers today compared with previous eras, it is salutary to recall that we still do not know as much about twin primes as Euclid knew about primes.

8 Further remarks

The historical development of the prime number theorem is traced in Landau [33]. The original papers of Chebyshev are available in [56]. Pintz [48] has given a simple proof of Chebyshev's result that $\pi(x) = x/(A \log x - B + o(1))$ implies $A = B = 1$.

There is an English translation of Riemann's memoir in Edwards [20]. Complex variable proofs of the prime number theorem, with error term, are contained in the books of Ayoub [4], Ellison and Ellison [21], and Patterson [47]. For a simple complex variable proof without error term, due to Newman (1980), see Zagier [63].

A proof with error term by the Wiener–Ikehara method is given in Čižek [12]. Wiener's general Tauberian theorem is proved in Rudin [52]. For its algebraic interpretation, see the resumé of Fourier analysis in [13]. The development of Selberg's method is surveyed in Diamond [18]. An elementary proof of the prime number theorem which is quite different from that of Selberg and Erdös has been given by Daboussi [15].

A clear account of Stieltjes integrals is given in Widder [62]. However, we do not use Stieltjes integrals in any essential way, but only for the formal convenience of treating integration by parts and summation by parts in the same manner. Widder's book also contains the Wiener–Ikehara proof of the prime number theorem.

By a theorem of S. Bernstein (1928), proved in Widder's book and also in Mattner [38], the hypotheses of Proposition 7 can be stated without reference to the function $\phi(x)$. Bernstein's theorem says that a real-valued function $F(\sigma)$ can be represented in the form

$$F(\sigma) = \int_0^\infty e^{-\sigma x} \, d\phi(x),$$

where $\phi(x)$ is a nondecreasing function for $x \geq 0$ and the integral is convergent for every $\sigma > 1$, if and only if $F(\sigma)$ has derivatives of all orders and

$$(-1)^k F^{(k)}(\sigma) \geq 0 \quad \text{for every } \sigma > 1 \quad (k = 0,1,2,\dots).$$

For the Poisson summation formula see, for example, Lasser [34] and Durán et al. [19]. There is a useful n-dimensional generalization, discussed more fully in §7 of Chapter XII, in which a sum over all points of a lattice is related to a sum over all points of the dual lattice. Further generalizations are mentioned in Chapter X.

More extended treatments of the gamma function are given in Andrews et al. [3] and Remmert [49].

More information about the Riemann zeta-function is given in the books of Patterson [47], Titchmarsh [57], and Karatsuba and Voronin [30]. For numerical data, see Rosser and Schoenfeld [50], van de Lune et al. [37] and Rumely [53].

For a proof that $\pi(x) - Li(x)$ changes sign infinitely often, see Diamond [17]. Estimates for values of x such that $\pi(x) > Li(x)$ are obtained by a technique due to Lehman [35]; for the most recent estimate, see Bays and Hudson [8].

For the pair correlation conjecture, see Montgomery [40], Goldston [24] and Odlyzko [45]. Random matrices are thoroughly discussed by Mehta [39]; for a nice introduction, see Tracy and Widom [58].

For Dedekind zeta functions see Stark [54], besides the books on algebraic number theory referred to in Chapter III. The prime ideal theorem is proved in Narkiewicz [44], for example. For consequences of the extended Riemann hypothesis, see Bach [5], Goldstein [23] and M.R.

Murty [41]. Many other generalizations of the zeta function are discussed in the article on zeta functions in [22].

Function fields are treated in the books of Chevalley [11] and Deuring [16]. The lengthy review of Chevalley's book by Weil in *Bull. Amer. Math. Soc.* **57** (1951), 384-398, is useful but over-critical. Even if geometric methods are better adapted for algebraic varieties of higher dimension, the algebraic methods available for curves are essentially simpler. Moreover it was the close analogy with number fields that suggested the possibility of a Riemann hypothesis for function fields. For a proof of the latter, see Bombieri [9]. For the Weil conjectures, see Weil [61] and Katz [32].

Stichtenoth [55] gives a good account of the theory of function fields with special emphasis on its applications to coding theory. For these applications, see also Goppa [26], Tsfasman *et al.* [60], and Tsfasman and Vladut [59]. Curves with a given genus which have the maximal number of $\mathbb{F}_q$-points are discussed by Cossidente *et al.* [14].

For introductions to Ramanujan's tau-function, see V.K. Murty [42] and Rankin's article (pp. 245-268) in Andrews *et al.* [2]. For analogues of the prime number theorem in the theory of dynamical systems, see Katok and Hasselblatt [31] and Parry and Pollicott [46]. Hadamard's pioneering study of geodesics on a surface of negative curvature and his proof of the prime number theorem are both reproduced in [27].

The 'equivalence' of Proposition 12 with the prime number theorem is proved in Ayoub [4]. A proof that the Riemann hypothesis is equivalent to $M(x) = O(x^\alpha)$ for every $\alpha > 1/2$ is contained in the book of Titchmarsh [57]. Good and Churchhouse's probabilistic conjectures appeared in [25]. For the central limit theorem and the law of the iterated logarithm see, for example, Adams [1], Kac [29], Bauer [7] and Loève [36].

Brun's theorem on twin primes is proved in Narkiewicz [43]. For numerical results, see Brent [10]. For conjectural asymptotic formulas, see Hardy and Littlewood [28] and Bateman and Horn [6]. There are several heuristic derivations of the twin prime formula, the most recent being Rubenstein [51]. It would be useful to try to analyse these heuristic derivations, so that the conclusion is seen as a consequence of precisely stated assumptions.

9 Selected references

[1] W.J. Adams, *The life and times of the central limit theorem*, Kaedmon, New York, 1974.

[2] G.E. Andrews, R.A. Askey, B.C. Berndt, K.G. Ramanathan and R.A. Rankin (ed.), *Ramanujan revisited*, Academic Press, London, 1988.

[3] G.E. Andrews, R. Askey and R. Roy, *Special functions*, Cambridge University Press, 1999.

[4] R. Ayoub, *An introduction to the analytic theory of numbers*, Math. Surveys no. 10, Amer. Math. Soc., Providence, 1963.

[5] E. Bach, Explicit bounds for primality testing and related problems, *Math. Comp.* **55** (1990), 353-380.

[6] P.T. Bateman and R.A. Horn, A heuristic asymptotic formula concerning the distribution of prime numbers, *Math. Comp.* **16** (1962), 363-367.

[7] H. Bauer, *Probability theory*, English transl. by R.B. Burckel, de Gruyter, Berlin, 1996.

[8] C. Bays and R.H. Hudson, A new bound for the smallest x with $\pi(x) > li(x)$, *Math. Comp.* **69** (1999), 1285-1296.

[9] E. Bombieri, Counting points on curves over finite fields (d'après S.A. Stepanov), *Séminaire Bourbaki vol. 1972/3, Exposés 418-435*, pp. 234-241, Lecture Notes in Mathematics **383** (1974), Springer-Verlag, Berlin.

[10] R.P. Brent, Irregularities in the distribution of primes and twin primes, *Math. Comp.* **29** (1975), 43-56.

[11] C. Chevalley, *Introduction to the theory of algebraic functions of one variable*, Math. Surveys no. 6, Amer. Math. Soc., New York, 1951.

[12] J. Čížek, On the proof of the prime number theorem, *Časopis Pěst. Mat.* **106** (1981), 395-401.

[13] W.A. Coppel, J.B. Fourier–On the occasion of his two hundredth birthday, *Amer. Math. Monthly* **76** (1969), 468-483.

[14] A. Cossidente, J.W.P. Hirschfeld, G. Korchmáros and F. Torres, On plane maximal curves, *Compositio Math.* **121** (2000), 163-181.

[15] H. Daboussi, Sur le théorème des nombres premiers, *C.R. Acad. Sci. Paris Sér. I* **298** (1984), 161-164.

[16] M. Deuring, *Lectures on the theory of algebraic functions of one variable*, Lecture Notes in Mathematics **314** (1973), Springer-Verlag, Berlin.

[17] H.G. Diamond, Changes of sign of $\pi(x) - li(x)$, *Enseign. Math.* **21** (1975), 1-14.

[18] H.G. Diamond, Elementary methods in the study of the distribution of prime numbers, *Bull. Amer. Math. Soc. (N.S.)* **7** (1982), 553-589.

[19] A.L. Durán, R. Estrada and R.P. Kanwal, Extensions of the Poisson summation formula, *J. Math. Anal. Appl.* **218** (1998), 581-606.

[20] H.M. Edwards, *Riemann's zeta function*, Academic Press, New York, 1974.

[21] W. Ellison and F. Ellison, *Prime numbers*, Wiley, New York, 1985.

[22] *Encyclopedic dictionary of mathematics* (ed. K. Ito), 2nd ed., Mathematical Society of Japan, MIT Press, Cambridge, Mass., 1987.

[23] L.J. Goldstein, Density questions in algebraic number theory, *Amer. Math. Monthly* **78** (1971), 342-351.

[24] D.A. Goldston, On the pair correlation conjecture for zeros of the Riemann zeta-function, *J. Reine Angew. Math.* **385** (1988), 24-40.

[25] I.J. Good and R.F. Churchhouse, The Riemann hypothesis and pseudorandom features of the Möbius sequence, *Math. Comp.* **22** (1968), 857-861.

[26] V.D. Goppa, Codes on algebraic curves, *Soviet Math. Dokl.* **24** (1981), 170-172.

[27] J. Hadamard, *Selecta*, Gauthier-Villars, Paris, 1935.

[28] G.H. Hardy and J.E. Littlewood, Some problems of partitio numerorum III, On the expression of a number as a sum of primes, *Acta Math.* **44** (1923), 1-70.

[29] M. Kac, *Statistical independence in probability, analysis and number theory*, Carus Mathematical Monograph **12**, Math. Assoc. of America, 1959.

[30] A.A. Karatsuba and S.M. Voronin, *The Riemann zeta-function*, English transl. by N. Koblitz, de Gruyter, Berlin, 1992.

[31] A. Katok and B. Hasselblatt, *Introduction to the modern theory of dynamical systems*, Cambridge University Press, Cambridge, 1995.

[32] N. Katz, An overview of Deligne's proof of the Riemann hypothesis for varieties over finite fields, *Mathematical developments arising from Hilbert problems*, Proc. Sympos. Pure Math. **28**, pp. 275-305, Amer. Math. Soc., Providence, 1976.

[33] E. Landau, *Handbuch der Lehre von der Verteilung der Primzahlen* (2 vols.), 2nd ed., Chelsea, New York, 1953.

[34] R. Lasser, *Introduction to Fourier series*, M. Dekker, New York, 1996.

[35] R.S. Lehman, On the difference $\pi(x) - li(x)$, *Acta Arith.* **11** (1966), 397-410.

[36] M. Loève, *Probability theory*, 4th ed. in 2 vols., Springer-Verlag, New York, 1978.

[37] J. van de Lune *et al.*, On the zeros of the Riemann zeta function in the critical strip IV, *Math. Comp.* **46** (1986), 667-681.

[38] L. Mattner, Bernstein's theorem, inversion formula of Post and Widder, and the uniqueness theorem for Laplace transforms, *Exposition. Math.* **11** (1993), 137-140.

[39] M.L. Mehta, *Random matrices*, 2nd ed., Academic Press, New York, 1991.

[40] H.L. Montgomery, The pair correlation of zeros of the zeta function, *Proc. Sympos. Pure Math.* **24**, pp. 181-193, Amer. Math. Soc., Providence, 1973.

[41] M. R. Murty, Artin's conjecture for primitive roots, *Math. Intelligencer* **10** (1988), no. 4, 59-67.

[42] V.K. Murty, Ramanujan and Harish-Chandra, *Math. Intelligencer* **15** (1993), no.2, 33-39.

[43] W. Narkiewicz, *Number theory*, World Scientific, Singapore, 1983.

[44] W. Narkiewicz, *Elementary and analytic theory of algebraic numbers*, 2nd ed., Springer-Verlag, Berlin, 1990.

[45] A.M. Odlyzko, On the distribution of spacings between zeros of the zeta function, *Math. Comp.* **48** (1987), 273-308.

[46] W. Parry and M. Pollicott, An analogue of the prime number theorem for closed orbits of Axiom A flows, *Ann. of Math.* **118** (1983), 573-591.

[47] S.J. Patterson, *An introduction to the theory of the Riemann zeta-function*, Cambridge University Press, Cambridge, 1988.

[48] J. Pintz, On Legendre's prime number formula, *Amer. Math. Monthly* **87** (1980), 733-735.

[49] R. Remmert, *Classical topics in complex function theory*, English transl. by L. Kay, Springer-Verlag, New York, 1998.

[50] J.B. Rosser and L. Schoenfeld, Approximate formulas for some functions of prime numbers, *Illinois J. Math.* **6** (1962), 64-94.

[51] M. Rubinstein, A simple heuristic proof of Hardy and Littlewood's conjecture B, *Amer. Math. Monthly* **100** (1993), 456-460.

[52] W. Rudin, *Functional analysis*, McGraw-Hill, New York, 1973.

[53] R. Rumely, Numerical computations concerning the ERH, *Math. Comp.* **61** (1993), 415-440.

[54] H.M. Stark, The analytic theory of algebraic numbers, *Bull. Amer. Math. Soc.* **81** (1975), 961-972.

[55] H. Stichtenoth, *Algebraic function fields and codes*, Springer-Verlag, Berlin, 1993.

[56] P.L. Tchebychef, *Oeuvres* (2 vols.), reprinted Chelsea, New York, 1962.

[57] E.C. Titchmarsh, *The theory of the Riemann zeta-function*, 2nd ed. revised by D.R. Heath-Brown, Clarendon Press, Oxford, 1986.

[58] C.A. Tracy and H. Widom, Introduction to random matrices, *Geometric and quantum aspects of integrable systems* (ed. G.F. Helminck), pp. 103-130, Lecture Notes in Physics **424**, Springer-Verlag, Berlin, 1993.

[59] M.A. Tsfasman and S.G. Vladut, *Algebraic-geometric codes*, Kluwer, Dordrecht, 1991.

[60] M.A. Tsfasman, S.G. Vladut and Th. Zink, Modular curves, Shimura curves, and Goppa codes, *Math. Nachr.* **109** (1982), 21-28.

[61] A. Weil, Number of solutions of equations in finite fields, *Bull. Amer. Math. Soc.* **55** (1949), 497-508.

[62] D.V. Widder, *The Laplace transform*, Princeton University Press, Princeton, 1941.

[63] D. Zagier, Newman's short proof of the prime number theorem, *Amer. Math. Monthly* **104** (1997), 705-708.

X
A character study

1 Primes in arithmetic progressions

Let a and m be integers with $1 \le a < m$. If a and m have a common divisor $d > 1$, then no term after the first of the arithmetic progression

$$a, a + m, a + 2m, \ldots \qquad (*)$$

is a prime. Legendre (1788) conjectured, and later (1808) attempted a proof, that *if a and m are relatively prime, then the arithmetic progression* (*) *contains infinitely many primes.*

If $a_1,\ldots,a_h$ are the positive integers less than m and relatively prime to m, and if $\pi_j(x)$ denotes the number of primes $\le x$ in the arithmetic progression

$$a_j, a_j + m, a_j + 2m, \ldots ,$$

then Legendre's conjecture can be stated in the form

$$\pi_j(x) \to \infty \ \text{ as } x \to \infty \ \ (j = 1,\ldots,h).$$

Legendre (1830) subsequently conjectured, and again gave a faulty proof, that

$$\pi_j(x)/\pi_k(x) \to 1 \ \text{ as } x \to \infty \ \text{ for all } j,k.$$

Since the total number $\pi(x)$ of primes $\le x$ satisfies

$$\pi(x) \ = \ \pi_1(x) + \ldots + \pi_h(x) + c,$$

where c is the number of different primes dividing m, Legendre's second conjecture is equivalent to

$$\pi_j(x)/\pi(x) \to 1/h \text{ as } x \to \infty \ \ (j = 1,\ldots,h).$$

Here $h = \varphi(m)$ is the number of positive integers less than m and relatively prime to m. If one assumes the truth of the prime number theorem, then the second conjecture is also equivalent to

$$\pi_j(x) \sim x/\varphi(m)\log x \quad (j = 1,...,\varphi(m)).$$

The validity of the second conjecture in this form is known as the *prime number theorem for arithmetic progressions*.

Legendre's first conjecture was proved by Dirichlet (1837) in an outstanding paper which combined number theory, algebra and analysis. His algebraic innovation was the use of *characters* to isolate the primes belonging to a particular residue class mod m. Legendre's second conjecture, which implies the first, was proved by de la Vallée Poussin (1896), again using characters, at the same time that he proved the ordinary prime number theorem.

Selberg (1949),(1950) has given proofs of both conjectures which avoid the use of complex analysis, but they are not very illuminating. The prime number theorem for arithmetic progressions will be proved here by an extension of the method used in the previous chapter to prove the ordinary prime number theorem.

For any integer a, with $1 \le a < m$ and $(a,m) = 1$, let

$$\pi(x;m,a) = \Sigma_{p \le x, p \equiv a \bmod m} \ 1.$$

Also, generalizing the definition of Chebyshev's functions in the previous chapter, put

$$\theta(x;m,a) = \Sigma_{p \le x, p \equiv a \bmod m} \ \log p, \quad \psi(x;m,a) = \Sigma_{n \le x, n \equiv a \bmod m} \ \Lambda(n).$$

Exactly as in the last chapter, we can show that the prime number theorem for arithmetic progressions,

$$\pi(x;m,a) \sim x/\varphi(m)\log x \quad \text{as } x \to \infty,$$

is equivalent to

$$\psi(x;m,a) \sim x/\varphi(m) \quad \text{as } x \to \infty.$$

It is in this form that the theorem will be proved.

2 Characters of finite abelian groups

Let G be an abelian group with identity element e. A *character* of G is a function $\chi: G \to \mathbb{C}$ such that

(i) $\chi(ab) = \chi(a)\chi(b)$ for all $a,b \in G$,
(ii) $\chi(c) \ne 0$ for some $c \in G$.

Since $\chi(c) = \chi(ca^{-1})\chi(a)$, by (i), it follows from (ii) that $\chi(a) \neq 0$ for every $a \in G$. (Thus χ is a *homomorphism* of G into the multiplicative group $\mathbb{C}^\times$ of nonzero complex numbers.) Moreover, since $\chi(a) = \chi(a)\chi(e)$, we must have $\chi(e) = 1$. Since $\chi(a)\chi(a^{-1}) = \chi(e)$, it follows that $\chi(a^{-1}) = \chi(a)^{-1}$.

The function $\chi_1 : G \to \mathbb{C}$ defined by $\chi_1(a) = 1$ for every $a \in G$ is obviously a character of G, the *trivial character* (also called the *principal* character!). Moreover, for any character χ of G, the function $\chi^{-1} : G \to \mathbb{C}$ defined by $\chi^{-1}(a) = \chi(a)^{-1}$ is also a character of G. Furthermore, if χ' and χ'' are characters of G, then the function $\chi'\chi'' : G \to \mathbb{C}$ defined by $\chi'\chi''(a) = \chi'(a)\chi''(a)$ is a character of G. Since

$$\chi_1\chi = \chi, \quad \chi'\chi'' = \chi''\chi', \quad \chi(\chi'\chi'') = (\chi\chi')\chi'',$$

it follows that the set $\hat{G}$ of all characters of G is itself an abelian group, the *dual group* of G, with the trivial character as identity element.

Suppose now that the group G is finite, of order g say. Then $\chi(a)$ is a g-th root of unity for every $a \in G$, since $a^g = e$ and hence

$$\chi(a)^g = \chi(a^g) = \chi(e) = 1.$$

It follows that $|\chi(a)| = 1$ and $\chi^{-1}(a) = \overline{\chi(a)}$. Thus we will sometimes write $\overline{\chi}$ instead of χ^{-1}.

PROPOSITION 1 *The dual group $\hat{G}$ of a finite abelian group G is a finite abelian group of the same order. Moreover, if $a \in G$ and $a \neq e$, then $\chi(a) \neq 1$ for some $\chi \in \hat{G}$.*

Proof Let g denote the order of G. Suppose first that G is a cyclic group, generated by the element c. Then any character χ of G is uniquely determined by the value $\chi(c)$, which is a g-th root of unity. Conversely if $\omega_j = e^{2\pi ij/g}$ ($0 \leq j < g$) is a g-th root of unity, then the functions $\chi^{(j)} : G \to \mathbb{C}$ defined by $\chi^{(j)}(c^k) = \omega_j^k$ are distinct characters of G and $\chi^{(1)}(c^k) \neq 1$ for $1 \leq k < g$. It follows that the proposition is true when G is cyclic. The general case can be reduced to this by using the fact (see §4 of Chapter III) that any finite abelian group is a direct product of cyclic groups. However, it can also be treated directly in the following way.

We use induction on g and suppose that G is not cyclic. Let H be a maximal proper subgroup of G and let h be the order of H. Let $a \in G \setminus H$ and let r be the least positive integer such that $b = a^r \in H$. Since G is generated by H and a, and $a^n \in H$ if and only if r divides n, each $x \in G$ can be uniquely expressed in the form

$$x = a^k y,$$

where $y \in H$ and $0 \leq k < r$. Hence $g = rh$.

If χ is any character of G, its restriction to H is a character ψ of H. Moreover χ is uniquely determined by ψ and the value $\chi(a)$, since

$$\chi(a^k y) = \chi(a)^k \psi(y).$$

Since $\chi(a)^r = \psi(b)$ is a root of unity, $\omega = \chi(a)$ is a root of unity such that $\omega^r = \psi(b)$.

Conversely, it is easily verified that, for each character ψ of H and for each of the r roots of unity ω such that $\omega^r = \psi(b)$, the function $\chi: G \to \mathbb{C}$ defined by $\chi(a^k y) = \omega^k \psi(y)$ is a character of G. Since H has exactly h characters by the induction hypothesis, it follows that G has exactly $rh = g$ characters. It remains to show that if $a^k y \neq e$, then $\chi(a^k y) \neq 1$ for some χ. But if $\omega^k \psi(y) = 1$ for all ω, then $k = 0$; hence $y \neq e$ and $\chi(y) = \psi(y) \neq 1$ for some ψ, by the induction hypothesis. $\square$

PROPOSITION 2 *Let G be a finite abelian group of order g and $\hat{G}$ its dual group. Then*

(i) $\displaystyle \sum_{a \in G} \chi(a) = \begin{cases} g & \text{if } \chi = \chi_1, \\ 0 & \text{if } \chi \neq \chi_1. \end{cases}$

(ii) $\displaystyle \sum_{\chi \in \hat{G}} \chi(a) = \begin{cases} g & \text{if } a = e, \\ 0 & \text{if } a \neq e. \end{cases}$

Proof Put

$$S = \sum_{a \in G} \chi(a).$$

Since it is obvious that $S = g$ if $\chi = \chi_1$, we assume $\chi \neq \chi_1$. Then $\chi(b) \neq 1$ for some $b \in G$. Since ab runs through all elements of G at the same time as a,

$$\chi(b)S = \sum_{a \in G} \chi(a)\chi(b) = \sum_{a \in G} \chi(ab) = S.$$

Since $\chi(b) \neq 1$, it follows that $S = 0$.

Now put

$$T = \sum_{\chi \in \hat{G}} \chi(a).$$

Evidently $T = g$ if $a = e$ since, by Proposition 1, $\hat{G}$ also has order g. Thus we now assume $a \neq e$. By Proposition 1 also, for some $\psi \in \hat{G}$ we have $\psi(a) \neq 1$. Since $\chi\psi$ runs through all elements of $\hat{G}$ at the same time as χ,

$$\psi(a)T = \sum_{\chi \in \hat{G}} \chi(a)\psi(a) = \sum_{\chi \in \hat{G}} \chi\psi(a) = T.$$

Since $\psi(a) \neq 1$, it follows that $T = 0$. $\square$

Since the product of two characters is again a character, and since $\overline{\psi}$ is the inverse of the character ψ, Proposition 2(i) can be stated in the apparently more general form

(i)' $\displaystyle\sum_{a \in G} \chi(a)\overline{\psi}(a) = \begin{cases} g & \text{if } \chi = \psi, \\ 0 & \text{if } \chi \neq \psi. \end{cases}$

Similarly, since $\overline{\chi}(b) = \chi(b^{-1})$, Proposition 2(ii) can be stated in the form

(ii)' $\displaystyle\sum_{\chi \in \hat{G}} \chi(a)\overline{\chi}(b) = \begin{cases} g & \text{if } a = b, \\ 0 & \text{if } a \neq b. \end{cases}$

The relations (i)' and (ii)' are known as the *orthogonality relations*, for the characters and elements respectively, of a finite abelian group.

3 Proof of the prime number theorem for arithmetic progressions

The finite abelian group in which we are interested is the multiplicative group $\mathbb{Z}_{(m)}^{\times}$ of integers relatively prime to m, where $m > 1$ will be fixed from now on. The group $G_m = \mathbb{Z}_{(m)}^{\times}$ has order $\varphi(m)$, where $\varphi(m)$ denotes as usual the number of positive integers less than m and relatively prime to m.

A *Dirichlet character* mod m is defined to be a function $\chi: \mathbb{Z} \to \mathbb{C}$ with the properties

(i) $\chi(ab) = \chi(a)\chi(b)$ for all $a,b \in \mathbb{Z}$,

(ii) $\chi(a) = \chi(b)$ if $a \equiv b \bmod m$,

(iii) $\chi(a) \neq 0$ if and only if $(a,m) = 1$.

Any character χ of G_m can be extended to a Dirichlet character mod m by putting $\chi(a) = 0$ if $a \in \mathbb{Z}$ and $(a,m) \neq 1$. Conversely, on account of (ii), any Dirichlet character mod m uniquely determines a character of G_m.

To illustrate the definition, here are some examples of Dirichlet characters. In each case we set $\chi(a) = 0$ if $(a,m) \neq 1$.

(I) $m = p$ is an odd prime and $\chi(a) = (a/p)$ if $p \nmid a$, where (a/p) is the Legendre symbol;

(II) $m = 4$ and $\chi(a) = 1$ or -1 according as $a \equiv 1$ or $-1 \bmod 4$;

(III) $m = 8$ and $\chi(a) = 1$ or -1 according as $a \equiv \pm 1$ or $\pm 3 \bmod 8$.

We now return to the general case. By the results of the previous section we have

$$\sum_{n=1}^{m} \chi(n) = \begin{cases} \varphi(m) & \text{if } \chi = \chi_1, \\ 0 & \text{if } \chi \neq \chi_1, \end{cases}$$

and

$$\sum_{\chi} \chi(a) = \begin{cases} \varphi(m) & \text{if } a \equiv 1 \bmod m, \\ 0 & \text{otherwise}, \end{cases}$$

where χ runs through all Dirichlet characters mod m. Furthermore

$$\sum_{n=1}^{m} \chi(n) \overline{\psi}(n) = \begin{cases} \varphi(m) & \text{if } \chi = \psi, \\ 0 & \text{if } \chi \neq \psi, \end{cases}$$

and

$$\sum_{\chi} \chi(a) \overline{\chi}(b) = \begin{cases} \varphi(m) & \text{if } (a,m) = 1 \text{ and } a \equiv b \bmod m, \\ 0 & \text{otherwise}. \end{cases}$$

LEMMA 3 *If* $\chi \neq \chi_1$ *is a Dirichlet character* mod m *then, for any positive integer N,*

$$\left| \sum_{n=1}^{N} \chi(n) \right| \leq \varphi(m)/2.$$

Proof Any positive integer N can be written in the form $N = qm + r$, where $q \geq 0$ and $1 \leq r \leq m$. Since $\chi(a) = \chi(b)$ if $a \equiv b \bmod m$, we have

$$\sum_{n=1}^{N} \chi(n) = \left(\sum_{n=1}^{m} + \sum_{n=m+1}^{2m} + \ldots + \sum_{n=(q-1)m+1}^{qm} \right)\chi(n) + \sum_{n=qm+1}^{qm+r} \chi(n)$$

$$= q \sum_{n=1}^{m} \chi(n) + \sum_{n=1}^{r} \chi(n).$$

But $\sum_{n=1}^{m} \chi(n) = 0$, since $\chi \neq \chi_1$. Hence

$$\sum_{n=1}^{N} \chi(n) = \sum_{n=1}^{r} \chi(n) = - \sum_{n=r+1}^{m} \chi(n).$$

Since $|\chi(n)| = 1$ or 0 according as $(n,m) = 1$ or $(n,m) \neq 1$, and since $\varphi(m)$ is the number of positive integers $n \leq m$ such that $(n,m) = 1$, the result follows. □

With each Dirichlet character χ, there is associated a *Dirichlet L-function*

$$L(s,\chi) = \sum_{n=1}^{\infty} \chi(n)/n^s.$$

Since $|\chi(n)| \leq 1$ for all n, the series is absolutely convergent for $\sigma := \Re s > 1$. We are going to show that if $\chi \neq \chi_1$, then the series is also convergent for $\sigma > 0$. (It does not converge if $\sigma \leq 0$, since then $|\chi(n)/n^s| \geq 1$ for infinitely many n.)

Put

Then
$$H(x) = \sum_{n \le x} \chi(n).$$

$$\sum_{n \le x} \chi(n) n^{-s} = \int_{1-}^{x+} t^{-s} dH(t)$$

$$= H(x) x^{-s} + s \int_1^x H(t) t^{-s-1} dt.$$

Since $H(x)$ is bounded, by Lemma 3, on letting $x \to \infty$ we obtain

$$L(s,\chi) = s \int_1^\infty H(t) t^{-s-1} dt \text{ for } \sigma > 0.$$

Moreover the integral on the right is uniformly convergent in any half-plane $\sigma \ge \delta$, where $\delta > 0$, and hence $L(s,\chi)$ is a holomorphic function for $\sigma > 0$.

The following discussion of Dirichlet L-functions and the prime number theorem for arithmetic progressions runs parallel to that of the Riemann ζ-function and the ordinary prime number theorem in the previous chapter. Consequently we will be more brief.

PROPOSITION 4 $L(s,\chi) = \prod_p (1 - \chi(p)p^{-s})^{-1}$ *for* $\sigma > 1$, *where the product is taken over all primes* p.

Proof The property $\chi(ab) = \chi(a)\chi(b)$ for all $a,b \in \mathbb{N}$ enables the proof of Euler's product formula for $\zeta(s)$ to be carried over to the present case. For $\sigma > 0$ we have

$$(1 - \chi(p)p^{-s})^{-1} = 1 + \chi(p)p^{-s} + \chi(p^2)p^{-2s} + \chi(p^3)p^{-3s} + \dots$$

and hence for $\sigma > 1$

$$\prod_{p \le x} (1 - \chi(p)p^{-s})^{-1} = \sum_{n \le N_x} \chi(n) n^{-s},$$

where N_x is the set of all positive integers whose prime factors are all $\le x$. Letting $x \to \infty$, we obtain the result. $\square$

It follows at once that

$$L(s,\chi_1) = \zeta(s) \prod_{p|m} (1 - p^{-s})$$

and that, for any Dirichlet character χ, $L(s,\chi) \ne 0$ for $\sigma > 1$.

PROPOSITION 5 $-L'(s,\chi)/L(s,\chi) = \sum_{n=1}^\infty \chi(n)\Lambda(n)/n^s$ *for* $\sigma > 1$.

Proof The series $\omega(s,\chi) = \sum_{n=1}^\infty \chi(n)\Lambda(n)n^{-s}$ converges absolutely and uniformly in any half-plane $\sigma \ge 1 + \varepsilon$, where $\varepsilon > 0$. Moreover, as in the proof of Proposition IX.6,

$$L(s,\chi)\omega(s,\chi) = \sum_{j=1}^\infty \chi(j)j^{-s} \sum_{k=1}^\infty \chi(k)\Lambda(k)k^{-s} = \sum_{n=1}^\infty n^{-s} \sum_{jk=n} \chi(j)\chi(k)\Lambda(k)$$

$$= \sum_{n=1}^\infty n^{-s}\chi(n) \sum_{d|n} \Lambda(d) = \sum_{n=1}^\infty n^{-s} \chi(n) \log n = -L'(s,\chi). \quad \square$$

As in the proof of Proposition IX.6, we can also prove directly that $L(s,\chi) \neq 0$ for $\sigma > 1$, and thus make the proof of the prime number theorem for arithmetic progressions independent of Proposition 4.

The following general result, due to Landau (1905), considerably simplifies the subsequent argument (and has other applications).

PROPOSITION 6 *Let $\phi(x)$ be a nondecreasing function for $x \geq 0$ such that the integral*

$$f(s) = \int_0^\infty e^{-sx} \, d\phi(x) \tag{†}$$

is convergent for $\Re s > \beta$. Thus f is holomorphic in this half-plane. If the definition of f can be extended so that it is holomorphic on the real segment $(\alpha,\beta]$, then the integral in (†) is convergent also for $\Re s > \alpha$. Thus f is actually holomorphic, and (†) holds, in this larger half-plane.

Proof Since f is holomorphic at β, we can choose $\delta > 0$ so that f is holomorphic in the disc $|s - (\beta + \delta)| < 2\delta$. Thus its Taylor series converges in this disc. But for $\Re s > \beta$ the n-th derivative of f is given by

$$f^{(n)}(s) = (-1)^n \int_0^\infty e^{-sx} \, x^n \, d\phi(x).$$

Hence, for any σ such that $\beta - \delta < \sigma < \beta + \delta$,

$$f(\sigma) = \sum_{n=0}^\infty (\sigma - \beta - \delta)^n \, f^{(n)}(\beta + \delta)/n!$$

$$= \sum_{n=0}^\infty (\sigma - \beta - \delta)^n \, (-1)^n \int_0^\infty e^{-(\beta+\delta)x} \, x^n \, d\phi(x)/n!$$

$$= \sum_{n=0}^\infty \int_0^\infty e^{-(\beta+\delta)x} \, (\beta + \delta - \sigma)^n x^n/n! \, d\phi(x).$$

Since the integrands are non-negative, we can interchange the orders of summation and integration, obtaining

$$f(\sigma) = \int_0^\infty e^{-(\beta+\delta)x} \sum_{n=0}^\infty (\beta + \delta - \sigma)^n x^n/n! \, d\phi(x)$$

$$= \int_0^\infty e^{-(\beta+\delta)x} \, e^{(\beta+\delta-\sigma)x} \, d\phi(x)$$

$$= \int_0^\infty e^{-\sigma x} \, d\phi(x).$$

Thus the integral in (†) converges for real $s > \beta - \delta$.

Let γ be the greatest lower bound of all real $s \in (\alpha,\beta)$ for which the integral in (†) converges. Then the integral in (†) is also convergent for $\Re s > \gamma$ and defines there a holomorphic function. Since this holomorphic function coincides with $f(s)$ for $\Re s > \beta$, it follows that (†) holds for $\Re s > \gamma$. Moreover $\gamma = \alpha$, since if $\gamma > \alpha$ we could replace β by γ in the preceding argument and thus obtain a contradiction to the definition of γ. $\square$

The punch-line is the following proposition:

PROPOSITION 7 $L(1 + it,\chi) \neq 0$ *for every real t and every* $\chi \neq \chi_1$.

Proof Assume on the contrary that $L(1 + i\alpha,\chi) = 0$ for some real α and some $\chi \neq \chi_1$. Then also $L(1 - i\alpha,\overline{\chi}) = 0$. If we put

$$f(s) = \zeta^2(s)L(s + i\alpha,\chi)L(s - i\alpha,\overline{\chi}),$$

then f is holomorphic and nonzero for $\sigma > 1$. Furthermore f is holomorphic on the real segment $[1/2,1]$, since the double pole of $\zeta^2(s)$ at $s = 1$ is cancelled by the zeros of the other two factors. By logarithmic differentiation we obtain, for $\sigma > 1$,

$$-f'(s)/f(s) = -2\zeta'(s)/\zeta(s) - L'(s + i\alpha,\chi)/L(s + i\alpha,\chi) - L'(s - i\alpha,\overline{\chi})/L(s - i\alpha,\overline{\chi})$$

$$= 2\sum_{n=1}^{\infty} \Lambda(n)n^{-s} + \sum_{n=1}^{\infty} \chi(n)\Lambda(n)n^{-s-i\alpha} + \sum_{n=1}^{\infty} \overline{\chi}(n)\Lambda(n)n^{-s+i\alpha}$$

$$= \sum_{n=2}^{\infty} c_n n^{-s},$$

where

$$c_n = \{2 + \chi(n)n^{-i\alpha} + \overline{\chi}(n)n^{i\alpha}\}\Lambda(n) = 2\{1 + \Re(\chi(n)n^{-i\alpha})\}\Lambda(n).$$

Since $|\chi(n)| \leq 1$ and $|n^{-i\alpha}| = 1$, it follows that $c_n \geq 0$ for all $n \geq 2$. If we put

$$g(s) = \sum_{n=2}^{\infty} c_n n^{-s}/\log n,$$

then $g'(s) = f'(s)/f(s)$ for $\sigma > 1$ and so the derivative of $e^{-g(s)}f(s)$ is

$$\{f'(s) - g'(s)f(s)\}e^{-g(s)} = 0.$$

Thus $f(s) = Ce^{g(s)}$, where C is a constant. In fact $C = 1$, since $g(\sigma) \to 0$ and $f(\sigma) \to 1$ as $\sigma \to +\infty$. Since $g(s)$ is the sum of an absolutely convergent Dirichlet series with nonnegative coefficients, so also are the powers $g^k(s)$ ($k = 2,3,...$). Hence also

$$f(s) = e^{g(s)} = 1 + g(s) + g^2(s)/2! + ... = \sum_{n=1}^{\infty} a_n n^{-s} \text{ for } \sigma > 1,$$

where $a_n \geq 0$ for every n. It follows from Proposition 6 that the series $\sum_{n=1}^{\infty} a_n n^{-\sigma}$ must actually converge with sum $f(\sigma)$ for $\sigma \geq 1/2$. We will show that this leads to a contradiction.

Take $n = p^2$, where p is a prime. Then, by the manner of its formation,

$$a_n \geq c_n/\log n + c_p{}^2/2 \log^2 p$$

$$= \{2 + \chi(p)^2 p^{-2i\alpha} + \overline{\chi}(p)^2 p^{2i\alpha}\}/2 + \{2 + \chi(p)p^{-i\alpha} + \overline{\chi}(p)p^{i\alpha}\}^2/2$$

$$= 2 - \chi(p)\overline{\chi}(p) + \{1 + \chi(p)p^{-i\alpha} + \overline{\chi}(p)p^{i\alpha}\}^2 \geq 1,$$

since $|\chi(p)| \leq 1$. Hence

$$f(1/2) = \sum_{n=1}^{\infty} a_n/n^{1/2} \geq \sum_{n=p^2} a_n/n^{1/2} \geq \sum_p 1/p.$$

Since $\sum_p 1/p$ diverges, this is a contradiction. $\square$

PROPOSITION 8 $\sum_{n \leq x} \chi_1(n)\Lambda(n) \sim x$, $\sum_{n \leq x} \chi(n)\Lambda(n) = o(x)$ if $\chi \neq \chi_1$.

Proof For any Dirichlet character χ, put

$$g(s) = -\zeta'(s)/\zeta(s) - L'(s,\chi)/2L(s,\chi) - L'(s,\overline{\chi})/2L(s,\overline{\chi}),$$

$$h(s) = -\zeta'(s)/\zeta(s) - L'(s,\chi)/2iL(s,\chi) + L'(s,\overline{\chi})/2iL(s,\overline{\chi}).$$

For $\sigma = \Re s > 1$ we have

$$g(s) = \sum_{n=1}^{\infty} \{1 + \Re\chi(n)\}\Lambda(n)n^{-s},$$

$$h(s) = \sum_{n=1}^{\infty} \{1 + \Im\chi(n)\}\Lambda(n)n^{-s}.$$

If $\chi \neq \chi_1$ then, by Proposition 7, $g(s) - 1/(s-1)$ and $h(s) - 1/(s-1)$ are holomorphic for $\Re s \geq 1$. Since the coefficients of the Dirichlet series for $g(s)$ and $h(s)$ are nonnegative, it follows from Ikehara's theorem (Theorem IX.9) that

$$\sum_{n \leq x} \{1 + \Re\chi(n)\}\Lambda(n) \sim x,$$

$$\sum_{n \leq x} \{1 + \Im\chi(n)\}\Lambda(n) \sim x.$$

On the other hand, if $\chi = \chi_1$ then $g(s) - 2/(s-1)$ and $h(s) - 1/(s-1)$ are holomorphic for $\Re s \geq 1$, from which we obtain in the same way

$$\sum_{n \leq x} \{1 + \chi_1(n)\}\Lambda(n) \sim 2x,$$

$$\sum_{n \leq x} \Lambda(n) \sim x.$$

The result follows. $\square$

The prime number theorem for arithmetic progressions can now be deduced immediately. For, by the orthogonality relations and Proposition 8, if $1 \leq a < m$ and $(a,m) = 1$, then

$$\psi(x;m,a) \; = \; \sum_{n \leq x, \; n \equiv a \bmod m} \Lambda(n)$$

$$= \; \sum_{\chi} \overline{\chi}(a) \sum_{n \leq x} \chi(n) \Lambda(n) \, / \varphi(m)$$

$$\sim \; x/\varphi(m).$$

It is also possible to obtain error bounds in the prime number theorem for arithmetic progressions of the same type as those in the ordinary prime number theorem. For example, it may be shown that for each $\alpha > 0$,

$$\psi(x;m,a) \; = \; x/\varphi(m) + O(x/\log^{\alpha}x),$$

$$\pi(x;m,a) \; = \; Li(x)/\varphi(m) + O(x/\log^{\alpha}x),$$

where the constants implied by the O-symbols depend on α, but not on m or a.

In the same manner as for the Riemann zeta-function $\zeta(s)$ it may be shown that the Dirichlet L-function $L(s,\chi)$ satisfies a functional equation, provided χ is a primitive character. (Here a Dirichlet character χ mod m is *primitive* if for each proper divisor d of m there exists an integer $a \equiv 1 \bmod d$ with $(a,m) = 1$ and $\chi(a) \neq 1$.) Explicitly, if χ is a primitive character mod m and if one puts

$$\Lambda(s,\chi) \; = \; (m/\pi)^{s/2} \, \Gamma((s + \delta)/2) L(s,\chi),$$

where $\delta = 0$ or 1 according as $\chi(-1) = 1$ or -1, then

$$\Lambda(1 - s, \overline{\chi}) \; = \; \varepsilon_{\chi} \Lambda(s,\chi),$$

where

$$\varepsilon_{\chi} \; = \; i^{-\delta} \, m^{-1/2} \sum_{k=1}^{m} \overline{\chi}(k) e^{2\pi i k/m}.$$

It follows from the functional equation that $|\varepsilon_{\chi}| = 1$. Indeed, by taking complex conjugates we obtain, for real s,

$$\Lambda(1 - s,\chi) \; = \; \overline{\varepsilon}_{\chi} \Lambda(s, \overline{\chi})$$

and hence, on replacing s by $1 - s$,

$$\Lambda(s,\chi) \; = \; \overline{\varepsilon}_{\chi} \Lambda(1 - s, \overline{\chi}) \; = \; \varepsilon_{\chi} \overline{\varepsilon}_{\chi} \Lambda(s,\chi).$$

The extended Riemann hypothesis implies that no Dirichlet L-function $L(s,\chi)$ has a zero in the half-plane $\Re s > 1/2$, since $f(s) = \prod_{\chi} L(s,\chi)$ is the Dedekind zeta-function of the algebraic number field $K = \mathbb{Q}(e^{2\pi i/m})$. Hence it may be shown that if the extended Riemann hypothesis holds, then

$$\psi(x;m,a) \; = \; x/\varphi(m) + O(x^{1/2} \log^2 x)$$

and

$$\pi(x;m,a) = Li(x)/\varphi(m) + O(x^{1/2} \log x),$$

where the constants implied by the O-symbols are independent of m and a. Assuming the extended Riemann hypothesis, Bach and Sorenson (1996) have shown that, for any a,m with $1 \le a < m$ and $(a,m) = 1$, the least prime $p \equiv a \bmod m$ satisfies $p < 2(m \log m)^2$.

Without any hypothesis, Linnik (1944) proved that there exists an absolute constant L such that the least prime in any arithmetic progression $a, a + m, a + 2m,...$, where $1 \le a < m$ and $(a,m) = 1$, does not exceed m^L if m is sufficiently large. Heath-Brown (1992) has shown that one can take any $L > 11/2$.

4 Representations of arbitrary finite groups

The problem of extending the character theory of finite abelian groups to arbitrary finite groups was proposed by Dedekind and solved by Frobenius (1896). Simplifications were afterwards found by Frobenius himself, Burnside and Schur (1905). We will follow Schur's treatment, which is distinguished by its simplicity. It turns out that for nonabelian groups the concept of 'representation' is more fundamental than that of 'character'.

A *representation* of a group G is a mapping ρ of G into the set of all linear transformations of a finite-dimensional vector space V over the field $\mathbb{C}$ of complex numbers which preserves products, i.e.

$$\rho(st) = \rho(s)\rho(t) \quad \text{for all } s,t \in G, \tag{1}$$

and maps the identity element of G into the identity transformation of V: $\rho(e) = I$. The dimension of the vector space V is called the *degree* of the representation (although 'dimension' would be more natural).

It follows at once from (1) that

$$\rho(s)\rho(s^{-1}) = \rho(s^{-1})\rho(s) = I.$$

Thus, for every $s \in G$, $\rho(s)$ is an invertible linear transformation of V and $\rho(s^{-1}) = \rho(s)^{-1}$. (Hence a representation of G is a *homomorphism* of G into the group $GL(V)$ of all invertible linear transformations of V.)

Any group has a *trivial representation* of degree 1 in which every element of the group is mapped into the scalar 1.

Also, with any group G of finite order g a representation of degree g may be defined in the following way. Let $s_1,...,s_g$ be an enumeration of the elements of G and let $e_1,...,e_g$ be a basis for a g-dimensional vector space V over $\mathbb{C}$. We define a linear transformation $A(s_i)$ of V by its action on the basis elements:

$$A(s_i)e_j = e_k \quad \text{if } s_i s_j = s_k.$$

Then, for all $s,t \in G$,

$$A(s^{-1})A(s) = I, \quad A(st) = A(s)A(t).$$

Thus the mapping $\rho_R: s_i \to A(s_i)$ is a representation of G, known as the *regular representation*.

By choosing a basis for the vector space we can reformulate the preceding definitions in terms of matrices. A representation of a group G is then a product-preserving map $s \to A(s)$ of G into the group of all $n \times n$ non-singular matrices of complex numbers. The positive integer n is the degree of the representation. However, we must regard two matrix representations $s \to A(s)$ and $s \to B(s)$ as *equivalent* if one is obtained from the other simply by changing the basis of the vector space, i.e. if there exists a non-singular matrix T such that

$$T^{-1}A(s)T = B(s) \quad \text{for every } s \in G.$$

It is easily verified that if $s \to A(s)$ is a matrix representation of degree n of a group G, then $s \to A(s^{-1})^t$ (the transpose of $A(s^{-1})$) is a representation of the same degree, the *contragredient representation*. Furthermore, $s \to \det A(s)$ is a representation of degree 1.

Again, if $\rho: s \to A(s)$ and $\sigma: s \to B(s)$ are matrix representations of a group G, of degrees m and n respectively, then the Kronecker product mapping

$$s \to A(s) \otimes B(s)$$

is also a representation of G, of degree mn, since

$$(A(s) \otimes B(s))(A(t) \otimes B(t)) = A(st) \otimes B(st).$$

We will call this representation simply the *product* of the representations ρ and σ, and denote it by $\rho \otimes \sigma$.

The basic problem of representation theory is to determine all possible representations of a given group. As we will see, all representations may in fact be built up from certain 'irreducible' ones.

Let ρ be a representation of a group G by linear transformations of a vector space V. If a subspace U of V is *invariant* under G, i.e. if

$$\rho(s)U \subseteq U \quad \text{for every } s \in G,$$

then the restrictions to U of the given linear transformations provide a representation ρ_U of G by linear transformations of the vector space U. If it happens that there exists another subspace W invariant under G such that V is the direct sum of U and W, i.e. $V = U + W$ and $U \cap W = \{0\}$, then the representation ρ is completely determined by the representations ρ_U and ρ_W and will be said simply to be their *sum*.

A representation ρ of a group G by linear transformations of a vector space V is said to be *irreducible* if no nontrivial proper subspace of V is invariant under G, and *reducible* otherwise. Evidently any representation of degree 1 is irreducible.

A matrix representation $s \to A(s)$, of degree n, of a group G is reducible if it is equivalent to a representation in which all matrices have the block form

$$\begin{pmatrix} P(s) & Q(s) \\ 0 & R(s) \end{pmatrix},$$

where $P(s)$ is a square matrix of order m, $0 < m < n$. Then $s \to P(s)$ and $s \to R(s)$ are representations of G of degrees m and $n - m$ respectively. The given representation is the sum of these representations if there exists a non-singular matrix T such that

$$T^{-1}A(s)T = \begin{pmatrix} P(s) & 0 \\ 0 & R(s) \end{pmatrix} \quad \text{for every } s \in G.$$

The following theorem of Maschke (1899) reduces the problem of finding all representations of a *finite* group to that of finding all irreducible representations.

PROPOSITION 9 *Every representation of a finite group is (equivalent to) a sum of irreducible representations.*

Proof We give a constructive proof due to Schur. Let $s \to A(s)$, where

$$A(s) = \begin{pmatrix} P(s) & Q(s) \\ 0 & R(s) \end{pmatrix},$$

be a reducible representation of a group G of finite order g. Since the mapping $s \to A(s)$ preserves products, we have

$$P(st) = P(s)P(t), \quad R(st) = R(s)R(t), \quad Q(st) = P(s)Q(t) + Q(s)R(t). \tag{2}$$

The non-singular matrix

$$T = \begin{pmatrix} I & M \\ 0 & I \end{pmatrix}$$

satisfies

$$\begin{pmatrix} P(t) & Q(t) \\ 0 & R(t) \end{pmatrix} T = T \begin{pmatrix} P(t) & 0 \\ 0 & R(t) \end{pmatrix} \tag{3}$$

if and only if

$$MR(t) = P(t)M + Q(t).$$

Take

$$M = g^{-1} \sum_{s \in G} Q(s)R(s^{-1}).$$

Then, by (2),

$$P(t)M = g^{-1} \sum_{s \in G} \{Q(ts) - Q(t)R(s)\}R(s^{-1})$$

$$= g^{-1} \sum_{s \in G} Q(ts)R(s^{-1}t^{-1})R(t) - Q(t) = MR(t) - Q(t),$$

and hence (3) holds.

Thus the given reducible representation $s \to A(s)$ is the sum of two representations $s \to P(s)$ and $s \to R(s)$ of lower degree. The result follows by induction on the degree. □

Maschke's original proof of Proposition 9 depended on showing that every representation of a finite group is equivalent to a representation by *unitary* matrices. We briefly sketch the argument. Let $\rho: s \to A(s)$ be a representation of a finite group G by linear transformations of a finite-dimensional vector space V. We may suppose V equipped with a positive definite inner product (u,v). It is easily verified that

$$(u,v)_G = g^{-1} \sum_{t \in G} (A(t)u, A(t)v)$$

is also a positive definite inner product on V and that it is invariant under G, i.e.

$$(A(s)u, A(s)v)_G = (u,v)_G \quad \text{for every } s \in G.$$

If U is a subspace of V which is invariant under G, and if $U^\perp$ is the subspace consisting of all vectors $v \in V$ such that $(u,v)_G = 0$ for every $u \in U$, then $U^\perp$ is also invariant under G and V is the direct sum of U and $U^\perp$. Thus ρ is the sum of its restrictions to U and $U^\perp$.

The basic result for irreducible representations is *Schur's lemma*, which comes in two parts:

PROPOSITION 10 (i) *Let $s \to A_1(s)$ and $s \to A_2(s)$ be irreducible representations of a group G by linear transformations of the vector spaces V_1 and V_2. If there exists a linear transformation $T \neq 0$ of V_1 into V_2 such that*

$$TA_1(s) = A_2(s)T \quad \text{for every } s \in G,$$

then the spaces V_1 and V_2 have the same dimension and T is invertible, so that the representations are equivalent.

(ii) *Let $s \to A(s)$ be an irreducible representation of a group G by linear transformations of a vector space V. A linear transformation T of V has the property*

$$TA(s) = A(s)T \quad \text{for every } s \in G \tag{4}$$

if and only if $T = \lambda I$ for some $\lambda \in \mathbb{C}$.

Proof (i) The image of V_1 under T is a subspace of V_2 which is invariant under the second representation. Since $T \neq 0$ and the representation is irreducible, it must be the whole space: $TV_1 = V_2$. On the other hand, those vectors in V_1 whose image under T is 0 form a subspace of V_1 which is invariant under the first representation. Since $T \neq 0$ and the representation is irreducible, it must contain only the zero vector. Hence distinct vectors of V_1 have distinct images in V_2 under T. Thus T is a one-to-one mapping of V_1 onto V_2.

(ii) By the fundamental theorem of algebra, there exists a complex number λ such that $\det(\lambda I - T) = 0$. Hence $T - \lambda I$ is not invertible. But if T has the property (4), so does $T - \lambda I$. Therefore $T - \lambda I = 0$, by (i) with $A_1 = A_2$. It is obvious that, conversely, (4) holds if $T = \lambda I$. $\square$

COROLLARY 11 *Every irreducible representation of an abelian group is of degree* 1.

Proof By Proposition 10 (ii) all elements of the group must be represented by scalar multiples of the identity transformation. But such a representation is irreducible only if its degree is 1. $\square$

5 Characters of arbitrary finite groups

By definition, the *trace* of an $n \times n$ matrix $A = (\alpha_{ij})$ is the sum of its main diagonal elements:

$$\operatorname{tr} A = \sum_{i=1}^{n} \alpha_{ii}.$$

It is easily verified that, for any $n \times n$ matrices A, B and any scalar λ, we have

$$\operatorname{tr}(A + B) = \operatorname{tr} A + \operatorname{tr} B, \quad \operatorname{tr}(\lambda A) = \lambda \operatorname{tr} A, \quad \operatorname{tr}(AB) = \operatorname{tr}(BA), \quad \operatorname{tr}(A \otimes B) = (\operatorname{tr} A)(\operatorname{tr} B).$$

Let $\rho: s \to A(s)$ be a matrix representation of a group G. By the *character* of the representation ρ we mean the mapping $\chi: G \to \mathbb{C}$ defined by

$$\chi(s) = \text{tr } A(s).$$

Since tr $(T^{-1}AT)$ = tr (ATT^{-1}) = tr A, equivalent representations have the same character. The significance of characters stems from the converse, which will be proved below.

Clearly the character χ of a representation ρ is a *class function*, i.e.

$$\chi(st) = \chi(ts) \quad \text{for all } s,t \in G.$$

The degree n of the representation ρ is determined by its character χ, since $A(e) = I_n$ and hence $\chi(e) = n$.

If the representation ρ is the sum of two representations ρ' and ρ'', the corresponding characters χ,χ',χ'' evidently satisfy

$$\chi(s) = \chi'(s) + \chi''(s) \quad \text{for every } s \in G.$$

On the other hand, if the representation ρ is the product of the representations ρ' and ρ'', then

$$\chi(s) = \chi'(s)\chi''(s) \quad \text{for every } s \in G.$$

Thus the set of all characters of a group is closed under addition and multiplication. The character of an irreducible representation will be called simply an *irreducible character*.

Let G be a group and ρ a representation of G of degree n with character χ. If s is an element of G of finite order m, then by restriction ρ defines a representation of the cyclic group generated by s. By Proposition 9 and Corollary 11, this representation is equivalent to a sum of representations of degree 1. Thus if S is the matrix representing s, there exists an invertible matrix T such that

$$T^{-1}ST = \text{diag } [\omega_1,...,\omega_n]$$

is a diagonal matrix. Moreover, since

$$T^{-1}S^kT = \text{diag } [\omega_1{}^k,...,\omega_n{}^k],$$

$\omega_1,...,\omega_n$ are all m-th roots of unity. Thus

$$\chi(s) = \omega_1 + ... + \omega_n$$

is a sum of n m-th roots of unity. Since the inverse of a root of unity ω is its complex conjugate $\overline{\omega}$, it follows that

$$\chi(s^{-1}) = \omega_1{}^{-1} + ... + \omega_n{}^{-1} = \overline{\chi(s)}.$$

Now let G be a group of finite order g, and let $\rho: s \to A(s)$ and $\sigma: s \to B(s)$ be irreducible matrix representations of G of degrees n and m respectively. For any $n \times m$ matrix C, form the matrix

$$T = \sum_{s \in G} A(s) C B(s^{-1}).$$

Since ts runs through the elements of G at the same time as s,

$$A(t)T = TB(t) \quad \text{for every } t \in G.$$

Therefore, by Schur's lemma, $T = O$ if ρ is not equivalent to σ and $T = \lambda I$ if $\rho = \sigma$. In particular, take C to be any one of the mn matrices which have a single entry 1 and all other entries 0. Then if $A = (\alpha_{ij})$, $B = (\beta_{kl})$, we get

$$\sum_{s \in G} \alpha_{ij}(s) \beta_{kl}(s^{-1}) = \begin{cases} 0 & \text{if } \rho, \sigma \text{ are inequivalent,} \\ \lambda_{jk}\delta_{il} & \text{if } \rho = \sigma, \end{cases}$$

where $\delta_{il} = 1$ or 0 according as $i = l$ or $i \neq l$ ('Kronecker delta'). Since for $(\alpha_{ij}) = (\beta_{ij})$ the left side is unchanged when i is interchanged with k and j with l, we must have $\lambda_{jk} = \lambda\delta_{jk}$. To determine λ set $i = l, j = k$ and sum with respect to k. Since the matrices representing s and s^{-1} are inverse, we get $g1 = n\lambda$. Thus

$$\sum_{s \in G} \alpha_{ij}(s) \alpha_{kl}(s^{-1}) = \begin{cases} g/n & \text{if } j = k \text{ and } i = l, \\ 0 & \text{otherwise.} \end{cases}$$

If μ, ν run through an index set for the inequivalent irreducible representations of G, then the relations which have been obtained can be rewritten in the form

$$\sum_{s \in G} \alpha_{ij}{}^{(\mu)}(s) \alpha_{kl}{}^{(\nu)}(s^{-1}) = \begin{cases} g/n_\mu & \text{if } \mu = \nu, j = k, i = l, \\ 0 & \text{otherwise.} \end{cases} \tag{5}$$

The *orthogonality relations* (5) for the irreducible matrix elements have several corollaries:

(i) *The functions* $\alpha_{ij}{}^{(\mu)}: G \to \mathbb{C}$ *are linearly independent.*

For suppose there exist $\lambda_{ij}{}^{(\mu)} \in \mathbb{C}$ such that

$$\sum_{i,j,\mu} \lambda_{ij}{}^{(\mu)} \alpha_{ij}{}^{(\mu)}(s) = 0 \quad \text{for every } s \in G.$$

Multiplying by $\alpha_{kl}{}^{(\nu)}(s^{-1})$ and summing over all $s \in G$, we get $(g/n_\nu)\lambda_{lk}{}^{(\nu)} = 0$. Hence every coefficient $\lambda_{lk}{}^{(\nu)}$ vanishes.

(ii)
$$\sum_{s\in G}\chi_\mu(s)\chi_\nu(s^{-1}) = g\delta_{\mu\nu}. \tag{6}$$

This follows from (5) by setting $i = j$, $k = l$ and summing over j, l.

(iii) *The irreducible characters χ_μ are linearly independent.*

In fact (iii) follows from (6) in the same way that (i) follows from (5).

The *orthogonality relations* (6) for the irreducible characters enable us to decompose a given representation ρ into irreducible representations. For if $\rho = \oplus m_\mu\rho_\mu$ is a direct sum decomposition of ρ into irreducible components ρ_μ, where the coefficients m_μ are non-negative integers, and if ρ has character χ, then

$$\chi(s) = \sum_\mu m_\mu\chi_\mu(s).$$

Multiplying by $\chi_\nu(s^{-1})$ and summing over all $s \in G$, we deduce from (6) that

$$g^{-1}\sum_{s\in G}\chi(s)\chi_\nu(s^{-1}) = m_\nu. \tag{7}$$

Thus the multiplicities m_ν are uniquely determined by the character χ of the representation ρ. It follows that *two representations are equivalent if and only if they have the same character.*

In the same way we find

$$g^{-1}\sum_{s\in G}\chi(s)\chi(s^{-1}) = \sum_\mu m_\mu^2. \tag{8}$$

Hence *a representation ρ with character χ is irreducible if and only if*

$$g^{-1}\sum_{s\in G}\chi(s)\chi(s^{-1}) = 1.$$

The procedure for decomposing a representation into its irreducible components may be applied, in particular, to the regular representation. Evidently the $g\times g$ matrix representing an element s has all its main diagonal elements 0 if $s \neq e$ and all its main diagonal elements 1 if $s = e$. Thus the character χ_R of the regular representation ρ_R is given by

$$\chi_R(e) = g, \quad \chi_R(s) = 0 \text{ if } s \neq e.$$

Since $\chi_\nu(e) = n_\nu$ is the degree of the ν-th irreducible representation, it follows from (7) that $m_\nu = n_\nu$. Thus *every irreducible representation is contained in the direct sum decomposition of the regular representation, and moreover each occurs as often as its degree.*

It follows that

$$\sum_\mu n_\mu^2 = g, \quad \sum_\mu n_\mu\chi_\mu(s) = 0 \text{ if } s \neq e. \tag{9}$$

Thus the total number of functions $\alpha_{ij}{}^{(\mu)}$ is $\sum_\mu n_\mu{}^2 = g$. Therefore, since they are linearly independent, *every function* $\phi: G \to \mathbb{C}$ *is a linear combination of functions* $\alpha_{ij}{}^{(\mu)}$ *occurring in irreducible matrix representations.*

We show next that *every class function* $\phi: G \to \mathbb{C}$ *is a linear combination of irreducible characters* χ_μ. By what we have just proved $\phi = \sum_\mu \phi_\mu$, where

$$\phi_\mu = \sum_{i,j=1}^{n_\mu} \lambda_{ij}{}^{(\mu)} \alpha_{ij}{}^{(\mu)}$$

and $\lambda_{ij}{}^{(\mu)} \in \mathbb{C}$. But $\phi(st) = \phi(ts)$ and

$$\phi_\mu(st) = \sum_{i,j,k} \lambda_{ik}{}^{(\mu)} \alpha_{ij}{}^{(\mu)}(s) \alpha_{jk}{}^{(\mu)}(t), \quad \phi_\mu(ts) = \sum_{i,j,k} \lambda_{kj}{}^{(\mu)} \alpha_{ki}{}^{(\mu)}(t) \alpha_{ij}{}^{(\mu)}(s).$$

Since the functions $\alpha_{ij}{}^{(\mu)}$ are linearly independent, we must have

$$\sum_k \lambda_{ik}{}^{(\mu)} \alpha_{jk}{}^{(\mu)}(t) = \sum_k \lambda_{kj}{}^{(\mu)} \alpha_{ki}{}^{(\mu)}(t).$$

If we denote by $T^{(\mu)}$ the transpose of the matrix $(\lambda_{ik}{}^{(\mu)})$, we can rewrite this in the form

$$A^{(\mu)}(t) T^{(\mu)} = T^{(\mu)} A^{(\mu)}(t).$$

Therefore, by Schur's lemma, $T^{(\mu)} = \lambda_\mu I_{n_\mu}$ and hence $\phi_\mu = \lambda_\mu \chi_\mu$. Thus $\phi = \sum_\mu \lambda_\mu \chi_\mu$.

Two elements u,v of a group G are said to be *conjugate* if $v = s^{-1}us$ for some $s \in G$. It is easily verified that conjugacy is an equivalence relation. Consequently G is the union of pairwise disjoint subsets, called *conjugacy classes*, such that two elements belong to the same subset if and only if they are conjugate. The inverses of all elements in a conjugacy class again form a conjugacy class, the *inverse class*.

In this terminology a function $\phi: G \to \mathbb{C}$ is a class function if and only if $\phi(u) = \phi(v)$ whenever u and v belong to the same conjugacy class. Thus the number of linearly independent class functions is just the number of conjugacy classes in G. Since the characters χ_μ form a basis for the class functions, it follows that *the number of inequivalent irreducible representations is equal to the number of conjugacy classes in the group.*

If a group of order g has r conjugacy classes then, by (9), $g = n_1{}^2 + \ldots + n_r{}^2$. Since it is abelian if and only if every conjugacy class contains exactly one element, i.e. if and only if $r = g$, it follows that *a finite group is abelian if and only if every irreducible representation has degree 1.*

Let $\mathscr{C}_1,\ldots,\mathscr{C}_r$ be the conjugacy classes of the group G and let h_k be the number of elements in $\mathscr{C}_k$ ($k = 1,\ldots,r$). Changing notation, we will now denote by χ_{ik} the common value of the character of all elements in the k-th conjugacy class in the i-th irreducible representation. Then, since $\chi(s^{-1}) = \overline{\chi(s)}$, the orthogonality relations (6) can be rewritten in the form

$$g^{-1} \sum_{j=1}^{r} h_j \chi_{ij} \overline{\chi_{kj}} = \begin{cases} 1 & \text{if } i = k, \\ 0 & \text{if } i \neq k. \end{cases} \tag{10}$$

Thus the $r \times r$ matrices $A = (\chi_{ik})$, $B = (g^{-1} h_i \overline{\chi_{ki}})$ satisfy $AB = I$. Therefore also $BA = I$, i.e.

$$\sum_{j=1}^{r} \overline{\chi_{ji}} \chi_{jk} = \begin{cases} g/h_k & \text{if } i = k, \\ 0 & \text{if } i \neq k. \end{cases} \tag{11}$$

It may be noted that h_k divides g since, for any $s_k \in \mathscr{C}_k$, g/h_k is the order of the subgroup formed by all elements of G which commute with s_k. We are going to show finally that *the degree of any irreducible representation divides the order of the group.*

Any representation $\rho\colon s \to A(s)$ of a finite group G may be extended by linearity to the set of all linear combinations of elements of G:

$$\rho(\textstyle\sum_{s \in G} \alpha_s s) = \sum_{s \in G} \alpha_s A(s).$$

In particular, let C_k denote the sum of all elements in the k-th conjugacy class $\mathscr{C}_k$ of G. For any $t, u \in G$,

$$u^{-1} s_k u t = t(t^{-1} u^{-1} s_k u t)$$

and hence

$$\rho(C_k) A(t) = \sum_{s \in \mathscr{C}_k} A(st) = \sum_{s \in \mathscr{C}_k} A(ts) = A(t) \rho(C_k).$$

If $\rho = \rho_i$ is an irreducible representation, it follows from Schur's lemma that $\rho_i(C_k) = \lambda_{ik} I_{n_i}$. Moreover, since

$$\operatorname{tr} \rho_i(C_k) = h_k \chi_{ik},$$

where h_k again denotes the number of elements in $\mathscr{C}_k$, we must have $\lambda_{ik} = h_k \chi_{ik}/n_i$. Now let

$$C = \sum_{k=1}^{r} (g/h_k) C_k C_{k'},$$

where $\mathscr{C}_{k'}$ is the conjugacy class inverse to $\mathscr{C}_k$. (Otherwise expressed, $C = \sum_{s,t \in G} sts^{-1} t^{-1}$). Then $\rho_i(C) = \gamma_i I_{n_i}$, where

$$\gamma_i = \sum_{k=1}^{r} (g/h_k) \lambda_{ik} \overline{\lambda_{ik}} = (g/n_i^2) \sum_{k=1}^{r} h_k \chi_{ik} \overline{\chi_{ik}} = (g/n_i)^2,$$

by (10). If $\rho_R(C)$ is the matrix representing C in the regular representation, it follows that there exists an invertible matrix T such that $T^{-1} \rho_R(C) T$ is a diagonal matrix, consisting of the matrices $(g/n_i)^2 I_{n_i}$, repeated n_i times, for every i. In particular, $(g/n_i)^2$ is a root of the characteristic polynomial $\phi(\lambda) = \det (\lambda I_g - \rho_R(C))$ for every i. But $\rho_R(C)$ is a matrix with integer entries and hence the polynomial $\phi(\lambda) = \lambda^g + a_1\lambda^{g-1} + ... + a_g$ has integer coefficients $a_1,...,a_g$. The following lemma, already proved in Proposition II.16 but reproved for convenience of reference here, now implies that $(g/n_i)^2$ is an integer and hence that n_i divides g.

LEMMA 12   *If* $\phi(\lambda) = \lambda^n + a_1\lambda^{n-1} + ... + a_n$ *is a monic polynomial with integer coefficients* $a_1,...,a_n$ *and* r *a rational number such that* $\phi(r) = 0$, *then* r *is an integer.*

Proof   We can write $r = b/c$, where b and c are relatively prime integers and $c > 0$. Then

$$b^n + a_1 b^{n-1} c + ... + a_n c^n = 0$$

and hence c divides b^n. Since c and b have no common prime factor, this implies $c = 1$. $\blacksquare$

If we apply the preceding argument to C_k, rather than to C, we see that there exists an invertible matrix T_k such that $T_k^{-1}\rho_R(C_k)T_k$ is a diagonal matrix, consisting of the matrices $(h_k\chi_{ik}/n_i)I_{n_i}$ repeated n_i times, for every i. Thus $h_k\chi_{ik}/n_i$ is a root of the characteristic polynomial $\phi_k(\lambda) = \det (\lambda I_g - \rho_R(C_k))$. Since this is a monic polynomial with integer coefficients, it follows that $h_k\chi_{ik}/n_i$ *is an algebraic integer.*

6   Induced representations and examples

Let H be a subgroup of finite index n of a group G, i.e. G is the disjoint union of n left cosets of H:

$$G = s_1 H \cup ... \cup s_n H.$$

Also, let there be given a representation $\sigma: t \to A(t)$ of H by linear transformations of a vector space V. The representation $\tilde{\sigma}: s \to \tilde{A}(s)$ of G *induced* by the given representation σ of H is defined in the following way:

Take the vector space $\tilde{V}$ to be the direct sum of n subspaces V_i, where V_i consists of all formal products $s_i \cdot v$ ($v \in V$) with the rules of combination

$$s_i \cdot (v + v') = s_i \cdot v + s_i \cdot v', \quad s_i \cdot (\lambda v) = \lambda(s_i \cdot v).$$

Then we set

$$\tilde{A}(s)s_i \cdot v = s_j \cdot A(t)v,$$

where t and s_j are determined from s and s_i by requiring that $t = s_j^{-1}ss_i \in H$. The degree of the induced representation of G is thus n times the degree of the original representation of H.

With respect to a given basis of V let $A(t)$ now denote the matrix representing $t \in H$ and put $A(s) = O$ if $s \in G \setminus H$. If one adopts corresponding bases for each of the subspaces V_i, then the matrix $\tilde{A}(s)$ representing $s \in G$ in the induced representation is the block matrix

$$
\tilde{A}(s) = \begin{pmatrix}
A(s_1^{-1}ss_1) & A(s_1^{-1}ss_2) & ... & A(s_1^{-1}ss_n) \\
A(s_2^{-1}ss_1) & A(s_2^{-1}ss_2) & ... & A(s_2^{-1}ss_n) \\
... & ... & ... & ... \\
A(s_n^{-1}ss_1) & A(s_n^{-1}ss_2) & ... & A(s_n^{-1}ss_n)
\end{pmatrix}.
$$

Evidently each row and each column contains exactly one nonzero block. It should be noted also that a different choice of coset representatives $s_i' = s_it_i$, where $t_i \in H$ $(i = 1,...,n)$, yields an equivalent representation, since

$$
\begin{pmatrix}
A(t_1)^{-1} & ... & 0 \\
... & ... & ... \\
0 & ... & A(t_n)^{-1}
\end{pmatrix}
\tilde{A}(s)
\begin{pmatrix}
A(t_1) & ... & 0 \\
... & ... & ... \\
0 & ... & A(t_n)
\end{pmatrix}
=
\begin{pmatrix}
A(s_1'^{-1}ss_1') & ... & A(s_1'^{-1}ss_n') \\
... & ... & ... \\
A(s_n'^{-1}ss_1') & ... & A(s_n'^{-1}ss_n')
\end{pmatrix}.
$$

Furthermore, changing the order of the cosets corresponds to performing the same permutation on the rows and columns of $\tilde{A}(s)$, and thus also yields an equivalent representation.

It follows that if ψ is the character of the original representation σ of H, then the character $\tilde{\psi}$ of the induced representation $\tilde{\sigma}$ of G is given by

$$
\tilde{\psi}(s) = \sum_{i=1}^{n} \psi(s_i^{-1}ss_i),
$$

where we set $\psi(s) = 0$ if $s \notin H$. If H is of finite order h, this can be rewritten in the form

$$
\tilde{\psi}(s) = h^{-1} \sum_{u \in G} \psi(u^{-1}su), \tag{12}
$$

since $\psi(t^{-1}s_i^{-1}ss_it) = \psi(s_i^{-1}ss_i)$ if $t \in H$.

From any representation of a group G we can also obtain a representation of a subgroup H simply by restricting the given representation to H. We will say that the representation of H is *deduced* from that of G. There is a remarkable reciprocity between induced and deduced representations, discovered by Frobenius (1898):

PROPOSITION 13 *Let* $\rho: s \to A(s)$ *be an irreducible representation of the finite group* G *and* $\sigma: t \to B(t)$ *an irreducible representation of the subgroup* H. *Then the number of times that* σ *occurs in the representation of* H *deduced from the representation* ρ *of* G *is equal to the number of times that* ρ *occurs in the representation of G induced by the representation* σ *of H.*

Proof Let χ denote the character of the representation ρ of G and ψ the character of the representation σ of H. By (7), the number of times that ρ occurs in the complete reduction of the induced representation $\tilde{\sigma}$ is

$$g^{-1} \sum_{s \in G} \tilde{\psi}(s)\chi(s^{-1}) = (gh)^{-1} \sum_{s,u \in G} \psi(u^{-1}su)\chi(s^{-1}).$$

If we put $u^{-1}s^{-1}u = t$, $u^{-1} = v$, then $s^{-1} = v^{-1}tv$ and (t,v) runs through all elements of $G \times G$ at the same time as (s,u). Therefore

$$g^{-1} \sum_{s \in G} \tilde{\psi}(s)\chi(s^{-1}) = (gh)^{-1} \sum_{t,v \in G} \chi(v^{-1}tv)\psi(t^{-1})$$

$$= h^{-1} \sum_{t \in G} \chi(t)\psi(t^{-1}) = h^{-1} \sum_{t \in H} \chi(t)\psi(t^{-1}),$$

which is the number of times that σ occurs in the complete reduction of the restriction of ρ to H. □

COROLLARY 14 *Each irreducible representation of a finite group G is contained in a representation induced by some irreducible representation of a given subgroup H.* □

A simple, but still significant, application of these results is to the case where the order of the subgroup H is half that of the whole group G. The subgroup H is then necessarily *normal* (as defined in Chapter I, §7) since, for any $v \in G \setminus H$, the elements of $G \setminus H$ form both a single left coset vH and a single right coset Hv. Hence if $s \to A(s)$ is a representation of H, then so also is $s \to A(v^{-1}sv)$, its *conjugate representation*. Since $v^2 \in H$, the conjugate of the conjugate is equivalent to the original representation. Evidently a representation is irreducible if and only if its conjugate representation is irreducible.

On the other hand G has a nontrivial character λ of degree 1, defined by

$$\lambda(s) = 1 \text{ or } -1 \text{ according as } s \in H \text{ or } s \notin H.$$

If χ is an irreducible character of G, then the character $\chi\lambda$ of the product representation is also irreducible, since

$$1 = g^{-1} \sum_{s \in G} \chi(s)\chi(s^{-1}) = \sum_{s \in G} \chi(s)\lambda(s)\chi(s^{-1})\lambda(s^{-1}).$$

Evidently χ and $\chi\lambda$ have the same degree.

If ψ_i is the character of an irreducible representation of H, we will denote by ψ_i^v the character of its conjugate representation. Thus

$$\psi_i{}^v(s) = \psi_i(v^{-1}sv).$$

The representation and its conjugate are equivalent if and only if $\psi_i{}^v(s) = \psi_i(s)$ for every $s \in H$.

Consider now the induced representation $\tilde{\psi}_i$ of G. Since H is a normal subgroup, it follows from (12) that

$$\tilde{\psi}_i(s) = \tilde{\psi}_i{}^v(s) = 0 \quad \text{if } s \in G \setminus H,$$
$$\tilde{\psi}_i(s) = \tilde{\psi}_i{}^v(s) = \psi_i(s) + \psi_i{}^v(s) \quad \text{if } s \in H.$$

Hence $\tilde{\psi}_i = \tilde{\psi}_i{}^v$ and

$$\sum_{s \in G} \tilde{\psi}_i(s) \tilde{\psi}_i(s^{-1}) = \sum_{s \in H} \{\psi_i(s) + \psi_i{}^v(s)\}\{\psi_i(s^{-1}) + \psi_i{}^v(s^{-1})\}$$

$$= \sum_{s \in H} \psi_i(s)\psi_i(s^{-1}) + \sum_{s \in H} \psi_i{}^v(s)\psi_i{}^v(s^{-1})$$

$$+ \sum_{s \in H} \{\psi_i(s)\psi_i{}^v(s^{-1}) + \psi_i(s^{-1})\psi_i{}^v(s)\}.$$

Consequently, by the orthogonality relations for H,

$$\sum_{s \in G} \tilde{\psi}_i(s) \tilde{\psi}_i(s^{-1}) = 2h + 2\sum_{s \in H} \psi_i(s)\psi_i{}^v(s^{-1}).$$

If ψ_i and $\psi_i{}^v$ are inequivalent, the second term on the right vanishes and we obtain

$$\sum_{s \in G} \tilde{\psi}_i(s) \tilde{\psi}_i(s^{-1}) = g.$$

Thus the induced representation $\tilde{\psi}_i$ of G is irreducible, its degree being twice that of ψ_i.

On the other hand, if ψ_i and $\psi_i{}^v$ are equivalent, then

$$\sum_{s \in G} \tilde{\psi}_i(s) \tilde{\psi}_i(s^{-1}) = 2g.$$

If $\tilde{\psi}_i = \sum_j m_j \chi_j$ is the decomposition of $\tilde{\psi}_i$ into irreducible characters χ_j of G, it follows from (8) that $\sum_j m_j{}^2 = 2$. This implies that $\tilde{\psi}_i$ decomposes into two inequivalent irreducible characters of G, say $\tilde{\psi}_i = \chi_k + \chi_l$. We will show that in fact $\chi_l = \chi_k \lambda$.

If $\chi_k(s) = 0$ for all $s \notin H$, then

$$\sum_{s \in H} \chi_k(s)\chi_k(s^{-1}) = \sum_{s \in G} \chi_k(s)\chi_k(s^{-1}) = g = 2h$$

and hence, by the same argument as that just used, the restriction of χ_k to H decomposes into two inequivalent irreducible characters of H. Since the restriction of $\tilde{\psi}_i$ to H is $2\psi_i$, this is a contradiction. We conclude that $\chi_k(s) \neq 0$ for some $s \notin H$, i.e. $\chi_k \lambda \neq \chi_k$. Since χ_k occurs once in the decomposition of $\tilde{\psi}_i$, and $\tilde{\psi}_i(s) = 0$ if $s \notin H$,

$$1 = g^{-1} \sum_{s \in G} \tilde{\psi}_i(s) \chi_k(s^{-1})$$

$$= g^{-1} \sum_{s \in H} \tilde{\psi}_i(s) \chi_k(s^{-1})$$

$$= g^{-1} \sum_{s \in H} \tilde{\psi}_i(s) \chi_k(s^{-1}) \lambda(s^{-1})$$

$$= g^{-1} \sum_{s \in G} \tilde{\psi}_i(s) \chi_k(s^{-1}) \lambda(s^{-1}).$$

Thus $\chi_k \lambda$ also occurs once in the decomposition of $\tilde{\psi}_i$, and since $\chi_k \lambda \neq \chi_k$ we must have $\chi_k \lambda = \chi_l$.

In the relation $\sum_i \psi_i(1)^2 = h$, partition the sum into a sum over pairs of distinct conjugate characters and a sum over self-conjugate characters:

$$\Sigma'\{\psi_i(1)^2 + \psi_i{}^\nu(1)^2\} + \Sigma''\psi_i(1)^2 = h.$$

Then for the corresponding characters of G we have

$$\Sigma' \tilde{\psi}_i(1)^2 + \Sigma''\{\chi_k(1)^2 + \chi_l(1)^2\} = 2\Sigma'\{\psi_i(1)^2 + \psi_i{}^\nu(1)^2\} + 2\Sigma''\psi_i(1)^2 = 2h = g.$$

Since, by Corollary 14, each irreducible character of G appears in the sum on the left, it follows from (9) that each occurs exactly once. Thus we have proved

PROPOSITION 15 *Let the finite group G have a subgroup H of half its order. Then each pair of distinct conjugate characters of H yields by induction a single irreducible character of G of twice the degree, whereas each self-conjugate character of H yields by induction two distinct irreducible characters of G of the same degree, which coincide on H and differ in sign on G\H. The irreducible characters of G thus obtained are all distinct, and every irreducible character of G is obtained in this way.* ☐

We will now use Proposition 15 to determine the irreducible characters of several groups of mathematical and physical interest. Let $\mathscr{S}_n$ denote the *symmetric* group consisting of all permutations of the set $\{1,2,...,n\}$, $\mathscr{A}_n$ the *alternating* group consisting of all even permutations, and C_n the *cyclic* group consisting of all cyclic permutations. Thus $\mathscr{S}_n$ has order $n!$, $\mathscr{A}_n$ has order $n!/2$ and C_n has order n.

The irreducible characters of the abelian group $\mathscr{A}_3 = C_3$ are all of degree 1 and can be arranged as a table in the following way, where ω is a primitive cube root of unity, say $\omega = e^{2\pi i/3} = (-1 + i\sqrt{3})/2$.

$\mathscr{A}_3$

	e	(123)	(132)
ψ_1	1	1	1
ψ_2	1	ω	ω^2
ψ_3	1	ω^2	ω

The group $\mathscr{S}_3$ contains $\mathscr{A}_3$ as a subgroup of index 2. The elements of $\mathscr{S}_3$ form three conjugacy classes: $\mathscr{C}_1$ containing only the identity element e, $\mathscr{C}_2$ containing the three elements (12),(13),(23) of order 2, and $\mathscr{C}_3$ containing the two elements (123),(132) of order 3. The irreducible character ψ_1 of $\mathscr{A}_3$ is self-conjugate and yields two irreducible characters of $\mathscr{S}_3$ of degree 1, the trivial character χ_1 and the sign character $\chi_2 = \chi_1 \lambda$. The irreducible characters ψ_2, ψ_3 of $\mathscr{A}_3$ are conjugate and yield a single irreducible character χ_3 of $\mathscr{S}_3$ of degree 2. Thus we obtain the character table:

$\mathscr{S}_3$

	$\mathscr{C}_1$	$\mathscr{C}_2$	$\mathscr{C}_3$
χ_1	1	1	1
χ_2	1	-1	1
χ_3	2	0	-1

The elements of $\mathscr{A}_4$ form four conjugacy classes: $\mathscr{C}_1$ containing only the identity element e, $\mathscr{C}_2$ containing the three elements $t_1 = (12)(34)$, $t_2 = (13)(24)$, $t_3 = (14)(23)$ of order 2, $\mathscr{C}_3$ containing four elements of order 3, namely c, ct_1, ct_2, ct_3, where $c = (123)$, and $\mathscr{C}_4$ containing the remaining four elements of order 3, namely $c^2, c^2t_1, c^2t_2, c^2t_3$. Moreover $N = \mathscr{C}_1 \cup \mathscr{C}_2$ is a normal subgroup of order 4, $H = \{e, c, c^2\}$ is a cyclic subgroup of order 3, and

$$\mathscr{A}_4 = HN, \quad H \cap N = \{e\}.$$

If χ is a character of degree 1 of H, then a character ψ of degree 1 of $\mathscr{A}_4$ is defined by

$$\psi(hn) = \chi(h) \quad \text{for all } h \in H, n \in N.$$

Since H is isomorphic to $\mathscr{A}_3$, we obtain in this way three characters ψ_1, ψ_2, ψ_3 of $\mathscr{A}_4$ of degree 1. Since $\mathscr{A}_4$ has order 12, and $12 = 1 + 1 + 1 + 9$, the remaining irreducible character ψ_4 of $\mathscr{A}_4$ has degree 3. The character table of $\mathscr{A}_4$ can be completed by means of the orthogonality relations (11) and has the following form, where again $\omega = (-1 + i\sqrt{3})/2$.

$$\mathscr{A}_4$$

$|\mathscr{C}|$	1	3	4	4
$\mathscr{C}$	$\mathscr{C}_1$	$\mathscr{C}_2$	$\mathscr{C}_3$	$\mathscr{C}_4$
ψ_1	1	1	1	1
ψ_2	1	1	ω	ω^2
ψ_3	1	1	ω^2	ω
ψ_4	3	-1	0	0

The group $\mathscr{S}_4$ contains $\mathscr{A}_4$ as a subgroup of index 2 and $v = (12) \in \mathscr{S}_4 \setminus \mathscr{A}_4$. The elements of $\mathscr{S}_4$ form five conjugacy classes: $\mathscr{C}_1$ containing only the identity element e, $\mathscr{C}_2$ containing six transpositions (jk) $(1 \le j < k \le 4)$, $\mathscr{C}_3$ containing the three elements of order 2 in $\mathscr{A}_4$, $\mathscr{C}_4$ containing eight elements of order 3, and $\mathscr{C}_5$ containing six elements of order 4.

The self-conjugate character ψ_1 of $\mathscr{A}_4$ yields two characters of $\mathscr{S}_4$ of degree 1, the trivial character χ_1 and the sign character $\chi_2 = \chi_1 \lambda$; the pair of conjugate characters ψ_2, ψ_3 of $\mathscr{A}_4$ yields an irreducible character χ_3 of $\mathscr{S}_4$ of degree 2; and the self-conjugate character ψ_4 of $\mathscr{A}_4$ yields two irreducible characters χ_4, χ_5 of $\mathscr{S}_4$ of degree 3. The rows of the character table corresponding to χ_4, χ_5 must have the form

$$3 \quad x \quad z \quad w \quad y$$
$$3 \quad -x \quad z \quad w \quad -y$$

and from the orthogonality relations (11) we obtain $z = -1$, $w = 0$, $xy = -1$. From the orthogonality relations (10) we further obtain $x + y = 0$. Hence $x^2 = 1$ and the complete character table is

$$\mathscr{S}_4$$

$|\mathscr{C}|$	1	6	3	8	6
$\mathscr{C}$	$\mathscr{C}_1$	$\mathscr{C}_2$	$\mathscr{C}_3$	$\mathscr{C}_4$	$\mathscr{C}_5$
χ_1	1	1	1	1	1
χ_2	1	-1	1	1	-1
χ_3	2	0	2	-1	0
χ_4	3	1	-1	0	-1
χ_5	3	-1	-1	0	1

The physical significance of these groups derives from the fact that $\mathscr{A}_4$ (resp. $\mathscr{S}_4$) is isomorphic to the group of all rotations (resp. orthogonal transformations) of $\mathbb{R}^3$ which map a regular tetrahedron onto itself. Similarly $\mathscr{A}_3$ (resp. $\mathscr{S}_3$) is isomorphic to the group of all plane rotations (resp. plane rotations and reflections) which map an equilateral triangle onto itself.

An important property of induced representations was proved by R. Brauer (1953): each character of a finite group is a linear combination with integer coefficients (not necessarily non-negative) of characters induced from characters of elementary subgroups. Here a group is said to be *elementary* if it is the direct product of a group whose order is a power of a prime and a cyclic group whose order is not divisible by that prime.

It may be deduced without difficulty from Brauer's theorem that, if G is a finite group and m the least common multiple of the orders of its elements, then (as had long been conjectured) any irreducible representation of G is equivalent to a representation in the field $\mathbb{Q}(e^{2\pi i/m})$. Green (1955) has shown that Brauer's theorem is actually best possible: if each character of a finite group G is a linear combination with integer coefficients of characters induced from characters of subgroups belonging to some family $\mathscr{F}$, then each elementary subgroup of G is contained in a conjugate of some subgroup in $\mathscr{F}$.

7 Applications

Character theory has turned out to be an invaluable tool in the study of abstract groups. We illustrate this by two results of Burnside (1904) and Frobenius (1901). It is remarkable, first that these applications were found so soon after the development of character theory and secondly that, one century later, there are still no proofs known which do not use character theory.

LEMMA 16 *If* $\rho: s \to A(s)$ *is a representation of degree n of a finite group G, then the character χ of ρ satisfies*

$$|\chi(s)| \leq n \quad \text{for any } s \in G.$$

Moreover, equality holds for some s if and only if $A(s) = \omega I_n$, where $\omega \in \mathbb{C}$.

Proof If $s \in G$ has order m, there exists an invertible matrix T such that

$$T^{-1}A(s)T = \text{diag } [\omega_1,...,\omega_n],$$

where $\omega_1,...,\omega_n$ are m-th roots of unity. Hence $\chi(s) = \omega_1 + ... + \omega_n$ and

$$|\chi(s)| \le |\omega_1| + \dots + |\omega_n| = n.$$

Moreover $|\chi(s)| = n$ only if $\omega_1,\dots,\omega_n$ all lie on the same ray through the origin and hence only if they are all equal, since they lie on the unit circle. But then $A(s) = \omega I_n$. $\blacksquare$

The *kernel* of the representation ρ is the set K_ρ of all $s \in G$ for which $\rho(s) = I_n$. Evidently K_ρ is a normal subgroup of G. By Lemma 16, K_ρ may be characterized as the set of all $s \in G$ such that $\chi(s) = n$.

LEMMA 17 *Let* $\rho: s \to A(s)$ *be an irreducible representation of degree n of a finite group G, with character χ, and let $\mathscr{C}$ be a conjugacy class of G containing h elements. If h and n are relatively prime then, for any $s \in \mathscr{C}$, either $\chi(s) = 0$ or $A(s) = \omega I_n$ for some $\omega \in \mathbb{C}$.*

Proof Since h and n are relatively prime, there exist integers a,b such that $ah + bn = 1$. Then

$$\chi(s)/n = ah\chi(s)/n + b\chi(s).$$

Since $h\chi(s)/n$ and $\chi(s)$ are algebraic integers, it follows that $\chi(s)/n$ is an algebraic integer. We may assume that $|\chi(s)| < n$, since otherwise the result follows from Lemma 16.

Suppose s has order m. If $(k,m) = 1$, then the conjugacy class containing s^k also has cardinality h and thus $\chi(s^k)/n$ is an algebraic integer, by what we have already proved. Hence

$$\alpha = \prod_k \chi(s^k)/n,$$

where k runs through all positive integers less than m and relatively prime to m, is also an algebraic integer. But $\chi(s^k) = f(\omega^k)$, where ω is a primitive m-th root of unity and

$$f(x) = x^{r_1} + \dots + x^{r_n}$$

for some non-negative integers $r_1,\dots,r_n$ less than m. Thus α is a symmetric function of the primitive roots ω^k. Since the *cyclotomic polynomial*

$$\Phi_n(x) = \prod_k (x - \omega^k)$$

has integer coefficients, it follows that $\alpha \in \mathbb{Q}$. Consequently, by Lemma 12, $\alpha \in \mathbb{Z}$.

But $|\alpha| < 1$, since $|\chi(s)| < n$ and $|\chi(s^k)| \le n$ for every k. Hence $\alpha = 0$, and thus $\chi(s^k) = 0$ for some k with $(k,m) = 1$. If $g(x)$ is the monic polynomial in $\mathbb{Q}[x]$ of least positive degree such that $g(\omega^k) = 0$, then any polynomial in $\mathbb{Q}[x]$ with ω^k as a root must be divisible by $g(x)$. Since we showed in Chapter II, §5 that the cyclotomic polynomial $\Phi_n(x)$ is irreducible over the field $\mathbb{Q}$, it follows that $g(x) = \Phi_n(x)$ and that $\Phi_n(x)$ divides $f(x)$. Hence also $\chi(s) = f(\omega) = 0$. $\blacksquare$

Before stating the next result we recall from Chapter I, §7 that a group is said to be *simple* if it contains more than one element and has no nontrivial proper normal subgroup.

PROPOSITION 18 *If a finite group G has a conjugacy class $\mathscr{C}$ of cardinality p^a, for some prime p and positive integer a, then G is not a simple group.*

Proof If $s \in \mathscr{C}$ then, by (9),

$$\sum_{\mu} n_{\mu} \chi_{\mu}(s) = 0.$$

Assume the notation chosen so that χ_1 is the character of the trivial representation. If $\chi_{\mu}(s) = 0$ for every $\mu > 1$ for which p does not divide n_{μ}, then the displayed equation has the form $1 + p\zeta = 0$, where ζ is an algebraic integer. Since $-1/p$ is not an integer, this contradicts Lemma 12. Consequently, by Lemma 17, for some $v > 1$ we must have $A^{(v)}(s) = \omega I_{n_v}$, where $\omega \in \mathbb{C}$. The set K_v of all elements of G which are represented by the identity transformation in the v-th irreducible representation is a normal subgroup of G. Moreover $K_v \neq \{e\}$, since K_v contains all elements $u^{-1}s^{-1}us$, and $K_v \neq G$, since $v > 1$. Thus G is not simple. $\square$

COROLLARY 19 *If G is a group of order $p^a q^b$, where p,q are distinct primes and a,b nonnegative integers such that $a + b > 1$, then G is not simple.*

Proof Let $\mathscr{C}_1,...,\mathscr{C}_r$ be the conjugacy classes of G, with $\mathscr{C}_1 = \{e\}$, and let h_k be the cardinality of $\mathscr{C}_k$ $(k = 1,...,r)$. Then h_k divides the order g of G and

$$g = h_1 + ... + h_r.$$

Suppose first that $h_j = 1$ for some $j > 1$. Then $\mathscr{C}_j = \{s_j\}$, where s_j commutes with every element of G. Thus the cyclic group H generated by s_j is a normal subgroup of G. Then G is not simple even if $H = G$, since $a + b > 1$ and any proper subgroup of a cyclic group is normal.

Suppose next that $h_k \neq 1$ for every $k > 1$. If G is simple then, by Proposition 18, q divides h_k for every $k > 1$. Since q divides g, it follows that q divides $h_1 = 1$, which is a contradiction. $\square$

It has been shown by Kazarin (1990) that the normal subgroup generated by the elements of the conjugacy class $\mathscr{C}$ in Proposition 18 is *solvable*. Although no proof of Burnside's Proposition 18 is known which does not use character theory, Goldschmidt (1970) and Matsuyama (1973) have given a rather intricate proof of the important Corollary 19 which is purely group theoretic.

The restriction to *two* distinct primes in the statement of Corollary 19 is essential, since the alternating group $\mathscr{A}_5$ of order $60 = 2^2 \cdot 3 \cdot 5$ is simple. It follows at once from Corollary 19, by induction on the order, that any finite group whose order is divisible by at most two distinct primes is *solvable*. P. Hall (1928/1937) has used Corollary 19 to show that a finite group G of order g is solvable if and only if G has a subgroup H of order h for every factorization $g = p^a h$, where $a > 0$ and p is a prime not dividing h.

The second application of group characters, due to Frobenius, has the following statement:

PROPOSITION 20 *If the finite group G has a nontrivial proper subgroup H such that*

$$x^{-1}Hx \cap H = \{e\} \quad \text{for every } x \in G \setminus H,$$

then G contains a normal subgroup N such that G is the semidirect product of H and N, i.e.

$$G = NH, \quad H \cap N = \{e\}.$$

Proof Obviously $x^{-1}Hx = y^{-1}Hy$ if $y \in Hx$ and the hypotheses imply that $x^{-1}Hx \cap y^{-1}Hy = \{e\}$ if $y \notin Hx$. If g,h are the orders of G,H respectively, it follows that the number of distinct conjugate subgroups $x^{-1}Hx$ (including H itself) is $n = g/h$. Furthermore the number of elements of G which belong to some conjugate subgroup is $n(h-1) + 1 = g - (n-1)$. Thus the set S of elements of G which do not belong to any conjugate subgroup has cardinality $n - 1$.

Let ψ_μ be the character of an irreducible representation of H and $\tilde{\psi}_\mu$ the character of the induced representation of G. By (12) and the hypotheses,

$$\tilde{\psi}_\mu(e) = n\psi_\mu(e), \quad \tilde{\psi}_\mu(s) = 0 \text{ if } s \in S, \quad \tilde{\psi}_\mu(s) = \psi_\mu(s) \text{ if } s \in H \setminus e.$$

For any fixed μ, form the class function

$$\chi = \tilde{\psi}_\mu - \psi_\mu(e)\{\tilde{\psi}_1 - \chi_1\},$$

where ψ_1 and χ_1 are the characters of the trivial representations of H and G respectively. Then χ is a *generalized character* of G, i.e. $\chi = \sum_\nu m_\nu \chi_\nu$ is a linear combination of irreducible characters χ_ν with integral, but not necessarily non-negative, coefficients m_ν. Moreover

$$\chi(e) = \psi_\mu(e), \quad \chi(s) = \psi_\mu(e) \text{ if } s \in S, \quad \chi(s) = \psi_\mu(s) \text{ if } s \in H \setminus e.$$

Hence

$$\sum_{s \in H \setminus e} \chi(s)\chi(s^{-1}) = \sum_{s \in H \setminus e} \psi_\mu(s)\psi_\mu(s^{-1}) = h - \psi_\mu(e)^2.$$

Since S has cardinality $n - 1$, it follows that

$$\sum_{s \in G} \chi(s)\chi(s^{-1}) = n\{h - \psi_\mu(e)^2\} + \psi_\mu(e)^2 + (n-1)\psi_\mu(e)^2 = g.$$

But the formula (8) holds also for generalized characters. Since $\chi(e) > 0$, we conclude that χ is in fact an irreducible character of G. Thus we have an irreducible representation of degree $\chi(e)$ in which the matrices representing elements of S have trace $\chi(e)$. The elements of S must therefore be represented by the unit matrix, i.e. they belong to the kernel K_μ of the representation.

On the other hand, for any $t \in H \setminus e$ we have

$$\sum_\mu \psi_\mu(e)\psi_\mu(t) = 0$$

and hence $\psi_\mu(t) \neq \psi_\mu(e)$ for some μ. Thus the intersection of the kernels K_μ for varying μ contains just the elements of S and e. Since K_μ is a normal subgroup of G, it follows that $N = S \cup \{e\}$ is also a normal subgroup. Furthermore, since $H \cap N = \{e\}$, HN has cardinality $hn = g$ and hence $HN = G$. $\square$

A finite group G which satisfies the hypotheses of Proposition 20 is said to be a *Frobenius group*. The subgroup H is said to be a *Frobenius complement* and the normal subgroup N a *Frobenius kernel*. It is readily shown that a finite permutation group is a Frobenius group if and only if it is transitive and no element except the identity fixes more than one symbol. Another characterization follows from Proposition 20: a finite group G is a Frobenius group if and only if it has a nontrivial proper normal subgroup N such that, if $x \in N$ and $x \neq e$, then $xy \neq yx$ for all $y \in G \setminus N$.

Frobenius groups are of some general significance and much is known about their structure. It is easily seen that h divides $n - 1$, so that the subgroups H and N have relatively prime orders. It has been shown by Thompson (1959) that the normal subgroup N is a direct product of groups of prime power order. The structure of H is known even more precisely through the work of Burnside (1901) and others.

Applications of group characters of quite a different kind arise in the study of molecular vibrations. We describe one such application within classical mechanics, due to Wigner (1930). However, there are further applications within quantum mechanics, e.g. to the determination of the possible spectral lines in the Raman scattering of light by a substance whose molecules have a particular symmetry group.

A basic problem of classical mechanics deals with the *small oscillations* of a system of particles about an equilibrium configuration. The equations of motion have the form

$$B \ddot{x} + C x = 0, \tag{13}$$

where $x \in \mathbb{R}^n$ is a vector of generalized coordinates and B,C are positive definite real symmetric matrices. In fact the kinetic energy is $(1/2)\,\dot{x}^t B\,\dot{x}$ and, as a first approximation for x near 0, the potential energy is $(1/2)x^t Cx$.

Since B and C are positive definite, there exists (see Chapter V, §4) a non-singular matrix T such that

$$T^t BT = I, \quad T^t CT = D,$$

where D is a diagonal matrix with positive diagonal elements. By the linear transformation $x = Ty$ the equations of motion are brought to the form

$$\ddot{y} + Dy = 0.$$

These 'decoupled' equations can be solved immediately: if

$$y = (\eta_1,...,\eta_n)^t, \quad D = \mathrm{diag}\ [\omega_1^2,...,\omega_n^2],$$

with $\omega_k > 0$ $(k = 1,...,n)$, then

$$\eta_k = \alpha_k \cos \omega_k t + \beta_k \sin \omega_k t,$$

where α_k, β_k $(k = 1,...,n)$ are arbitrary constants of integration. Hence there exist vectors $a_k, b_k \in \mathbb{R}^n$ such that every solution of (13) is a linear combination of solutions of the form

$$a_k \cos \omega_k t, \quad b_k \sin \omega_k t \quad (k = 1,...,n),$$

the so-called *normal modes* of oscillation. The eigenvalues of the matrix $B^{-1}C$ are the squares of the *normal frequencies* $\omega_1,...,\omega_n$.

An important example is the system of particles formed by a molecule of N atoms. Since the displacement of each atom from its equilibrium position is specified by three coordinates, the internal configuration of the molecule without regard to its position and orientation in space may be specified by $n = 3N - 6$ internal coordinates. The determination of the corresponding normal frequencies $\omega_1,...,\omega_n$ may be a formidable task even for moderate values of N. However, the problem is considerably reduced by taking advantage of the symmetry of the molecule.

A *symmetry operation* is an isometry of $\mathbb{R}^3$ which sends the equilibrium position of any atom into the equilibrium position of an atom of the same type. The set of all symmetry operations is clearly a group under composition, the *symmetry group* of the molecule.

For example, the methane molecule CH_4 has four hydrogen atoms at the vertices of a regular tetrahedron and a carbon atom at the centre, from which it follows that the symmetry group of CH_4 is isomorphic to $\mathscr{S}_4$. Similarly, the ammonia molecule NH_3 has three hydrogen atoms and a nitrogen atom at the four vertices of a regular tetrahedron, and hence the symmetry group of NH_3 is isomorphic to $\mathscr{S}_3$.

We return now to the general case. If G is the symmetry group of the molecule, then to each $s \in G$ there corresponds a linear transformation $A(s)$ of the configuration space $\mathbb{R}^n$. Moreover the map $\rho: s \to A(s)$ is a representation of G. Since the kinetic and potential energies are unchanged by a symmetry operation, we have

$$A(s)^t B A(s) = B, \quad A(s)^t C A(s) = C \quad \text{for every } s \in G.$$

It follows that

$$B^{-1} C A(s) = A(s) B^{-1} C \quad \text{for every } s \in G.$$

Assume the notation chosen so that the distinct ω's are $\omega_1,...,\omega_p$ and ω_k occurs m_k times in the sequence $\omega_1,...,\omega_n$ ($k = 1,...,p$). Thus $n = m_1 + ... + m_p$. If V_k is the set of all $v \in \mathbb{R}^n$ such that

$$B^{-1} C v = \omega_k^2 v,$$

then V_k is an m_k-dimensional subspace of $\mathbb{R}^n$ ($k = 1,...,p$) and $\mathbb{R}^n$ is the direct sum of $V_1,...,V_p$. Moreover each eigenspace V_k is invariant under $A(s)$ for every $s \in G$. Hence, by Maschke's theorem (which holds also for representations in a real vector space), V_k is a direct sum of real-irreducible invariant subspaces. It follows that there exists a real non-singular matrix T such that, for every $s \in G$,

$$T^{-1} A(s) T = \begin{pmatrix} A_1(s) & 0 & ... & 0 \\ 0 & A_2(s) & ... & 0 \\ ... & ... & ... & ... \\ 0 & 0 & ... & A_q(s) \end{pmatrix},$$

where $s \to A_k(s)$ is a real-irreducible representation of G, of degree n_k say ($k = 1,...,q$), and

$$T^{-1} B^{-1} C T = \begin{pmatrix} \lambda_1 I_{n_1} & 0 & ... & 0 \\ 0 & \lambda_2 I_{n_2} & ... & 0 \\ ... & ... & ... & ... \\ 0 & 0 & ... & \lambda_q I_{n_q} \end{pmatrix}.$$

If the real-irreducible representations $s \to A_k(s)$ ($k = 1,...,q$) are also complex-irreducible, then their degrees and multiplicities can be found by character theory. Thus by decomposing the representation ρ of G into its irreducible components we can determine the degeneracy of the normal frequencies.

We will not consider here the modifications needed when some real-irreducible component is not also complex-irreducible. Also, it should be noted that it may happen 'accidentally' that $\lambda_j = \lambda_k$ for some $j \neq k$.

As a simple illustration of the preceding discussion we consider the ammonia molecule NH_3. Its internal configuration may be described by the six internal coordinates r_1, r_2, r_3 and $\alpha_{23}, \alpha_{31}, \alpha_{12}$, where r_j is the change from its equilibrium value of the distance from the nitrogen atom to the j-th hydrogen atom, and α_{jk} is the change from its equilibrium value of the angle between the rays joining the nitrogen atom to the j-th and k-th hydrogen atoms.

We will determine the character χ of the corresponding representation ρ of the symmetry group $\mathscr{S}_3$. In the notation of the character table previously given for $\mathscr{S}_3$, there is an element $s \in \mathscr{C}_3$ for which the symmetry operation $A(s)$ cyclically permutes r_1, r_2, r_3 and $\alpha_{23}, \alpha_{31}, \alpha_{12}$. Consequently $\chi(s) = 0$ if $s \in \mathscr{C}_3$. Also, there is an element $t \in \mathscr{C}_2$ for which the symmetry operation $A(t)$ interchanges r_1 with r_2 and α_{23} with α_{31}, but fixes r_3 and α_{12}. Consequently $\chi(t) = 2$ if $t \in \mathscr{C}_2$. Since it is obvious that $\chi(e) = 6$, this determines χ and we adjoin it to the character table of $\mathscr{S}_3$:

$|\mathscr{C}|$	1	3	2
$\mathscr{C}$	$\mathscr{C}_1$	$\mathscr{C}_2$	$\mathscr{C}_3$
χ_1	1	1	1
χ_2	1	-1	1
χ_3	2	0	-1
χ	6	2	0

Decomposing the character χ into its irreducible components by means of (7), we obtain $\chi = 2\chi_1 + 2\chi_3$. Since the irreducible representations of $\mathscr{S}_3$ are all real, this means that the configuration space $\mathbb{R}^6$ is the direct sum of four irreducible invariant subspaces, two of dimension 1 and two of dimension 2. Knowing what to look for, we may verify that the one-dimensional subspaces spanned by $r_1 + r_2 + r_3$ and $\alpha_{23} + \alpha_{31} + \alpha_{12}$ are invariant. Also, the two-dimensional subspace formed by all vectors $\mu_1 r_1 + \mu_2 r_2 + \mu_3 r_3$ with $\mu_1 + \mu_2 + \mu_3 = 0$ is invariant and irreducible, and so is the two-dimensional subspace formed by all vectors $\nu_1 \alpha_{23} + \nu_2 \alpha_{31} + \nu_3 \alpha_{12}$ with $\nu_1 + \nu_2 + \nu_3 = 0$. Hence we can find a real non-singular matrix T such that

$$T^{-1}B^{-1}CT = \begin{pmatrix} \lambda_1 I_1 & 0 & 0 & 0 \\ 0 & \lambda_2 I_1 & 0 & 0 \\ 0 & 0 & \lambda_3 I_2 & 0 \\ 0 & 0 & 0 & \lambda_4 I_2 \end{pmatrix}.$$

This shows that the ammonia molecule NH_3 has two nondegenerate normal frequencies and two doubly degenerate normal frequencies.

8 Generalizations

During the past century the character theory of finite groups has been extensively generalized to infinite groups with a topological structure. It may be helpful to give an overview here, without proofs, of this vast development. The reader wishing to pursue some particular topic may consult the references at the end of the chapter.

A *topological group* is a group G with a topology such that the map $(s,t) \to st^{-1}$ of $G \times G$ into G is continuous. Throughout the following discussion we will assume that G is a topological group which, as a topological space, is *locally compact and Hausdorff*, i.e. any two distinct points are contained in open sets whose closures are disjoint compact sets. (A closed set E in a topological space is *compact* if each open cover of E has a finite subcover. In a metric space this is consistent with the definition of sequential compactness in Chapter I, §4.)

Let $\mathscr{C}_0(G)$ denote the set of all continuous functions $f \colon G \to \mathbb{C}$ such that $f(s) = 0$ for all s outside some compact subset of G (which may depend on f). A map $M \colon \mathscr{C}_0(G) \to \mathbb{C}$ is said to be a *nonnegative linear functional* if

(i) $M(f_1 + f_2) = M(f_1) + M(f_2)$ for all $f_1, f_2 \in \mathscr{C}_0(G)$,

(ii) $M(\lambda f) = \lambda M(f)$ for all $\lambda \in \mathbb{C}$ and $f \in \mathscr{C}_0(G)$,

(iii) $M(f) \geq 0$ if $f(s) \geq 0$ for every $s \in G$.

It is said to be a *left* (resp. *right*) *Haar integral* if, in addition, it is nontrivial, i.e. $M(f) \neq 0$ for some $f \in \mathscr{C}_0(G)$, and left (resp. right) invariant, i.e.

(iv) $M(_t f) = M(f)$ for every $t \in G$ and $f \in \mathscr{C}_0(G)$, where $_t f(s) = f(t^{-1}s)$, (resp. $M(f_t) = M(f)$ for every $t \in G$ and $f \in \mathscr{C}_0(G)$, where $f_t(s) = f(st)$).

It was shown by Haar (1933) that a left Haar integral exists on any locally compact group; it was later shown to be uniquely determined apart from a positive multiplicative constant. By defining $M^*(f) = M(f^*)$, where $f^*(s) = f(s^{-1})$ for every $s \in G$, it follows that a right Haar integral also exists and is uniquely determined apart from a positive multiplicative constant.

The notions of left and right Haar integral obviously coincide if the group G is abelian, and it may be shown that they also coincide if G is compact or is a semi-simple Lie group.

We now restrict attention to the case of a left Haar integral. It is easily seen that

$$M(\bar{f}) = \overline{M(f)},$$

where $\bar{f}(s) = \overline{f(s)}$ for every $s \in G$. If we set $(f,g) = M(f\bar{g})$, then the usual inner product properties hold:

$$(f_1 + f_2, g) = (f_1, g) + (f_2, g),$$
$$(\lambda f, g) = \lambda(f, g),$$
$$(f, g) = \overline{(g, f)},$$
$$(f, f) \geq 0, \text{ with equality only if } f \equiv 0.$$

By the *Riesz representation theorem*, there is a unique *positive measure* μ on the σ-algebra $\mathcal{M}$ generated by the compact subsets of G (cf. Chapter XI, §3) such that $\mu(K)$ is finite for every compact set $K \subseteq G$, $\mu(E)$ is the supremum of $\mu(K)$ over all compact $K \subseteq E$ for each $E \in \mathcal{M}$, and

$$M(f) = \int_G f \, d\mu \quad \text{for every } f \in \mathcal{C}_0(G).$$

The measure μ is necessarily left invariant:

$$\mu(E) = \mu(sE) \quad \text{for all } E \in \mathcal{M} \text{ and } s \in G,$$

where $sE = \{sx : x \in E\}$.

For $p = 1$ or 2, let $L^p(G)$ denote the set of all μ-measurable functions $f : G \to \mathbb{C}$ such that

$$\int_G |f|^p \, d\mu < \infty.$$

The definition of M can be extended to $L^1(G)$ by setting

$$M(f) = \int_G f \, d\mu,$$

and the inner product can be extended to $L^2(G)$ by setting

$$(f, g) = \int_G f\bar{g} \, d\mu.$$

Moreover, with this inner product $L^2(G)$ is a *Hilbert space*. If we define the *convolution product* $f*g$ of $f,g \in L^1(G)$ by

$$f*g\,(s)\ =\ \int_G f(st)g(t^{-1})\ \mathrm{d}\mu(t),$$

then $L^1(G)$ is a *Banach algebra* and

$$M(f*g)\ =\ M(f)M(g)\ \text{ for all } f,g \in L^1(G).$$

A *unitary representation* of G in a Hilbert space $\mathcal{H}$ is a map ρ of G into the set of all linear transformations of $\mathcal{H}$ which maps the identity element e of G into the identity transformation of $\mathcal{H}$:

$$\rho(e)\ =\ I,$$

which preserves not only products in G:

$$\rho(st)\ =\ \rho(s)\rho(t)\ \text{ for all } s,t \in G,$$

but also inner products in $\mathcal{H}$:

$$(\rho(s)u,\rho(s)v)\ =\ (u,v)\ \text{ for all } s \in G \text{ and all } u,v \in \mathcal{H},$$

and for which the map $(s,v) \to \rho(s)v$ of $G \times \mathcal{H}$ into $\mathcal{H}$ is continuous (or, equivalently, for which the map $s \to (\rho(s)v,v)$ of G into $\mathbb{C}$ is continuous at e for every $v \in \mathcal{H}$).

For example, any locally compact group G has a unitary representation ρ in $L^2(G)$, its *regular representation*, defined by

$$(\rho(t)f)(s)\ =\ f(t^{-1}s)\ \text{ for all } f \in L^2(G) \text{ and all } s,t \in G.$$

If ρ is a unitary representation of G in a Hilbert space $\mathcal{H}$, and if a closed subspace V of $\mathcal{H}$ is invariant under $\rho(s)$ for every $s \in G$, then so also is its orthogonal complement $V^{\perp}$. The representation ρ is said to be *irreducible* if the only closed subspaces of $\mathcal{H}$ which are invariant under $\rho(s)$ for every $s \in G$ are $\mathcal{H}$ and $\{0\}$. It has been shown by Gelfand and Raikov (1943) that, for any locally compact group G and any $s \in G \setminus e$, there is an irreducible unitary representation ρ of G with $\rho(s) \neq I$.

Consider now the case in which the locally compact group G is abelian. Then any irreducible unitary representation of G is one-dimensional. Hence if we define a *character* of G to be a continuous function $\chi\colon G \to \mathbb{C}$ such that

(i) $\chi(st) = \chi(s)\chi(t)$ for all $s,t \in G$,

(ii) $|\chi(s)| = 1$ for every $s \in G$,

then every irreducible unitary representation is a character, and vice versa.

If multiplication and inversion of characters are defined pointwise, then the set $\hat{G}$ of all characters of G is again an abelian group, the *dual group* of G. Moreover, we can put a topology on $\hat{G}$ by defining a subset of $\hat{G}$ to be open if it is a union of sets of the form

$$N(\psi,\varepsilon,K) = \{\chi \in \hat{G}: |\chi(s)/\psi(s) - 1| < \varepsilon \text{ for all } s \in K\},$$

where $\psi \in \hat{G}, \varepsilon > 0$ and K is a compact subset of G. Then $\hat{G}$ is not only abelian, but also a locally compact topological group.

For each fixed $s \in G$, the map $\hat{s}: \chi \to \chi(s)$ is a character of $\hat{G}$. Moreover the map $s \to \hat{s}$ is one-to-one, by the theorem of Gelfand and Raikov, and every character of $\hat{G}$ is obtained in this way. In fact the *duality theorem* of Pontryagin and van Kampen (1934/5) states that G is isomorphic and homeomorphic to the dual group of $\hat{G}$.

The *Fourier transform* of a function $f \in L^1(G)$ is the function $\hat{f}: \hat{G} \to \mathbb{C}$ defined by

$$\hat{f}(\chi) = \int_G f(s)\overline{\chi(s)}\, d\mu(s),$$

where μ is the Haar measure on G. If $f_1, f_2 \in L^1(G) \cap L^2(G)$, then $\hat{f}_1, \hat{f}_2 \in L^2(\hat{G})$ and, with a suitable fixed normalization of the Haar measure $\hat{\mu}$ on $\hat{G}$,

$$(f_1, f_2)_G = (\hat{f}_1, \hat{f}_2)_{\hat{G}}.$$

Furthermore, the map $f \to \hat{f}$ can be uniquely extended to a unitary map of $L^2(G)$ onto $L^2(\hat{G})$. This generalizes *Plancherel's theorem* for Fourier integrals on the real line.

If $f = g * h$, where $g, h \in L^1(G)$, then $f \in L^1(G)$ and

$$\hat{f}(\chi) = \hat{g}(\chi)\hat{h}(\chi) \text{ for every } \chi \in \hat{G}.$$

If, in addition, $g, h \in L^2(G)$, then $\hat{f} \in L^1(\hat{G})$ and, with the same choice as before for the Haar measure $\hat{\mu}$ on $\hat{G}$, the *Fourier inversion formula* holds:

$$f(s) = \int_{\hat{G}} \hat{f}(\chi)\chi(s)\, d\hat{\mu}(\chi).$$

The *Poisson summation formula* can also be extended to this general setting. Let H be a closed subgroup of G and let K denote the factor group G/H. If the Haar measures $\mu, \hat{\nu}$ on $H, \hat{K}$ are suitably chosen then, with appropriate hypotheses on $f \in L^1(G)$,

$$\int_H f(t)\, d\mu(t) = \int_{\hat{K}} \hat{f}(\psi)\, d\hat{\nu}(\psi).$$

We now give some examples (without spelling out the topologies). If $G = \mathbb{R}$ is the additive group of all real numbers, then its characters are the functions $\chi_t: \mathbb{R} \to \mathbb{C}$, with $t \in \mathbb{R}$, defined by

$$\chi_t(s) = e^{its}.$$

In this case G is isomorphic and homeomorphic to $\hat{G}$ itself under the map $t \to \chi_t$. The Haar integral of $f \in L^1(G)$ is the ordinary Lebesgue integral

$$M(f) = \int_{-\infty}^{\infty} f(s)\,ds,$$

the Fourier transform of f is

$$\hat{f}(t) = \int_{-\infty}^{\infty} f(s)e^{-its}\,ds,$$

and the Fourier inversion formula has the form

$$f(s) = (1/2\pi)\int_{-\infty}^{\infty} \hat{f}(t)e^{its}\,dt.$$

If $G = \mathbb{Z}$ is the additive group of all integers, then its characters are the functions $\chi_z : \mathbb{Z} \to \mathbb{C}$, with $z \in \mathbb{C}$ and $|z| = 1$, defined by

$$\chi_z(n) = z^n.$$

Thus $\hat{G}$ is the multiplicative group of all complex numbers of absolute value 1. The Haar integral of $f \in L^1(G)$ is

$$M(f) = \sum_{-\infty}^{\infty} f(n),$$

the Fourier transform of f is

$$\hat{f}(e^{i\phi}) = \sum_{-\infty}^{\infty} f(n)e^{-in\phi},$$

and the Fourier inversion formula has the form

$$f(n) = (1/2\pi)\int_0^{2\pi} \hat{f}(e^{i\phi})e^{in\phi}\,d\phi.$$

Thus the classical theories of Fourier integrals and Fourier series are just special cases. As another example, let $G = \mathbb{Q}_p$ be the additive group of all p-adic numbers. The characters in this case are the functions $\chi_t : \mathbb{Q}_p \to \mathbb{C}$, with $t \in \mathbb{Q}_p$, defined by

$$\chi_t(s) = e^{2\pi i\lambda(st)},$$

where $\lambda(x) = \sum_{j<0} x_j p^j$ if $x \in \mathbb{Q}_p$ is given by $x = \sum_{-\infty}^{\infty} x_j p^j$, $x_j \in \{0,1,...,p-1\}$ and $x_j = 0$ for all large $j < 0$. Also in this case G is isomorphic and homeomorphic to $\hat{G}$ itself under the map $t \to \chi_t$. If we choose the Haar measure on G so that the measure of the compact set $\mathbb{Z}_p$ of all p-adic integers is 1, then the same choice for $\hat{G}$ is the appropriate one for Plancherel's theorem and the Fourier inversion formula.

Consider next the case in which the group G is compact, but not necessarily abelian. In this case $\mathscr{C}_0(G)$ coincides with the set $\mathscr{C}(G)$ of all continuous functions $f\colon G \to \mathbb{C}$. The Haar integral is both left and right invariant, and we suppose it normalized so that the integral of the constant 1 has the value 1. Then the integral $M(f)$ of any $f \in \mathscr{C}(G)$, or $L^1(G)$, may be called the *invariant mean* of f.

It may be shown that if ρ is a unitary representation of a compact group G in a Hilbert space $\mathscr{H}$, then $\mathscr{H}$ may be represented as a direct sum $\mathscr{H} = \oplus_\alpha \mathscr{H}_\alpha$ of mutually orthogonal finite-dimensional invariant subspaces $\mathscr{H}_\alpha$ such that, for every α, the restriction of ρ to $\mathscr{H}_\alpha$ is irreducible.

In particular, any irreducible unitary representation of a compact group is finite-dimensional. Consequently it is possible to talk about matrix elements and traces, i.e. characters, of irreducible unitary representations. The orthogonality relations for matrix elements and for characters of irreducible representations of finite groups remain valid for irreducible unitary representations of compact groups if one replaces $g^{-1} \sum_{s\in G} f(s)$ by the invariant mean $M(f)$.

Furthermore, any function $f \in \mathscr{C}(G)$ can be uniformly approximated by finite linear combinations of matrix elements of irreducible unitary representations, and any class function $f \in \mathscr{C}(G)$ can be uniformly approximated by finite linear combinations of characters of irreducible unitary representations. Finally, in the direct sum decomposition of the regular representation into finite-dimensional irreducible unitary representations, each irreducible representation occurs as often as its dimension.

Thus the representation theory of compact groups is completely analogous to that of finite groups. Indeed we may regard the representation theory of finite groups as a special case, since any finite group is compact with the discrete topology and any representation is equivalent to a unitary representation.

An example of a compact group which is neither finite nor abelian is the group $G = SU(2)$ of all 2×2 unitary matrices with determinant 1. The elements of G have the form

$$g = \begin{pmatrix} \gamma & \delta \\ -\bar{\delta} & \bar{\gamma} \end{pmatrix},$$

where γ,δ are complex numbers such that $|\gamma|^2 + |\delta|^2 = 1$. Writing $\gamma = \xi_0 + i\xi_3$, $\delta = \xi_1 + i\xi_2$, we see that topologically $SU(2)$ is homeomorphic to the sphere

$$S^3 = \{x = (\xi_0,\xi_1,\xi_2,\xi_3) \in \mathbb{R}^4 \colon \xi_0{}^2 + \xi_1{}^2 + \xi_2{}^2 + \xi_3{}^2 = 1\}$$

and hence is compact and *simply-connected* (i.e. it is path-connected and any closed path can be continuously deformed to a point).

For any integer $n \geq 0$, let V_n denote the vector space of all polynomials $f(z_1,z_2)$ with complex coefficients which are homogeneous of degree n. Writing $z = (z_1,z_2)$, we have

$$zg = (\gamma z_1 - \bar{\delta} z_2, \delta z_1 + \bar{\gamma} z_2).$$

Hence if we define a linear transformation T_g of V_n by $(T_g f)(z) = f(zg)$, then $\rho_n : g \rightarrow T_g$ is a representation of $SU(2)$ in V_n. It may be shown that this representation is irreducible and is unitary with respect to the inner product

$$(\textstyle\sum_{k=0}^{n} \alpha_k z_1{}^k z_2{}^{n-k}, \sum_{k=0}^{n} \beta_k z_1{}^k z_2{}^{n-k}) = \sum_{k=0}^{n} k!(n-k)! \alpha_k \bar{\beta}_k.$$

Moreover, every irreducible representation of $SU(2)$ is equivalent to ρ_n for some $n \geq 0$.

To determine the character χ_n of ρ_n we observe that any $g \in G$ is conjugate in G to a diagonal matrix

$$t = \begin{pmatrix} e^{i\theta} & 0 \\ 0 & e^{-i\theta} \end{pmatrix},$$

where $\theta \in \mathbb{R}$. If $f_k(z_1,z_2) = z_1{}^k z_2{}^{n-k}$ $(0 \leq k \leq n)$, then

$$(T_t f_k)(z_1,z_2) = (e^{i\theta} z_1)^k (e^{-i\theta} z_2)^{n-k} = e^{i(2k-n)\theta} f_k(z_1,z_2).$$

Since the polynomials $f_0,...,f_n$ are a basis for V_n it follows that

$$\chi_n(g) = \chi_n(t) = \sum_{k=0}^{n} e^{i(2k-n)\theta}.$$

Thus $\chi_n(I) = n + 1$, $\chi_n(-I) = (-1)^n(n+1)$ and

$$\chi_n(g) = \{e^{i(n+1)\theta} - e^{-i(n+1)\theta}\}/\{e^{i\theta} - e^{-i\theta}\} = \sin(n+1)\theta / \sin\theta \quad \text{if } g \neq I, -I.$$

From this formula we can easily deduce the decomposition of the product representation $\rho_m \otimes \rho_n$ into irreducible components. Since

$$\chi_m(g)\chi_n(g) = (e^{in\theta} + e^{i(n-2)\theta} + ... + e^{-in\theta})\{e^{i(m+1)\theta} - e^{-i(m+1)\theta}\}/\{e^{i\theta} - e^{-i\theta}\}$$

$$= \chi_{m+n}(g) + \chi_{m+n-2}(g) + ... + \chi_{|m-n|}(g),$$

we have the *Clebsch–Gordan formula*

$$\rho_m \otimes \rho_n = \rho_{m+n} + \rho_{m+n-2} + ... + \rho_{|m-n|}.$$

This formula is the group-theoretical basis for the rule in atomic physics which determines the possible values of the angular momentum when two systems with given angular momenta are coupled.

The complex numbers γ, δ with $|\gamma|^2 + |\delta|^2 = 1$ which specify the matrix $g \in SU(2)$ can be uniquely expressed in the form

$$\gamma = e^{i(\psi+\varphi)/2} \cos \theta/2, \quad \delta = e^{i(\psi-\varphi)/2} \sin \theta/2,$$

where $0 \le \theta \le \pi$, $0 \le \varphi < 2\pi$, $-2\pi \le \psi < 2\pi$. The invariant mean of any continuous function $f\colon SU(2) \to \mathbb{C}$ is then given by

$$M(f) = (1/16\pi^2) \int_{-2\pi}^{2\pi} \int_0^{2\pi} \int_0^{\pi} f(\theta,\varphi,\psi)\sin \theta \; d\theta \; d\varphi \; d\psi.$$

Another example of a compact group which is neither finite nor abelian is the group $SO(3)$ of all 3×3 real orthogonal matrices with determinant 1. The representations of $SO(3)$ may actually be obtained from those of $SU(2)$, since the two groups are intimately related. This was already shown in §6 of Chapter I, but another version of the proof will now be given.

The set V of all 2×2 matrices v which are skew-Hermitian and have zero trace,

$$v = \begin{pmatrix} \alpha & \beta \\ -\bar{\beta} & \bar{\alpha} \end{pmatrix}, \quad \text{where } \Re\alpha = 0,$$

is a three-dimensional real vector space which may be identified with $\mathbb{R}^3$ by writing $\alpha = i\xi_3$, $\beta = \xi_1 + i\xi_2$. Any $g \in G = SU(2)$ defines a linear transformation $T_g\colon v \to gvg^{-1}$ of $\mathbb{R}^3$. Moreover T_g is an orthogonal transformation, since if

$$T_g v = v_1 = \begin{pmatrix} \alpha_1 & \beta_1 \\ -\bar{\beta}_1 & \bar{\alpha}_1 \end{pmatrix}$$

then, by the product rule for determinants,

$$|\alpha_1|^2 + |\beta_1|^2 = |\alpha|^2 + |\beta|^2.$$

Hence $\det T_g = \pm 1$. In fact, since T_g is a continuous function of g and $SU(2)$ is connected, we must have $\det T_g = \det T_e = 1$ for every $g \in G$. Thus $T_g \in SO(3)$. Since $T_{gh} = T_g T_h$, the map $g \to T_g$ is a representation of G.

Every element of $SO(3)$ is represented in this way, since

$$\text{if } g_\varphi = \begin{pmatrix} e^{-i\varphi/2} & 0 \\ 0 & e^{i\varphi/2} \end{pmatrix} \text{ then } T_{g_\varphi} = B_\varphi = \begin{pmatrix} \cos\varphi & \sin\varphi & 0 \\ -\sin\varphi & \cos\varphi & 0 \\ 0 & 0 & 1 \end{pmatrix},$$

$$\text{if } h_\theta = \begin{pmatrix} \cos\theta/2 & -\sin\theta/2 \\ \sin\theta/2 & \cos\theta/2 \end{pmatrix} \text{ then } T_{h_\theta} = C_\theta = \begin{pmatrix} 1 & 0 & 0 \\ 0 & \cos\theta & \sin\theta \\ 0 & -\sin\theta & \cos\theta \end{pmatrix},$$

and every $A \in SO(3)$ can be expressed as a product $A = B_\psi C_\theta B_\varphi$, where φ, θ, ψ are *Euler's angles*.

If $T_g = I_3$ is the identity matrix, i.e. if $gv = vg$ for every $v \in V$, then $g = \pm I_2$, since any 2×2 matrix which commutes with both the matrices

$$\begin{pmatrix} 0 & 1 \\ -1 & 0 \end{pmatrix}, \begin{pmatrix} i & 0 \\ 0 & -i \end{pmatrix}$$

must be a scalar multiple of the identity matrix. It follows that $SO(3)$ *is isomorphic to the factor group* $SU(2)/\{\pm I_2\}$.

These examples, and higher-dimensional generalizations, can be treated systematically by the theory of Lie groups. A *Lie group* is a group G with the structure of a finite-dimensional real analytic manifold such that the map $(x,y) \to xy^{-1}$ of $G \times G$ into G is real analytic.

Some examples of Lie groups are

(i) a *Euclidean space* $\mathbb{R}^n$ under vector addition;

(ii) an *n-dimensional torus* (or *n-torus*) $\mathbb{T}^n$, i.e. the direct product of n copies of the multiplicative group $\mathbb{T}^1$ of all complex numbers of absolute value 1;

(iii) the *general linear group* $GL(n)$ of all real nonsingular $n \times n$ matrices under matrix multiplication;

(iv) the *orthogonal group* $O(n)$ of all matrices $X \in GL(n)$ such that $X^t X = I_n$;

(v) the *unitary group* $U(n)$ of all complex $n \times n$ matrices X such that $X^*X = I_n$, where X^* is the conjugate transpose of X; ($U(n)$ may be viewed as a subgroup of $GL(2n)$)

(vi) the *unitary symplectic group* $Sp(n)$ of all quaternion $n \times n$ matrices X such that $X^*X = I_n$, where X^* is the conjugate transpose of X. ($Sp(n)$ may be viewed as a subgroup of $GL(4n)$)

The definition implies that any Lie group is a locally compact topological group. The fifth Paris problem of Hilbert (1900) asks for a characterization of Lie groups among all topological groups. A complete solution was finally given by Gleason, Montgomery and Zippin (1953): a topological group can be given the structure of a Lie group if and only if it is *locally Euclidean*, i.e. there is a neighbourhood of the identity which is homeomorphic to $\mathbb{R}^n$ for some n.

The advantage of Lie groups over arbitrary topological groups is that, by replacing them by their Lie algebras, they can be studied by the methods of *linear* analysis.

A real (resp. complex) *Lie algebra* is a finite-dimensional real (resp. complex) vector space L with a map $(u,v) \to [u,v]$ of $L \times L$ into L, which is linear in u and in v and has the properties

(i) $[v,v] = 0$ for every $v \in L$,
(ii) $[u,[v,w]] + [v,[w,u]] + [w,[u,v]] = 0$ for all $u,v,w \in L$.

It follows from (i) and the linearity of the bracket product that

$$[u,v] + [v,u] = 0 \text{ for all } u,v \in L.$$

An example of a real (resp. complex) Lie algebra is the vector space $\mathfrak{gl}(n,\mathbb{R})$ (resp. $\mathfrak{gl}(n,\mathbb{C})$) of all $n \times n$ real (resp. complex) matrices X with $[X,Y] = XY - YX$. Other examples are easily constructed as subalgebras.

A *Lie subalgebra* of a Lie algebra L is a vector subspace M of L such that $u \in M$ and $v \in M$ imply $[u,v] \in M$. Some Lie subalgebras of $\mathfrak{gl}(n,\mathbb{C})$ are

(i)   the set A_n of all $X \in \mathfrak{gl}(n + 1,\mathbb{C})$ with tr $X = 0$,
(ii) the set B_n of all $X \in \mathfrak{gl}(2n + 1,\mathbb{C})$ such that $X^t + X = 0$,
(iii) the set C_n of all $X \in \mathfrak{gl}(2n,\mathbb{C})$ such that $X^t J + JX = 0$, where

$$J = \begin{pmatrix} 0 & I_n \\ -I_n & 0 \end{pmatrix},$$

(iv) the set D_n of all $X \in \mathfrak{gl}(2n,\mathbb{C})$ such that $X^t + X = 0$.

The manifold structure of a Lie group G implies that with each $s \in G$ there is associated a real vector space, the *tangent space* at s. The group structure of the Lie group G implies that the tangent space at the identity e of G is a real Lie algebra, which will be denoted by $L(G)$. For example, if $G = GL(n)$ then $L(G) = \mathfrak{gl}(n,\mathbb{R})$. The properties of Lie groups are mirrored by those of their Lie algebras in the following way.

For every real Lie algebra L, there is a simply-connected Lie group $\tilde{G}$ such that $L(\tilde{G}) = L$. Moreover, $\tilde{G}$ is uniquely determined up to isomorphism by L. A connected Lie group G has $L(G) = L$ if and only if G is isomorphic to a factor group $\tilde{G}/D$, where D is a discrete subgroup of the centre of $\tilde{G}$.

A *Lie subgroup* of a Lie group G is a real analytic submanifold H of G which is also a Lie group under the restriction to H of the group structure on G. It may be shown that a subgroup H of a Lie group G is a Lie subgroup if it is a closed subset of G, and is a connected Lie subgroup if and only if it is path-connected. Thus any closed subgroup of $GL(n)$ is a Lie group.

If H is a Lie subgroup of the Lie group G, then $L(H)$ is a Lie subalgebra of $L(G)$. Moreover, if M is a Lie subalgebra of $L(G)$, there is a unique connected Lie subgroup H of G such that $L(H) = M$.

If G_1,G_2 are Lie groups, then a map $f: G_1 \to G_2$ is a *Lie group homomorphism* if it is an analytic map, regarding G_1,G_2 as manifolds, and a homomorphism, regarding G_1,G_2 as groups. It may be shown that any continuous map $f: G_1 \to G_2$ which is a group homomorphism is actually a Lie group homomorphism. (It follows that a locally Euclidean topological group can be given the structure of a Lie group in only one way.)

If L_1,L_2 are Lie algebras, then a map $T: L_1 \to L_2$ is a *Lie algebra homomorphism* if it is linear and $T[u,v] = [Tu,Tv]$ for all $u,v \in L_1$. If G_1,G_2 are Lie groups and if $f: G_1 \to G_2$ is a Lie group homomorphism, then the derivative of f at the identity, $f'(e): L(G_1) \to L(G_2)$, is a Lie algebra homomorphism. Moreover, if G_1 is connected then distinct Lie group homomorphisms give rise to distinct Lie algebra homomorphisms, and if G_1 is simply-connected then every Lie algebra homomorphism $L(G_1) \to L(G_2)$ arises from some Lie group homomorphism. (In particular, the representations of a connected Lie group are determined by the representations of its Lie algebra.)

A Lie algebra L is *abelian* if $[u,v] = 0$ for all $u,v \in L$. A connected Lie group is abelian if and only if its Lie algebra is abelian. Since the Euclidean space $\mathbb{R}^n$ is a simply-connected Lie group with an n-dimensional abelian Lie algebra, it follows that any n-dimensional connected abelian Lie group is isomorphic to a direct product $\mathbb{R}^{n-k} \times \mathbb{T}^k$ (where $\mathbb{T}^k$ is a k-torus) for some k such that $0 \le k \le n$.

An *ideal* of a Lie algebra L is a vector subspace M of L such that $u \in L$ and $v \in M$ imply $[u,v] \in M$. A connected Lie subgroup H of a connected Lie group G is a normal subgroup if and only if $L(H)$ is an ideal of $L(G)$.

A Lie algebra L is *simple* if it has no ideals except $\{0\}$ and L and is not one-dimensional, and *semisimple* if it has no abelian ideal except $\{0\}$. It may be shown that a Lie algebra is semisimple if and only if it is the direct sum of finitely many ideals, each of which is a simple Lie algebra.

A Lie group is *semisimple* if it is connected and has no connected abelian normal Lie subgroup except $\{e\}$. It follows that a connected Lie group G is semisimple if and only if its Lie algebra $L(G)$ is semisimple.

We turn our attention now to compact Lie groups. It may be shown that a compact topological group can be given the structure of a Lie group if and only if it is finite-dimensional and locally connected. Furthermore, a compact Lie group is isomorphic to a closed subgroup of $GL(n)$ for some n. Other basic results are:

(i)   a compact Lie group, and even any compact topological group, has only finitely many connected components;

(ii)   a connected compact Lie group is abelian if and only if it is an n-torus $\mathbb{T}^n$ for some n;

(iii) a semisimple connected compact Lie group G has a finite centre. Moreover the simply-connected Lie group $\tilde{G}$ such that $L(\tilde{G}) = L(G)$ is not only semisimple but also compact;

(iv)   an arbitrary connected compact Lie group G has the form $G = ZH$, where Z,H are connected compact Lie subgroups, H is semisimple and Z is the component of the centre of G which contains the identity e.

These results essentially reduce the classification of arbitrary compact Lie groups to the classification of those which are semisimple and simply-connected. It may be shown that the latter are in one-to-one correspondence with the semisimple *complex* Lie algebras. Since a semisimple Lie algebra is a direct sum of finitely many simple Lie algebras, we are thus reduced to the classification of the simple complex Lie algebras. The miracle is that these can be completely enumerated: the non-isomorphic simple complex Lie algebras consist of the four infinite families A_n ($n \geq 1$), B_n ($n \geq 2$), C_n ($n \geq 3$), D_n ($n \geq 4$), of dimensions $n(n+2)$, $n(2n+1)$, $n(2n+1)$, $n(2n-1)$ respectively, and five *exceptional* Lie algebras G_2,F_4,E_6,E_7,E_8 of dimensions 14, 52, 78, 133, 248 respectively.

To the simple complex Lie algebra A_n corresponds the compact Lie group $SU(n + 1)$ of all matrices in $U(n + 1)$ with determinant 1; to B_n corresponds the compact Lie group $SO(2n + 1)$ of all matrices in $O(2n + 1)$ with determinant 1; to C_n corresponds the compact Lie group $Sp(n)$ (whose matrices all have determinant 1), and to D_n corresponds the compact Lie group $SO(2n)$ of all matrices in $O(2n)$ with determinant 1. The groups $SU(n)$ and $Sp(n)$ are simply-connected if $n \geq 2$, whereas $SO(n)$ is connected but has index 2 in its simply-connected covering group $Spin(n)$ if $n \geq 5$. The compact Lie groups corresponding to the five exceptional simple complex Lie algebras are all related to the algebra of *octonions* or Cayley numbers.

Space does not permit consideration here of the methods by which this classification has been obtained, although the methods are just as significant as the result. Indeed they provide a uniform approach to many problems involving the classical groups, giving explicit formulas for the invariant mean and for the characters of all irreducible representations. There is also a fascinating connection with *groups generated by reflections*.

The classification of arbitrary semisimple Lie groups reduces similarly to the classification of simple *real* Lie algebras, which have also been completely enumerated. The irreducible unitary representations of non-compact semisimple Lie groups have been extensively studied, notably by Harish-Chandra. However, the non-compact case is essentially more difficult than the compact, since any nontrivial representation is infinite-dimensional, and the results are still incomplete. Much of the motivation for this work has come from elementary particle physics where, in the original formulation of Wigner (1939), a particle (specified by its mass and spin) corresponds to an irreducible unitary representation of the inhomogeneous Lorentz group.

9 Further remarks

The history of Legendre's conjectures on primes in arithmetic progressions is described in Vol. I of Dickson [13]. Dirichlet's original proof is contained in [33], pp. 313-342. Although no simple general proof of Dirichlet's theorem is known, simple proofs have been given for the existence of infinitely many primes congruent to 1 mod m; see Sedrakian and Steinig [41].

If all arithmetic progressions $a, a + m,...$ with $(a,m) = 1$ contain a prime, then they all contain infinitely many, since for any $k > 1$ the arithmetic progression $a + m^k, a + 2m^k,...$ contains a prime.

It may be shown that any finite abelian group G is *isomorphic* to its dual group $\hat{G}$ (although not in a canonical way) by expressing G as a direct product of cyclic groups; see, for example, W. & F. Ellison [15].

In the final step of the proof of Proposition 7 we have followed Bateman [3]. Other proofs that $L(1,\chi) \neq 0$ for every $\chi \neq \chi_1$, which do not use Proposition 6, are given in Hasse [21]. The functional equation for Dirichlet L-functions was first proved by Hurwitz (1882). For proofs of some of the results stated at the end of §3, see Bach and Sorenson [1], Davenport [12], W. & F. Ellison [15] and Prachar [40]. Funakura [18] characterizes Dirichlet L-functions by means of their analytic properties.

The history of the theory of group representations and group characters is described in Curtis [10]. More complete expositions of the subject than ours are given by Serre [42], Feit [16], Huppert [27], and Curtis and Reiner [11]. The proof given here that the degree of an irreducible representation divides the order of the group is not Frobenius' original proof. It first appeared in a footnote of a paper by Schur (1904) on projective representations, where it is attributed to Frobenius. Zassenhaus [50] gives an interpretation in terms of *Casimir operators.*

A character-free proof of Corollary 19 is given in Gagen [19]. P. Hall's theorem is proved in Feit [16], for example. Frobenius groups are studied further in Feit [16] and Huppert [27].

For physical and chemical applications of group representations, see Cornwell [9], Janssen [29], Meijer [36], Birman [4] and Wilson *et al.* [48].

Dym and McKean [14] give an outward-looking introduction to the classical theory of Fourier series and integrals. The formal definition of a topological group is due to Schreier (1926). The Haar integral is discussed by Nachbin [37]. General introductions to abstract harmonic analysis are given by Weil [46], Loomis [34] and Folland [17]. More detailed information on topological groups and their representations is contained in Pontryagin [39], Hewitt and Ross [23] and Gurarii [20]. A simple proof that the additive group $\mathbb{Q}_p$ of all p-adic numbers is isomorphic to its dual group is given by Washington [45]. In the adelic approach to algebraic number theory this isomorphism lies behind the functional equation of the Riemann zeta function; see, for example, Lang [31].

For Hilbert's fifth problem, see Yang [49] and Hirschfeld [24]. The correspondence between Lie groups and Lie algebras was set up by Sophus Lie (1873-1893) in a purely local way, i.e. between neighbourhoods of the identity in the Lie group and of zero in the Lie algebra. Over half a century elapsed before the correspondence was made global by Cartan, Pontryagin and Chevalley. A basic property of solvable Lie algebras was established by Lie, but we owe to Killing (1888-1890) the remarkable classification of simple complex Lie algebras. Some gaps and inaccuracies in Killing's pioneering work were filled and corrected in the thesis of Cartan (1894). The classification of simple real Lie algebras is due to Cartan (1914). The representation theory of semisimple Lie algebras and compact semisimple Lie groups is the creation of Cartan (1913) and Weyl (1925-7). The introduction of groups generated by reflections is due to Weyl.

For the theory of Lie groups, see Chevalley [7], Warner [44], Varadarajan [43], Helgason [22] and Barut and Raczka [2]. The last reference also has information on representations of noncompact Lie groups and applications to quantum theory. The purely algebraic theory of Lie algebras is discussed by Jacobson [28] and Humphreys [25]. Niederle [38] gives a survey of the applications of the exceptional Lie algebras and Lie superalgebras in particle physics. Groups generated by reflections are treated by Humphreys [26], Bourbaki [5] and Kac [30], while Cohen [8] gives a useful overview.

The character theory of locally compact abelian groups, whose roots lie in Dirichlet's theorem on primes in arithmetic progressions, has given something back to number theory in the adelic approach to algebraic number fields; see the thesis of Tate, reproduced (pp. 305-347) in Cassels and Fröhlich [6], Lang [31] and Weil [47]. For a broad historical perspective and future plans, see Mackey [35] and Langlands [32].

10   Selected references

[1]   E. Bach and J. Sorenson, Explicit bounds for primes in residue classes, *Math. Comp.* **65** (1996), 1717-1735.

[2]   A.O. Barut and R. Raczka, *Theory of group representations and applications*, 2nd ed., Polish Scientific Publishers, Warsaw, 1986.

[3]   P.T. Bateman, A theorem of Ingham implying that Dirichlet's *L*-functions have no zeros with real part one, *Enseign. Math.* **43** (1997), 281-284.

[4]   J.L. Birman, *Theory of crystal space groups and lattice dynamics*, Springer-Verlag, Berlin, 1984.

[5]   N. Bourbaki, *Groupes et algèbres de Lie: Chapitres 4,5 et 6*, Masson, Paris, 1981.

[6]   J.W.S. Cassels and A. Fröhlich (ed.), *Algebraic number theory*, Academic Press, London, 1967.

[7]   C. Chevalley, *Theory of Lie groups I*, Princeton University Press, Princeton, 1946. [Reprinted, 1999]

[8]   A.M. Cohen, Coxeter groups and three related topics, *Generators and relations in groups and geometries* (ed. A. Barlotti *et al.*), pp. 235-278, Kluwer, Dordrecht, 1991.

[9]   J.F. Cornwell, *Group theory in physics*, 3 vols., Academic Press, London, 1984-1989.

[10]   C.W. Curtis, *Pioneers of representation theory: Frobenius, Burnside, Schur, and Brauer*, American Mathematical Society, Providence, R.I., 1999.

[11]   C.W. Curtis and I. Reiner, *Methods of representation theory*, 2 vols., Wiley, New York, 1990.

[12]   H. Davenport, *Multiplicative number theory*, 3rd ed. revised by H.L. Montgomery, Springer-Verlag, New York, 2000.

[13] L.E. Dickson, *History of the theory of numbers*, 3 vols., reprinted Chelsea, New York, 1966.

[14] H. Dym and H.P. McKean, *Fourier series and integrals*, Academic Press, Orlando, FL, 1972.

[15] W. Ellison and F. Ellison, *Prime numbers*, Wiley, New York, 1985.

[16] W. Feit, *Characters of finite groups*, Benjamin, New York, 1967.

[17] G.B. Folland, *A course in abstract harmonic analysis*, CRC Press, Boca Raton, FL, 1995.

[18] T. Funakura, On characterization of Dirichlet L-functions, *Acta Arith.* **76** (1996), 305-315.

[19] T.M. Gagen, *Topics in finite groups*, London Mathematical Society Lecture Note Series **16**, Cambridge University Press, 1976.

[20] V.P. Gurarii, *Group methods in commutative harmonic analysis*, English transl. by D. and S. Dynin, Encyclopaedia of Mathematical Sciences **25**, Springer-Verlag, Berlin, 1998.

[21] H. Hasse, *Vorlesungen über Zahlentheorie*, 2nd ed., Springer-Verlag, Berlin, 1964.

[22] S. Helgason, *Differential geometry, Lie groups and symmetric spaces*, Academic Press, New York, 1978.

[23] E. Hewitt and K.A. Ross, *Abstract harmonic analysis*, 2 vols., Springer-Verlag, Berlin, 1963/1970. [Corrected reprint of Vol. I, 1979]

[24] J. Hirschfeld, The nonstandard treatment of Hilbert's fifth problem, *Trans. Amer. Math. Soc.* **321** (1990), 379-400.

[25] J.E. Humphreys, *Introduction to Lie algebras and representation theory*, Springer-Verlag, New York, 1972.

[26] J.E. Humphreys, *Reflection groups and Coxeter groups*, Cambridge University Press, Cambridge, 1990.

[27] B. Huppert, *Character theory of finite groups,* de Gruyter, Berlin, 1998.

[28] N. Jacobson, *Lie algebras*, Interscience, New York, 1962.

[29] T. Janssen, *Crystallographic groups*, North-Holland, Amsterdam, 1973.

[30] V.G. Kac, *Infinite dimensional Lie Algebras*, corrected reprint of 3rd ed., Cambridge University Press, Cambridge, 1995.

[31] S. Lang, *Algebraic number theory*, 2nd ed., Springer-Verlag, New York, 1994.

[32] R.P. Langlands, Representation theory: its rise and its role in number theory, *Proceedings of the Gibbs symposium* (ed. D.G. Caldi and G.D. Mostow), pp. 181-210, Amer. Math. Soc., Providence, Rhode Island, 1990.

[33] G. Lejeune-Dirichlet, *Werke*, reprinted in one volume, Chelsea, New York, 1969.

[34] L.H. Loomis, *An introduction to abstract harmonic analysis*, Van Nostrand, New York, 1953.

[35] G.W. Mackey, Harmonic analysis as the exploitation of symmetry - a historical survey, *Bull. Amer. Math. Soc. (N.S.)* **3** (1980), 543-698. [Reprinted, with related articles, in G.W. Mackey, *The scope and history of commutative and noncommutative harmonic analysis*, American Mathematical Society, Providence, R.I., 1992]

[36] P.H. Meijer (ed.), *Group theory and solid state physics: a selection of papers*, Vol. 1, Gordon and Breach, New York, 1964.

[37] L. Nachbin, *The Haar integral*, reprinted, Krieger, Huntington, New York, 1976.

[38] J. Niederle, The unusual algebras and their applications in particle physics, *Czechoslovak J. Phys. B* **30** (1980), 1-22.

[39] L.S. Pontryagin, *Topological groups*, English transl. of 2nd ed. by A. Brown, Gordon and Breach, New York, 1966. [Russian original, 1954]

[40] K. Prachar, *Primzahlverteilung*, Springer-Verlag, Berlin, 1957.

[41] N. Sedrakian and J. Steinig, A particular case of Dirichlet's theorem on arithmetic progressions, *Enseign. Math.* **44** (1998), 3-7.

[42] J.-P. Serre, *Linear representations of finite groups*, Springer-Verlag, New York, 1977.

[43] V.S. Varadarajan, *Lie groups, Lie algebras and their representations*, corrected reprint, Springer-Verlag, New York, 1984.

[44] F.W. Warner, *Foundations of differentiable manifolds and Lie groups*, corrected reprint, Springer-Verlag, New York, 1983.

[45] L. Washington, On the self-duality of Q_p, *Amer. Math. Monthly* **81** (1974), 369-371.

[46] A. Weil, *L'integration dans les groupes topologiques et ses applications*, 2nd ed., Hermann, Paris, 1953.

[47] A. Weil, *Basic number theory*, 2nd ed., Springer-Verlag, Berlin, 1973.

[48] E.B. Wilson, J.C. Decius and P.C. Cross, *Molecular vibrations*, McGraw-Hill, New York, 1955.

[49] C.T. Yang, Hilbert's fifth problem and related problems on transformation groups, *Mathematical developments arising from Hilbert problems* (ed. F.E. Browder), pp. 142-146, Amer. Math. Soc., Providence, R.I., 1976.

[50] H. Zassenhaus, An equation for the degrees of the absolutely irreducible representations of a group of finite order, *Canad. J. Math.* **2** (1950), 166-167.

XI
Uniform distribution and ergodic theory

A trajectory of a system which is evolving with time may be said to be 'recurrent' if it keeps returning to any neighbourhood, however small, of its initial point, and 'dense' if it passes arbitrarily near to every point. It may be said to be 'uniformly distributed' if the proportion of time it spends in any region tends asymptotically to the ratio of the volume of that region to the volume of the whole space. In the present chapter these notions will be made precise and some fundamental properties derived. The subject of dynamical systems has its roots in mechanics, but we will be particularly concerned with its applications in number theory.

1 Uniform distribution

Before introducing our subject, we establish the following interesting result:

LEMMA 0 *Let $J = [a,b]$ be a compact interval and $f_n: J \to \mathbb{R}$ a sequence of nondecreasing functions. If $f_n(t) \to f(t)$ for every $t \in J$ as $n \to \infty$, where $f: J \to \mathbb{R}$ is a continuous function, then $f_n(t) \to f(t)$ uniformly on J.*

Proof Evidently f is also nondecreasing. Moreover, since J is compact, f is uniformly continuous on J. Thus, for any $\varepsilon > 0$, there is a subdivision $a = t_0 < t_1 < ... < t_m = b$ such that

$$f(t_k) - f(t_{k-1}) < \varepsilon \quad (k = 1,...,m).$$

We can choose a positive integer p so that, for all $n > p$,

$$\left| f_n(t_k) - f(t_k) \right| < \varepsilon \quad (k = 0,1,...,m).$$

If $t \in J$, then $t \in [t_{k-1},t_k]$ for some $k \in \{1,...,m\}$. Hence

$$f_n(t) - f(t) \leq f_n(t_k) - f(t_k) + f(t_k) - f(t_{k-1}) < 2\varepsilon$$

and similarly

$$f_n(t) - f(t) \geq f_n(t_{k-1}) - f(t_{k-1}) + f(t_{k-1}) - f(t_k) > -2\varepsilon.$$

Thus $|f_n(t) - f(t)| < 2\varepsilon$ for every $t \in J$ if $n > p$. $\blacksquare$

For any real number ξ, let $\lfloor \xi \rfloor$ denote again the greatest integer $\leq \xi$ and let

$$\{\xi\} = \xi - \lfloor \xi \rfloor$$

denote the *fractional part* of ξ. We are going to prove that, if ξ is irrational, then the sequence $(\{n\xi\})$ of the fractional parts of the multiples of ξ is *dense* in the unit interval $I = [0,1]$, i.e. every point of I is a limit point of the sequence.

It is sufficient to show that the points $z_n = e^{2\pi i n \xi}$ ($n = 1,2,...$) are dense on the unit circle. Since ξ is irrational, the points z_n are all distinct and $z_n \neq \pm 1$. Consequently they have a limit point on the unit circle. Thus, for any given $\varepsilon > 0$, there exist positive integers m,r such that

$$|z_{m+r} - z_m| < \varepsilon.$$

But

$$|z_{m+r} - z_m| = |z_r - 1| = |z_{n+r} - z_n| \text{ for every } n \in \mathbb{N}.$$

If we write $z_r = e^{2\pi i \theta}$, where $0 < \theta < 1$, then $z_{kr} = e^{2\pi i k\theta}$ ($k = 1,2,...$). Define the positive integer N by $1/(N+1) < \theta < 1/N$. Then the points $z_r, z_{2r},...,z_{Nr}$ follow one another in order on the unit circle and every point of the unit circle is distant less than ε from one of these points.

It may be asked if the sequence $(\{n\xi\})$ is not only dense in I, but also spends 'the right amount of time' in each subinterval of I. To make the question precise we introduce the following definition:

A sequence (ξ_n) of real numbers is said to be *uniformly distributed mod* 1 if, for all α, β with $0 \leq \alpha < \beta \leq 1$,

$$\varphi_{\alpha,\beta}(N)/N \to \beta - \alpha \quad \text{as } N \to \infty,$$

where $\varphi_{\alpha,\beta}(N)$ is the number of positive integers $n \leq N$ such that $\alpha \leq \{\xi_n\} < \beta$.

In this definition we need only require that $\varphi_{0,\alpha}(N)/N \to \alpha$ for every $\alpha \in (0,1)$, since

$$\varphi_{\alpha,\beta}(N) = \varphi_{0,\beta}(N) - \varphi_{0,\alpha}(N)$$

and hence

$$|\varphi_{\alpha,\beta}(N)/N - (\beta - \alpha)| \leq |\varphi_{0,\beta}(N)/N - \beta| + |\varphi_{0,\alpha}(N)/N - \alpha|.$$

It follows from Lemma 0, with $f_n(t) = \varphi_{0,t}(n)/n$ and $f(t) = t$, that the sequence (ξ_n) is uniformly distributed mod 1 if and only if

$$\varphi_{\alpha,\beta}(N)/N \to \beta - \alpha \quad \text{as } N \to \infty$$

uniformly for all α,β with $0 \le \alpha < \beta \le 1$.

It was first shown by Bohl (1909) that, if ξ is irrational, the sequence $(n\xi)$ is uniformly distributed mod 1 in the sense of our definition. Later Weyl (1914,1916) established this result by a less elementary, but much more general argument, which was equally applicable to multi-dimensional problems. The following two theorems, due to Weyl, replace the problem of showing that a sequence is uniformly distributed mod 1 by a more tractable analytic problem.

THEOREM 1 *A real sequence (ξ_n) is uniformly distributed mod 1 if and only if, for every function $f: I \to \mathbb{C}$ which is Riemann integrable,*

$$N^{-1}\sum_{n=1}^{N}f(\{\xi_n\}) \to \int_I f(t)\, dt \quad \text{as } N \to \infty. \tag{1}$$

Proof For any $\alpha,\beta \in I$ with $\alpha < \beta$, let $\chi_{\alpha,\beta}$ denote the *indicator function* of the interval $[\alpha,\beta)$, i.e.

$$\chi_{\alpha,\beta}(t) = 1 \quad \text{for } \alpha \le t < \beta,$$
$$= 0 \quad \text{otherwise.}$$

Since

$$\int_I \chi_{\alpha,\beta}(t)\, dt = \beta - \alpha,$$

the definition of uniform distribution can be rephrased by saying that the sequence (ξ_n) is uniformly distributed mod 1 if and only if, for all choices of α and β,

$$N^{-1}\sum_{n=1}^{N}\chi_{\alpha,\beta}(\{\xi_n\}) \to \int_I \chi_{\alpha,\beta}(t)\, dt \quad \text{as } N \to \infty.$$

Thus the sequence (ξ_n) is certainly uniformly distributed mod 1 if (1) holds for every Riemann integrable function f.

Suppose now that the sequence (ξ_n) is uniformly distributed mod 1. Then (1) holds not only for every function $f = \chi_{\alpha,\beta}$, but also for every finite linear combination of such functions, i.e. for every *step-function f*. But, for any real-valued Riemann integrable function f and any $\varepsilon > 0$, there exist step-functions f_1, f_2 such that

$$f_1(t) \le f(t) \le f_2(t) \quad \text{for every } t \in I$$

and

$$\int_I (f_2(t) - f_1(t))\, dt < \varepsilon.$$

Hence

$$N^{-1} \sum_{n=1}^{N} f(\{\xi_n\}) - \int_I f(t) \, dt \leq N^{-1} \sum_{n=1}^{N} f_2(\{\xi_n\}) - \int_I f_2(t) \, dt + \varepsilon$$

$$< 2\varepsilon \quad \text{for all large } N,$$

and similarly

$$N^{-1} \sum_{n=1}^{N} f(\{\xi_n\}) - \int_I f(t) \, dt > -2\varepsilon \quad \text{for all large } N.$$

Thus (1) holds when the Riemann integrable function f is real-valued and also, by linearity, when it is complex-valued. $\square$

A converse of Theorem 1 has been proved by de Bruijn and Post (1968): if a function $f: I \to \mathbb{C}$ has the property that

$$\lim_{N \to \infty} N^{-1} \sum_{n=1}^{N} f(\{\xi_n\})$$

exists for every sequence (ξ_n) which is uniformly distributed mod 1, then f is Riemann integrable.

In the statement of the next result, and throughout the rest of the chapter, we use the abbreviation

$$e(t) = e^{2\pi i t}.$$

In the proof of the next result we use the *Weierstrass approximation theorem*: any continuous function $f: I \to \mathbb{C}$ of period 1 is the uniform limit of a sequence (f_n) of trigonometric polynomials. In fact, as Fejér (1904) showed, one can take f_n to be the arithmetic mean $(S_0 + \dots + S_{n-1})/n$, where

$$S_m = S_m(x) := \sum_{h=-m}^{m} c_h \, e(hx)$$

is the m-th partial sum of the Fourier series for f. This yields the explicit formula

$$f_n(x) = \int_I K_n(x - t) f(t) \, dt,$$

where

$$K_n(u) = (\sin^2 n\pi u)/(n \sin^2 \pi u).$$

THEOREM 2 *A real sequence (ξ_n) is uniformly distributed mod 1 if and only if, for every integer $h \neq 0$,*

$$N^{-1} \sum_{n=1}^{N} e(h\xi_n) \to 0 \quad \text{as } N \to \infty. \tag{2}$$

Proof If the sequence (ξ_n) is uniformly distributed mod 1 then, by taking $f(t) = e(ht)$ in Theorem 1 we obtain (2) since, for every integer $h \neq 0$,

$$\int_I e(ht) \, dt = 0.$$

Conversely, suppose (2) holds for every nonzero integer h. Then, by linearity, for any trigonometric polynomial

$$g(t) = \sum_{h=-m}^{m} b_h \, e(ht)$$

we have

$$N^{-1} \sum_{n=1}^{N} g(\{\xi_n\}) \to b_0 = \int_I g(t) \, dt \quad \text{as } N \to \infty.$$

If f is a continuous function of period 1 then, by the Weierstrass approximation theorem, for any $\varepsilon > 0$ there exists a trigonometric polynomial $g(t)$ such that $|f(t) - g(t)| < \varepsilon$ for every $t \in I$. Hence

$$\left| N^{-1} \sum_{n=1}^{N} f(\{\xi_n\}) - \int_I f(t) \, dt \right|$$

$$\leq \left| N^{-1} \sum_{n=1}^{N} (f(\{\xi_n\}) - g(\{\xi_n\})) \right| + \left| N^{-1} \sum_{n=1}^{N} g(\{\xi_n\}) - \int_I g(t) \, dt \right| + \left| \int_I (g(t) - f(t)) \, dt \right|$$

$$< 2\varepsilon + \left| N^{-1} \sum_{n=1}^{N} g(\{\xi_n\}) - \int_I g(t) \, dt \right|$$

$$< 3\varepsilon \quad \text{for all large } N.$$

Thus (1) holds for every continuous function f of period 1.

Finally, if $\chi_{\alpha,\beta}$ is the function defined in the proof of Theorem 1 then, for any $\varepsilon > 0$, there exist continuous functions f_1, f_2 of period 1 such that

$$f_1(t) \leq \chi_{\alpha,\beta}(t) \leq f_2(t) \quad \text{for every } t \in I$$

and

$$\int_I (f_2(t) - f_1(t)) \, dt < \varepsilon,$$

from which it follows similarly that

$$N^{-1} \sum_{n=1}^{N} \chi_{\alpha,\beta}(\{\xi_n\}) \to \int_I \chi_{\alpha,\beta}(t) \, dt \quad \text{as } N \to \infty.$$

Thus the sequence (ξ_n) is uniformly distributed mod 1. $\square$

Weyl's criterion, as Theorem 2 is usually called, immediately implies Bohl's result:

PROPOSITION 3 *If ξ is irrational, the sequence $(n\xi)$ is uniformly distributed mod* 1.

Proof For any nonzero integer h,

$$e(h\xi) + e(2h\xi) + \ldots + e(Nh\xi) = (e((N+1)h\xi) - e(h\xi))/(e(h\xi) - 1).$$

Hence

$$\left| N^{-1} \sum_{n=1}^{N} e(hn\xi) \right| \leq 2|e(h\xi) - 1|^{-1} N^{-1},$$

and the result follows from Theorem 2. ∎

These results can be immediately extended to higher dimensions. A sequence (x_n) of vectors in $\mathbb{R}^d$ is said to be *uniformly distributed mod* 1 if, for all vectors $a = (\alpha_1,...,\alpha_d)$ and $b = (\beta_1,...,\beta_d)$ with $0 \le \alpha_k < \beta_k \le 1$ $(k = 1,...,d)$,

$$\varphi_{a,b}(N)/N \to \prod_{k=1}^d (\beta_k - \alpha_k) \quad \text{as } N \to \infty,$$

where $x_n = (\xi_n^{(1)},...,\xi_n^{(d)})$ and $\varphi_{a,b}(N)$ is the number of positive integers $n \le N$ such that $\alpha_k \le \{\xi_n^{(k)}\} < \beta_k$ for every $k \in \{1,...,d\}$. Let I^d be the set of all $x = (\xi^{(1)},...,\xi^{(d)})$ such that $0 \le \xi^{(k)}) \le 1$ $(k = 1,...,d)$ and, for an arbitrary vector $x = (\xi^{(1)},...,\xi^{(d)})$, put

$$\{x\} = (\{\xi^{(1)}\},...,\{\xi^{(d)}\}).$$

Then Theorems 1 and 2 have the following generalizations:

THEOREM 1′ *A sequence (x_n) of vectors in $\mathbb{R}^d$ is uniformly distributed mod 1 if and only if, for every function $f: I^d \to \mathbb{C}$ which is Riemann integrable,*

$$N^{-1}\sum_{n=1}^N f(\{x_n\}) \to \int_I ... \int_I f(t_1,...,t_d)\, dt_1...dt_d \quad \text{as } N \to \infty. \quad \square$$

THEOREM 2′ *A sequence (x_n) of vectors in $\mathbb{R}^d$ is uniformly distributed mod 1 if and only if, for every nonzero vector $m = (\mu_1,...,\mu_d) \in \mathbb{Z}^d$,*

$$N^{-1}\sum_{n=1}^N e(m \cdot x_n) \to 0 \quad \text{as } N \to \infty,$$

where $m \cdot x_n = \mu_1 \xi_n^{(1)} + ... + \mu_d \xi_n^{(d)}$. $\square$

Proposition 3 can also be generalized in the following way:

PROPOSITION 3′ *If $x = (\xi^{(1)},...,\xi^{(d)})$ is any vector in $\mathbb{R}^d$ such that $1,\xi^{(1)},...,\xi^{(d)}$ are linearly independent over the field $\mathbb{Q}$ of rational numbers, then the sequence (nx) is uniformly distributed mod 1.* ∎

In particular, the sequence $(\{nx\}) = (\{n\xi^{(1)}\},...,\{n\xi^{(d)}\})$ is dense in the d-dimensional unit cube if $1,\xi^{(1)},...,\xi^{(d)}$ are linearly independent over the field $\mathbb{Q}$ of rational numbers. This much weaker assertion had already been proved before Weyl by Kronecker (1884).

It is easily seen that the linear independence of $1,\xi^{(1)},...,\xi^{(d)}$ over the field $\mathbb{Q}$ of rational numbers is also necessary for the sequence $(\{nx\})$ to be dense in the d-dimensional unit cube and, *a fortiori*, for the sequence (nx) to be uniformly distributed mod 1. For if $1,\xi^{(1)},...,\xi^{(d)}$ are linearly dependent over $\mathbb{Q}$ there exists a nonzero vector $m = (\mu_1,...,\mu_d) \in \mathbb{Z}^d$ such that

$$m \cdot x = \mu_1 \xi^{(1)} + \dots + \mu_d \xi^{(d)} \in \mathbb{Z}.$$

It follows that each point of the sequence (nx) lies on some hyperplane $m \cdot y = h$, where $h \in \mathbb{Z}$. Without loss of generality, suppose $\mu_1 \neq 0$. Then no point of the d-dimensional unit cube which is sufficiently close to the point $(|2\mu_1|^{-1}, 0, \dots, 0)$ lies on such a hyperplane.

We now return to the one-dimensional case. Weyl used Theorem 2 to prove, not only Proposition 3, but also a deeper result concerning the uniform distribution of the sequence $(f(n))$, where f is a polynomial of any positive degree. We will derive Weyl's result by a more general argument due to van der Corput (1931), based on the following inequality:

LEMMA 4 *If* $\zeta_1, \dots, \zeta_N$ *are arbitrary complex numbers then, for any positive integer* $M \leq N$,

$$M^2 \left| \sum_{n=1}^{N} \zeta_n \right|^2 \leq M(M + N - 1) \sum_{n=1}^{N} |\zeta_n|^2 + 2(M + N - 1) \sum_{m=1}^{M-1} (M - m) \left| \sum_{n=1}^{N-m} \overline{\zeta_n} \zeta_{n+m} \right|.$$

Proof Put $\zeta_n = 0$ if $n \leq 0$ or $n > N$. Then it is easily verified that

$$M \sum_{n=1}^{N} \zeta_n = \sum_{h=1}^{M+N-1} \left(\sum_{k=0}^{M-1} \zeta_{h-k} \right).$$

Applying Schwarz's inequality (Chapter I, §4), we get

$$M^2 \left| \sum_{n=1}^{N} \zeta_n \right|^2 \leq (M + N - 1) \sum_{h=1}^{M+N-1} \left| \sum_{k=0}^{M-1} \zeta_{h-k} \right|^2$$

$$= (M + N - 1) \sum_{h=1}^{M+N-1} \sum_{j,k=0}^{M-1} \zeta_{h-k} \overline{\zeta_{h-j}}.$$

On the right side any term $|\zeta_n|^2$ occurs exactly M times, namely for $h - k = h - j = n$. A term $\overline{\zeta_n} \zeta_{n+m}$ or $\zeta_n \overline{\zeta_{n+m}}$, where $m > 0$, occurs only if $m < M$ and then it occurs exactly $M - m$ times. Thus the right side is equal to

$$M(M + N - 1) \sum_{n=1}^{N} |\zeta_n|^2 + (M + N - 1) \sum_{m=1}^{M-1} (M - m) \sum_{n=1}^{N-m} (\overline{\zeta_n} \zeta_{n+m} + \zeta_n \overline{\zeta_{n+m}}).$$

The lemma follows. □

COROLLARY 5 *If* (ξ_n) *is a real sequence such that, for each positive integer* m,

$$N^{-1} \sum_{n=1}^{N} e(\xi_{n+m} - \xi_n) \to 0 \quad as\ N \to \infty,$$

then

$$N^{-1} \sum_{n=1}^{N} e(\xi_n) \to 0 \quad as\ N \to \infty.$$

Proof By taking $\zeta_n = e(\xi_n)$ in Lemma 4 we obtain, for $1 \leq M \leq N$,

$N^{-2} |\sum_{n=1}^{N} e(\xi_n)|^2 \le$

$$(M + N - 1)M^{-1}N^{-1} + 2(M + N - 1)M^{-2}N^{-2} \sum_{m=1}^{M-1} (M - m) \left| \sum_{n=1}^{N-m} e(\xi_{n+m} - \xi_n) \right|.$$

Keeping M fixed and letting $N \to \infty$, we get

$$\overline{\lim}_{N \to \infty} N^{-2} |\sum_{n=1}^{N} e(\xi_n)|^2 \le M^{-1}.$$

But M can be chosen as large as we please. ◻

An immediate consequence is van der Corput's *difference theorem*:

PROPOSITION 6 *The real sequence (ξ_n) is uniformly distributed mod 1 if, for each positive integer m, the sequence $(\xi_{n+m} - \xi_n)$ is uniformly distributed mod 1.*

Proof If the sequences $(\xi_{n+m} - \xi_n)$ are uniformly distributed mod 1 then, by Theorem 2,

$$N^{-1} \sum_{n=1}^{N} e(h(\xi_{n+m} - \xi_n)) \to 0 \quad \text{as } N \to \infty$$

for all integers $h \ne 0, m > 0$. Replacing ξ_n by $h\xi_n$ in Corollary 5 we obtain, for all integers $h \ne 0$,

$$N^{-1} \sum_{n=1}^{N} e(h\xi_n) \to 0 \quad \text{as } N \to \infty.$$

Hence, by Theorem 2 again, the sequence (ξ_n) is uniformly distributed mod 1. ◻

The sequence $(n\xi)$, with ξ irrational, shows that we cannot replace 'if' by 'if and only if' in the statement of Proposition 6. Weyl's result will now be derived from Proposition 6:

PROPOSITION 7 *If*

$$f(t) = \alpha_r t^r + \alpha_{r-1} t^{r-1} + \dots + \alpha_0$$

is any polynomial with real coefficients α_k such that α_k is irrational for at least one $k > 0$, then the sequence $(f(n))$ is uniformly distributed mod 1.

Proof If $r = 1$, then the result holds by the same argument as in Proposition 3. We assume that $r > 1, \alpha_r \ne 0$ and the result holds for polynomials of degree less than r.

For any positive integer m,

$$g_m(t) = f(t + m) - f(t)$$

is a polynomial of degree $r - 1$ with leading coefficient $rm\alpha_r$. If α_r is irrational, then $rm\alpha_r$ is also irrational and hence, by the induction hypothesis, the sequence $(g_m(n))$ is uniformly distributed mod 1. Consequently, by Proposition 6, the sequence $(f(n))$ is also uniformly distributed mod 1.

Suppose next that the leading coefficient α_r is rational, and let α_s $(1 \leq s < r)$ be the coefficient nearest to it which is irrational. Then the coefficients of $t^{r-1},...,t^s$ of the polynomial $g_m(t)$ are rational, but the coefficient of t^{s-1} is irrational. If $s > 1$, it follows again from the induction hypothesis and Proposition 6 that the sequence $(f(n))$ is uniformly distributed mod 1.

Suppose finally that $s = 1$ and put

$$F(t) = \alpha_r t^r + \alpha_{r-1}t^{r-1} + ... + \alpha_2 t^2.$$

If $q > 0$ is a common denominator for the rational numbers $\alpha_2,...,\alpha_r$ then, for any integer $h \neq 0$ and any nonnegative integers j,k,

$$e(hF(jq + k)) = e(hF(k)).$$

Write $N = \ell q + k$, where $\ell = \lfloor N/q \rfloor$ and $0 \leq k < q$. Since $f(t) = F(t) + \alpha_1 t + \alpha_0$, we obtain

$$N^{-1}\sum_{n=0}^{N-1} e(hf(n)) = N^{-1}\sum_{k=0}^{q-1}\sum_{j=0}^{\ell-1} e(hf(jq + k)) + N^{-1}\sum_{n=\ell q}^{N} e(hf(n))$$

$$= N^{-1}\lfloor N/q \rfloor\sum_{k=0}^{q-1} e(hF(k))\sum_{j=0}^{\ell-1}\ell^{-1}e(h(jq\alpha_1+k\alpha_1+\alpha_0)) + N^{-1}\sum_{n=\ell q}^{N} e(hf(n)).$$

The last term tends to zero as $N \to \infty$, since the sum contains at most q terms, each of absolute value 1. By Theorem 2, each of the q inner sums in the first term also tends to zero as $N \to \infty$, because the result holds for $r = 1$. Hence, by Theorem 2 again, the sequence $(f(n))$ is uniformly distributed mod 1. $\square$

An interesting extension of Proposition 6 was derived by Korobov and Postnikov (1952):

PROPOSITION 8 *If, for every positive integer* m, *the sequence* $(\xi_{n+m} - \xi_n)$ *is uniformly distributed mod 1 then, for all integers* $q > 0$ *and* $r \geq 0$, *the sequence* (ξ_{qn+r}) *is uniformly distributed mod 1.*

Proof We may suppose $q > 1$, since the assertion follows at once from Proposition 6 if $q = 1$. By Theorem 2 it is enough to show that, for every integer $m \neq 0$,

$$S: = N^{-1}\sum_{n=1}^{N} e(m\xi_{qn+r}) \to 0 \quad \text{as } N \to \infty.$$

Since

$$q^{-1}\sum_{k=1}^{q} e(nk/q) = 1 \quad \text{if } n \equiv 0 \mod q,$$
$$= 0 \quad \text{if } n \not\equiv 0 \mod q,$$

we can write

$$S = (qN)^{-1} \sum_{n=1}^{qN} e(m\xi_{n+r}) \sum_{k=1}^{q} e(nk/q)$$
$$= (qN)^{-1} \sum_{k=1}^{q} \sum_{n=1}^{qN} e(m\eta_n^{(k)}),$$

where we have put

$$\eta_n^{(k)} = \xi_{n+r} + nk/mq.$$

By hypothesis, for every positive integer h, the sequence

$$\eta_{n+h}^{(k)} - \eta_n^{(k)} = \xi_{n+h+r} - \xi_{n+r} - hk/mq$$

is uniformly distributed mod 1. Hence $\eta_n^{(k)}$ is uniformly distributed mod 1, by Proposition 6. Thus, for each $k \in \{1,...,q\}$,

$$(qN)^{-1} \sum_{n=1}^{qN} e(m\eta_n^{(k)}) \to 0 \quad \text{as } N \to \infty,$$

and consequently also $S \to 0$ as $N \to \infty$. □

As an application of Proposition 8 we prove

PROPOSITION 9 *Let A be a d×d matrix of integers, no eigenvalue of which is a root of unity. If, for some $x \in \mathbb{R}^d$, the sequence $(A^n x)$ is uniformly distributed mod 1 then, for any integers $q > 0$ and $r \geq 0$, the sequence $(A^{qn+r}x)$ is also uniformly distributed mod 1.*

Proof It follows from Theorem 2′ that, for any nonzero vector $m \in \mathbb{Z}^d$, the scalar sequence $\xi_n = m \cdot A^n x$ is uniformly distributed mod 1. For any positive integer h, the sequence

$$\xi_{n+h} - \xi_n = m \cdot (A^h - I)A^n x = (A^h - I)^t m \cdot A^n x$$

has the same form as the sequence ξ_n, since the hypotheses ensure that $(A^h - I)^t m$ is a nonzero vector in $\mathbb{Z}^d$. Hence the sequence $\xi_{n+h} - \xi_n$ is uniformly distributed mod 1. Therefore, by Proposition 8, the sequence $\xi_{qn+r} = m \cdot A^{qn+r}x$ is uniformly distributed mod 1, and thus the sequence $A^{qn+r}x$ is uniformly distributed mod 1. □

It may be noted that the matrix A in Proposition 9 is necessarily nonsingular. For if $\det A = 0$, there exists a nonzero vector $z \in \mathbb{Z}^d$ such that $A^t z = 0$. Then, for any $x \in \mathbb{R}^d$ and any positive integer n, $e(z \cdot A^n x) = e((A^t)^n z \cdot x) = 1$. Thus $N^{-1} \sum_{n=1}^{N} e(z \cdot A^n x) = 1$ and hence, by Theorem 2′, the sequence $A^n x$ is not uniformly distributed mod 1.

Further examples of uniformly distributed sequences are provided by the following result, which is due to Fejér (c. 1924):

PROPOSITION 10 *Let* (ξ_n) *be a sequence of real numbers such that* $\eta_n := \xi_{n+1} - \xi_n$ *tends to zero monotonically as* $n \to \infty$. *Then* (ξ_n) *is uniformly distributed mod* 1 *if* $n|\eta_n| \to \infty$ *as* $n \to \infty$.

Proof By changing the signs of all ξ_n we may restrict attention to the case where the sequence (η_n) is strictly decreasing. For any real numbers α,β we have

$$
\begin{aligned}
|e(\alpha) - e(\beta) - 2\pi i(\alpha - \beta)e(\beta)| &= |e(\alpha - \beta) - 1 - 2\pi i(\alpha - \beta)| \\
&= 4\pi^2 \, |\textstyle\int_0^{\alpha-\beta}(\alpha - \beta - t)e(t)\,dt| \\
&\leq 4\pi^2 \, |\textstyle\int_0^{\alpha-\beta}(\alpha - \beta - t)\,dt| \\
&= 2\pi^2(\alpha - \beta)^2.
\end{aligned}
$$

If we take $\alpha = h\xi_{n+1}$ and $\beta = h\xi_n$, where h is any nonzero integer, this yields

$$
|e(h\xi_{n+1})/\eta_n - e(h\xi_n)/\eta_n - 2\pi i h e(h\xi_n)| \leq 2\pi^2 h^2 \eta_n
$$

and hence

$$
|e(h\xi_{n+1})/\eta_{n+1} - e(h\xi_n)/\eta_n - 2\pi i h e(h\xi_n)| \leq 1/\eta_{n+1} - 1/\eta_n + 2\pi^2 h^2 \eta_n.
$$

Taking $n = 1,\dots,N$ and adding, we obtain

$$
\begin{aligned}
|2\pi h \textstyle\sum_{n=1}^{N} e(h\xi_n)| &\leq 1/\eta_{N+1} + 1/\eta_1 + \textstyle\sum_{n=1}^{N}(1/\eta_{n+1} - 1/\eta_n) + 2\pi^2 h^2 \textstyle\sum_{n=1}^{N}\eta_n \\
&= 2/\eta_{N+1} + 2\pi^2 h^2 \textstyle\sum_{n=1}^{N}\eta_n.
\end{aligned}
$$

Thus

$$
N^{-1}|\textstyle\sum_{n=1}^{N} e(h\xi_n)| \leq (\pi|h|N\eta_{N+1})^{-1} + \pi|h|N^{-1}\textstyle\sum_{n=1}^{N}\eta_n.
$$

But the right side of this inequality tends to zero as $N \to \infty$, since $N\eta_N \to \infty$ and $\eta_N \to 0$. $\square$

By the mean value theorem, the hypotheses of Proposition 10 are certainly satisfied if $\xi_n = f(n)$, where f is a differentiable function such that $f'(t) \to 0$ monotonically as $t \to \infty$ and $t|f'(t)| \to \infty$ as $t \to \infty$. Consequently the sequence (an^α) is uniformly distributed mod 1 if $a \neq 0$ and $0 < \alpha < 1$, and the sequence $(a(\log n)^\alpha)$ is uniformly distributed mod 1 if $a \neq 0$ and $\alpha > 1$. By using van der Corput's difference theorem and an inductive argument starting from Proposition 10, it may be further shown that the sequence (an^α) is uniformly distributed mod 1 for any $a \neq 0$ and any $\alpha > 0$ which is not an integer.

It has been shown by Kemperman (1973) that 'if' may be replaced by 'if and only if' in the statement of Proposition 10. Consequently the sequence $(a(\log n)^\alpha)$ is not uniformly distributed mod 1 if $0 < \alpha \leq 1$.

The theory of uniform distribution has an application, and its origin, in astronomy. In his investigations on the secular perturbations of planetary orbits Lagrange (1782) was led to the problem of *mean motion*: if

$$z(t) = \sum_{k=1}^{n} \rho_k \, e(\omega_k t + \alpha_k),$$

where $\rho_k > 0$ and $\alpha_k, \omega_k \in \mathbb{R}$ ($k = 1,...,n$), does $t^{-1} \arg z(t)$ have a finite limit as $t \to +\infty$? It is assumed that $z(t)$ never vanishes and $\arg z(t)$ is then defined by continuity. (Zeros of $z(t)$ can be admitted by writing $z(t) = \rho(t)e(\phi(t))$, where $\rho(t)$ and $\phi(t)$ are continuous real-valued functions and $\rho(t)$ is required to change sign at a zero of $z(t)$ of odd multiplicity.)

In the astronomical application $\arg z(t)$ measures the longitude of the perihelion of the planetary orbit. Lagrange showed that the limit

$$\mu = \lim_{t \to +\infty} t^{-1} \arg z(t)$$

does exist when $n = 2$ and also, for arbitrary n, when some ρ_k exceeds the sum of all the others. The only planets which do not satisfy this second condition are Venus and Earth. Lagrange went on to say that, when neither of the two conditions was satisfied, the problem was "very difficult and perhaps impossible".

There was no further progress until the work of Bohl (1909), who took $n = 3$ and considered the non-Lagrangian case when there exists a triangle with sidelengths ρ_1, ρ_2, ρ_3. He showed that the limit μ exists if $\omega_1, \omega_2, \omega_3$ are linearly independent over the rational field $\mathbb{Q}$ and then $\mu = \lambda_1 \omega_1 + \lambda_2 \omega_2 + \lambda_3 \omega_3$, where $\pi\lambda_1, \pi\lambda_2, \pi\lambda_3$ are the angles of the triangle with sidelengths ρ_1, ρ_2, ρ_3. In the course of the proof he stated and proved Proposition 3 (without formulating the general concept of uniform distribution).

Using his earlier results on uniform distribution, Weyl (1938) showed that the limit μ exists if $\omega_1,...,\omega_n$ are linearly independent over the rational field $\mathbb{Q}$ and then

$$\mu = \lambda_1 \omega_1 + ... + \lambda_n \omega_n,$$

where $\lambda_k \geq 0$ ($k = 1,...,n$) and $\sum_{k=1}^{n} \lambda_k = 1$. The coefficients λ_k depend only on the ρ's, not on the α's or ω's, and there is even an explicit expression for λ_k, involving Bessel functions, which is derived from the theory of random walks.

Finally, it was shown by Jessen and Tornehave (1945) that the limit μ exists for arbitrary $\omega_k \in \mathbb{R}$.

2 Discrepancy

The *star discrepancy* of a finite set of points $\xi_1,...,\xi_N$ in the unit interval $I = [0,1]$ is defined to be

$$D_N^* = D_N^*(\xi_1,...,\xi_N) = \sup_{0<\alpha\le 1} |\varphi_\alpha(N)/N - \alpha|,$$

where $\varphi_\alpha(N) = \varphi_{0,\alpha}(N)$ denotes the number of positive integers $n \le N$ such that $0 \le \xi_n < \alpha$. Here we will omit the qualifier 'star', since we will not be concerned with any other type of discrepancy and the notation D_N^* should provide adequate warning.

It was discovered only in 1972, by Niederreiter, that the preceding definition may be reformulated in the following simple way:

PROPOSITION 11 *If $\xi_1,...,\xi_N$ are real numbers such that $0 \le \xi_1 \le ... \le \xi_N \le 1$, then*

$$D_N^* = D_N^*(\xi_1,...,\xi_N) = \max_{1\le k\le N} \max (|\xi_k - k/N|, |\xi_k - (k-1)/N|)$$

$$= (2N)^{-1} + \max_{1\le k\le N} |\xi_k - (2k-1)/2N|.$$

Proof Put $\xi_0 = 0$, $\xi_{N+1} = 1$. Since the distinct ξ_k with $0 \le k \le N + 1$ define a subdivision of the unit interval I, we have

$$D_N^* = \max_{k:\xi_\kappa<\xi_{\kappa+1}} \sup_{\xi_\kappa\le\alpha<\xi_{\kappa+1}} |\varphi_\alpha(N)/N - \alpha|$$

$$= \max_{k:\xi_\kappa<\xi_{\kappa+1}} \sup_{\xi_\kappa\le\alpha<\xi_{\kappa+1}} |k/N - \alpha|.$$

But the function $f_k(t) = |k/N - t|$ attains its maximum in the interval $\xi_k \le t \le \xi_{k+1}$ at one of the endpoints of this interval. Consequently

$$D_N^* = \max_{k:\xi_\kappa<\xi_{\kappa+1}} \max (|k/N - \xi_k|, |k/N - \xi_{k+1}|).$$

We are going to show that in fact

$$D_N^* = \max_{0\le k\le N} \max (|k/N - \xi_k|, |k/N - \xi_{k+1}|).$$

Suppose $\xi_k < \xi_{k+1} = \xi_{k+2} = ... = \xi_{k+r} < \xi_{k+r+1}$ for some $r \ge 2$. By applying the same reasoning as before to the function $g_k(t) = |t - \xi_{k+1}|$ we obtain, for $1 \le j < r$,

$$|(k+j)/N - \xi_{k+j}| = |(k+j)/N - \xi_{k+j+1}| = |(k+j)/N - \xi_{k+1}|$$

$$< \max (|k/N - \xi_{k+1}|, |(k+r)/N - \xi_{k+1}|)$$

$$= \max (|k/N - \xi_{k+1}|, |(k+r)/N - \xi_{k+r}|).$$

Since both terms in the last maximum appear in the expression already obtained for D_N^*, it follows that this expression is not altered by dropping the restriction to those k for which $\xi_k < \xi_{k+1}$.

Since $|0/N - \xi_0| = |N/N - \xi_{N+1}| = 0$, we can now also write

$$D_N^* = \max_{1 \le k \le N} \max \left(|k/N - \xi_k|, |(k-1)/N - \xi_k| \right).$$

The second expression for D_N^* follows immediately, since

$$\max \left(|k/N - \alpha|, |(k-1)/N - \alpha| \right) = |(k - 1/2)/N - \alpha| + 1/2N. \quad \square$$

COROLLARY 12 *If $\xi_1,...,\xi_N$ are real numbers such that $0 \le \xi_1 \le ... \le \xi_N \le 1$, then $D_N^* \ge (2N)^{-1}$. Moreover, equality holds if and only if $\xi_k = (2k - 1)/N$ for $k = 1,...,N$.* $\square$

Thus Proposition 11 says that the discrepancy of any set of N points of I is obtained by adding to its minimal value $1/2N$ the maximum deviation of the set from the unique minimizing set, when both sets are arranged in order of magnitude.

The next result shows that the discrepancy $D_N^*(\xi_1,...,\xi_N)$ is a continuous function of $\xi_1,...,\xi_N$.

PROPOSITION 13 *If $\xi_1,...,\xi_N$ and $\eta_1,...,\eta_N$ are two sets of N points of I, with discrepancies D_N^* and E_N^* respectively, then*

$$|D_N^* - E_N^*| \le \max_{1 \le k \le N} |\xi_k - \eta_k|.$$

Proof Let $x_1 \le ... \le x_N$ and $y_1 \le ... \le y_N$ be the two given sets rearranged in order of magnitude. It is enough to show that

$$\max_{1 \le k \le N} |x_k - y_k| \le \delta := \max_{1 \le k \le N} |\xi_k - \eta_k|,$$

since it then follows from Proposition 11 that

$$D_N^* \le \delta + E_N^*, \quad E_N^* \le \delta + D_N^*.$$

Assume, on the contrary, that $|x_k - y_k| > \delta$ for some k. Then either $x_k > y_k + \delta$ or $y_k > x_k + \delta$. Without loss of generality we restrict attention to the first case. By hypothesis, for each y_i with $1 \le i \le k$ there exists an x_{j_i} with $1 \le j_i \le N$ such that $|y_i - x_{j_i}| \le \delta$ and such that the subscripts j_i are distinct. Since $y_1 \le ... \le y_k$, it follows that

$$x_{j_i} \le y_i + \delta \le y_k + \delta < x_k.$$

But this is a contradiction, since there are at most $k - 1$ x's less than x_k. $\square$

We now show how the notion of discrepancy makes it possible to obtain estimates for the accuracy of various methods of numerical integration.

PROPOSITION 14 *If the function f satisfies the 'Lipschitz condition'*

$$|f(t_2) - f(t_1)| \leq L|t_2 - t_1| \quad \text{for all } t_1, t_2 \in I,$$

then for any finite set $\xi_1, ..., \xi_N \in I$ *with discrepancy* D_N^*,

$$\left| N^{-1} \sum_{n=1}^{N} f(\xi_n) - \int_I f(t) \, dt \right| \leq L D_N^*.$$

Proof Without loss of generality we may assume $\xi_1 \leq ... \leq \xi_N$. Writing

$$\int_I f(t) \, dt = \sum_{n=1}^{N} \int_{(n-1)/N}^{n/N} f(t) \, dt,$$

we obtain

$$\left| N^{-1} \sum_{n=1}^{N} f(\xi_n) - \int_I f(t) \, dt \right| \leq \sum_{n=1}^{N} \int_{(n-1)/N}^{n/N} |f(\xi_n) - f(t)| \, dt$$

$$\leq L \sum_{n=1}^{N} \int_{(n-1)/N}^{n/N} |\xi_n - t| \, dt.$$

But for $(n-1)/N \leq t \leq n/N$ we have

$$|\xi_n - t| \leq \max\left(|\xi_n - n/N|, |\xi_n - (n-1)/N|\right) \leq D_N^*,$$

by Proposition 11. The result follows. □

As Koksma (1942) first showed, Proposition 14 can be sharpened in the following way:

PROPOSITION 15 *If the function f has bounded variation on the unit interval I, with total variation V, then for any finite set* $\xi_1, ..., \xi_N \in I$ *with discrepancy* D_N^*,

$$\left| N^{-1} \sum_{n=1}^{N} f(\xi_n) - \int_I f(t) \, dt \right| \leq V D_N^*.$$

Proof Without loss of generality we may assume $\xi_1 \leq ... \leq \xi_N$ and we put $\xi_0 = 0$, $\xi_{N+1} = 1$. By integration and summation by parts we obtain

$$\sum_{n=0}^{N} \int_{\xi_n}^{\xi_{n+1}} (t - n/N) \, df(t) = \int_I t \, df(t) - N^{-1} \sum_{n=0}^{N} n(f(\xi_{n+1}) - f(\xi_n))$$

$$= [t f(t)]_0^1 - \int_I f(t) \, dt - f(1) + N^{-1} \sum_{n=0}^{N-1} f(\xi_{n+1})$$

$$= N^{-1} \sum_{n=1}^{N} f(\xi_n) - \int_I f(t) \, dt.$$

The result follows, since for $\xi_n \leq t \leq \xi_{n+1}$ we have

$$|t - n/N| \leq \max\left(|\xi_n - n/N|, |\xi_{n+1} - n/N|\right) \leq D_N^*. \quad \square$$

As an application of Proposition 15 we prove

PROPOSITION 16 *If* $\xi_1,...,\xi_N$ *are points of the unit interval I with discrepancy* D_N^* *then, for any integer* $h \neq 0$,

$$|N^{-1}\sum_{n=1}^{N} e(h\xi_n)| \leq 4|h|D_N^*.$$

Proof We can write

$$N^{-1}\sum_{n=1}^{N} e(h\xi_n) = \rho e(\alpha),$$

where $\rho \geq 0$ and $\alpha \in I$. Thus

$$\rho = N^{-1}\sum_{n=1}^{N} e(h\xi_n - \alpha).$$

Adding this relation to its complex conjugate, we obtain

$$\rho = N^{-1}\sum_{n=1}^{N} \cos 2\pi(h\xi_n - \alpha).$$

The result follows by applying Proposition 15 to the function $f(t) = \cos 2\pi(ht - \alpha)$, which has bounded variation on I with total variation $\int_I |f'(t)|\, dt = 4|h|$. $\square$

An inequality in the opposite direction to Proposition 16 was obtained by Erdös and Turan (1948) who showed that, for any positive integer m,

$$D_N^* \leq C(m^{-1} + \sum_{h=1}^{m} h^{-1}|N^{-1}\sum_{n=1}^{N} e(h\xi_n)|),$$

where the positive constant C is independent of m,N and the ξ's. Niederreiter and Philipp (1973) showed that one can take $C = 4$. Furthermore they generalized the result and simplified the proof.

The connection between these results and the theory of uniform distribution is close at hand. Let (ξ_n) be an arbitrary sequence of real numbers and let δ_N denote the discrepancy of the fractional parts $\{\xi_1\},...,\{\xi_N\}$. By the remark after the definition of uniform distribution in §1, *the sequence* (ξ_n) *is uniformly distributed mod* 1 *if and only if* $\delta_N \to 0$ as $N \to \infty$. It follows from Proposition 16 and the inequality of Erdös and Turan (in which m may be arbitrarily large) that $\delta_N \to 0$ as $N \to \infty$ if and only if, for every integer $h \neq 0$,

$$N^{-1}\sum_{n=1}^{N} e(h\xi_n) \to 0 \quad \text{as } N \to \infty.$$

This provides a new proof of Theorem 2. Furthermore, from bounds for the exponential sums we can obtain estimates for the rapidity with which δ_N tends to zero.

Propositions 14 and 15 show that in a formula for numerical integration the nodes (ξ_n) should be chosen to have as small a discrepancy as possible. For a given finite number N of nodes Corollary 12 shows how this can be achieved. In practice, however, one does not know in advance an appropriate choice of N, since universal error bounds may grossly overestimate the error in a specific case. Consequently it is also of interest to consider the problem of choosing an infinite sequence (ξ_n) of nodes so that the discrepancy δ_N of $\xi_1,...,\xi_N$ tends to zero as rapidly as possible when $N \to \infty$. There is a limit to what can be achieved in this way. W. Schmidt (1972), improving earlier results of van Aardenne-Ehrenfest (1949) and Roth (1954), showed that there exists an absolute constant $C > 0$ such that

$$\overline{\lim}_{N \to \infty} N\delta_N /\log N \geq C$$

for *every* infinite sequence (ξ_n). Kuipers and Niederreiter (1974) showed that a possible value for C was $(132 \log 2)^{-1} = 0.010..$ and Bejian (1979) sharpened this to $(24 \log 2)^{-1} = 0.060..$.

Schmidt's result is best possible, apart from the value of the constant. Ostrowski (1922) had already shown that for the sequence $(\{n\alpha\})$, where $\alpha \in (0,1)$ is irrational, one has

$$s^*(\alpha): = \overline{\lim}_{N \to \infty} N\delta_N /\log N < \infty$$

if in the continued fraction expansion

$$\alpha = [0;a_1,a_2,...] = \cfrac{1}{a_1 + \cfrac{1}{a_2 + ...}}$$

the partial quotients a_k are bounded. Dupain and Sós (1984) have shown that the minimum value of $s^*(\alpha)$, for all such α, is $(4 \log (1 + \sqrt{2}))^{-1} = 0.283...$ and the minimum is attained for $\alpha = \sqrt{2} - 1 = [0;2,2,...]$. Schoessengeier (1984) has proved that, for any irrational $\alpha \in (0,1)$, one has $N\delta_N = O(\log N)$ if and only if the partial quotients a_k satisfy $\sum_{k=1}^{n} a_k = O(n)$.

There are other low discrepancy sequences. Haber (1966) showed that, for a sequence (ξ_n) constructed by van der Corput (1935),

$$\overline{\lim}_{N \to \infty} N\delta_N /\log N = (3 \log 2)^{-1} = 0.481... .$$

van der Corput's sequence is defined in the following way: if $n - 1 = a_m 2^m + ... + a_1 2^1 + a_0$, where $a_k \in \{0,1\}$, then $\xi_n = a_0 2^{-1} + a_1 2^{-2} + ... + a_m 2^{-m-1}$. In other words, the expression for ξ_n in the base 2 is obtained from that for $n - 1$ by reflection in the 'decimal' point, a construction which is easily implemented on a computer. Various generalizations of this construction have been given, and Faure (1981) defined in this way a sequence (ξ_n) for which

$$\overline{\lim}_{N \to \infty} N\delta_N /\log N = (1919)(3454 \log 12)^{-1} = 0.223... .$$

Thus if C^* is the least upper bound for all admissible values of C in Schmidt's result then, by what has been said, $0.060... \leq C^* \leq 0.223...$. It is natural to ask: what is the exact value of C^*, and is there a sequence (ξ_n) for which it is attained?

The notion of discrepancy is easily extended to higher dimensions by defining the discrepancy of a finite set of vectors $x_1,...,x_N$ in the d-dimensional unit cube $I^d = I \times ... \times I$ to be

$$D_N^*(x_1,...,x_N) = \sup_{0 < \alpha_k \leq 1 \ (k=1,...,d)} |\varphi_a(N)/N - \alpha_1 \cdots \alpha_d|,$$

where $x_n = (\xi_n^{(1)},...,\xi_n^{(d)})$, $a = (\alpha_1,...,\alpha_d)$ and $\varphi_a(N)$ is the number of positive integers $n \leq N$ such that $0 \leq \xi_n^{(k)} < \alpha_k$ for every $k \in \{1,...,d\}$.

For $d > 1$ there is no simple reformulation of the definition analogous to Proposition 11, but many results do carry over. In particular, Proposition 15 was generalized and applied to the numerical evaluation of multiple integrals by Hlawka (1961/62). Indeed this application has greater value in higher dimensions, where other methods perform poorly.

For the application one requires a set of vectors $x_1,...,x_N \in I^d$ whose discrepancy $D_N^*(x_1,...,x_N)$ is small. A simple procedure for obtaining such a set, which is most useful when the integrand is smooth and has period 1 in each of its variables, is the method of 'good lattice points' introduced by Korobov (1959). Here, for a suitably chosen $g \in \mathbb{Z}^d$, one takes $x_n = \{(n-1)g/N\}$ $(n = 1,...,N)$. A result of Niederreiter (1986) implies that, for every $d \geq 2$ and every $N \geq 2$, one can choose g so that

$$ND_N^* \leq (1 + \log N)^d + d2^d.$$

The van der Corput sequence has also been generalized to any finite number of dimensions by Halton (1960). He defined an infinite sequence (x_n) of vectors in $\mathbb{R}^d$ for which

$$\overline{\lim}_{N \to \infty} N\delta_N / (\log N)^d < \infty.$$

It is conjectured that for each $d > 1$ (as for $d = 1$) there exists an absolute constant $C_d > 0$ such that

$$\overline{\lim}_{N \to \infty} N\delta_N / (\log N)^d \geq C_d$$

for every infinite sequence (x_n) of vectors in $\mathbb{R}^d$. However, the best known result remains that of Roth (1954), in which the exponent d is replaced by $d/2$.

3 Birkhoff's ergodic theorem

In statistical mechanics there is a procedure for calculating the physical properties of a system by simply averaging over all possible states of the system. To justify this procedure Boltzmann (1871) introduced what he later called the 'ergodic hypothesis'. In the formulation of Maxwell (1879) this says that "the system, if left to itself in its actual state of motion, will, sooner or later, pass through every phase which is consistent with the equation of energy". The word *ergodic*, coined by Boltzmann (1884), was a composite of the Greek words for 'energy' and 'path'. It was recognized by Poincaré (1894) that it was too much to ask that a path pass through every state on the same energy surface as its initial state, and he suggested instead that it pass arbitrarily close to every such state. Moreover, he observed that it would still be necessary to exclude certain exceptional initial states.

A breakthrough came with the work of G.D. Birkhoff (1931), who showed that Lebesgue measure was the appropriate tool for treating the problem. He established a deep and general result which says that, apart from a set of initial states of measure zero, there is a definite limiting value for the proportion of time which a path spends in any given measurable subset B of an energy surface X. The proper formulation for the ergodic hypothesis was then that this limiting value should coincide with the ratio of the measure of B to that of X, i.e. that 'the paths through almost all initial states should be uniformly distributed over arbitrary measurable sets'. It was not difficult to deduce that this was the case if and only if 'any invariant measurable subset of X either had measure zero or had the same measure as X'.

Birkhoff proved his theorem in the framework of classical mechanics and for *flows* with continuous time. We will prove his theorem in the abstract setting of probability spaces and for *cascades* with discrete time. The abstract formulation makes possible other applications, for which continuous time is not appropriate.

Let $\mathcal{B}$ be a σ-*algebra* of subsets of a given set X, i.e. a nonempty family of subsets of X such that

(B1) the complement of any set in $\mathcal{B}$ is again a set in $\mathcal{B}$,
(B2) the union of any finite or countable collection of sets in $\mathcal{B}$ is again a set in $\mathcal{B}$.

It follows that $X \in \mathcal{B}$, since $B \in \mathcal{B}$ implies $B^c := X \setminus B \in \mathcal{B}$ and $X = B \cup B^c$. Hence also $\varnothing = X^c \in \mathcal{B}$. Furthermore, the intersection of any finite or countable collection of sets in $\mathcal{B}$ is again a set in $\mathcal{B}$, since $\bigcap_n B_n = X \setminus (\bigcup_n B_n{}^c)$. Hence if $A, B \in \mathcal{B}$, then

$$B \setminus A = B \cap A^c \in \mathcal{B}$$

and the *symmetric difference*

$$A \Delta B: = (B \setminus A) \cup (A \setminus B) \in \mathcal{B}.$$

The family of all subsets of X is certainly a σ-algebra. Furthermore, the intersection of any collection of σ-algebras is again a σ-algebra. It follows that, for any family $\mathcal{A}$ of subsets of X, there is a σ-algebra $\sigma(\mathcal{A})$ which contains $\mathcal{A}$ and is contained in every σ-algebra which contains $\mathcal{A}$. We call $\sigma(\mathcal{A})$ the σ-algebra of subsets of X *generated* by $\mathcal{A}$.

Suppose $\mathcal{B}$ is a σ-algebra of subsets of X and a function $\mu: \mathcal{B} \to \mathbb{R}$ is defined such that

(Pr1) $\mu(B) \geq 0$ for every $B \in \mathcal{B}$,

(Pr2) $\mu(X) = 1$,

(Pr3) if (B_n) is a sequence of pairwise disjoint sets in $\mathcal{B}$, then $\mu(\bigcup_n B_n) = \Sigma_n \mu(B_n)$.

Then μ is said to be a *probability measure* and the triple $(X, \mathcal{B}, \mu)$ is said to be a *probability space*.

It is easily seen that the definition implies

(i) $\mu(\varnothing) = 0$,

(ii) $\mu(B^c) = 1 - \mu(B)$,

(iii) $\mu(A) \leq \mu(B)$ if $A, B \in \mathcal{B}$ and $A \subseteq B$,

(iv) $\mu(B_n) \to \mu(B)$ if (B_n) is a sequence of sets in $\mathcal{B}$ such that $B_1 \supseteq B_2 \supseteq \ldots$ and $B = \bigcap_n B_n$.

If a property of points in a probability space $(X, \mathcal{B}, \mu)$ holds for all $x \in B$, where $B \in \mathcal{B}$ and $\mu(B) = 1$, then the property is said to hold for (μ-)*almost all* $x \in X$, or simply *almost everywhere* (a.e.).

A function $f: X \to \mathbb{R}$ is *measurable* if, for every $\alpha \in \mathbb{R}$, the set $\{x \in X: f(x) < \alpha\}$ is in $\mathcal{B}$. Let $f: X \to [0, \infty)$ be measurable and for any partition $\mathcal{P}$ of X into finitely many pairwise disjoint sets $B_1, \ldots, B_n \in \mathcal{B}$, put

$$L_{\mathcal{P}}(f) = \Sigma_{k=1}^{n} f_k \mu(B_k),$$

where $f_k = \inf\{f(x): x \in B_k\}$. We say that f is *integrable* if

$$\int_X f \, d\mu: = \sup_{\mathcal{P}} L_{\mathcal{P}}(f) < \infty.$$

The set of all measurable functions $f: X \to \mathbb{R}$ such that $|f|$ is integrable is denoted by $L(X, \mathcal{B}, \mu)$.

A map $T: X \to X$ is said to be a *measure-preserving transformation* of the probability space $(X, \mathcal{B}, \mu)$ if, for every $B \in \mathcal{B}$, the set $T^{-1}B = \{x \in X: Tx \in B\}$ is again in $\mathcal{B}$ and $\mu(T^{-1}B) = \mu(B)$. This is equivalent to $\mu(TB) = \mu(B)$ for every $B \in \mathcal{B}$ if the measure-preserving transformation T is *invertible*, i.e. if T is bijective and $TB \in \mathcal{B}$ for every $B \in \mathcal{B}$.

However, we do not wish to restrict attention to the invertible case. Several important examples of measure-preserving transformations of probability spaces will be given in the next section.

Birkhoff's ergodic theorem, which is also known as the 'individual' or 'pointwise' ergodic theorem, has the following statement:

THEOREM 17 *Let T be a measure-preserving transformation of the probability space $(X,\mathcal{B},\mu)$. If $f \in L(X,\mathcal{B},\mu)$ then, for almost all $x \in X$, the limit*

$$f^*(x) = \lim_{n \to \infty} n^{-1} \sum_{k=0}^{n-1} f(T^k x)$$

exists and $f^*(Tx) = f^*(x)$. Moreover, $f^* \in L(X,\mathcal{B},\mu)$ and $\int_X f^* \, d\mu = \int_X f \, d\mu$.

Proof It is sufficient to prove the theorem for nonnegative functions, since we can write $f = f_+ - f_-$, where

$$f_+(x) = \max\{f(x),0\}, \quad f_-(x) = \max\{-f(x),0\},$$

and $f_+, f_- \in L(X,\mathcal{B},\mu)$.

Put

$$\overline{f}(x) = \overline{\lim}_{n \to \infty} n^{-1} \sum_{k=0}^{n-1} f(T^k x), \quad \underline{f}(x) = \underline{\lim}_{n \to \infty} n^{-1} \sum_{k=0}^{n-1} f(T^k x).$$

Then $\overline{f}$ and $\underline{f}$ are μ-measurable functions since, for any sequence (g_n),

$$\overline{\lim}_{n \to \infty} g_n(x) = \inf_m (\sup_{n \geq m} g_n(x)), \quad \underline{\lim}_{n \to \infty} g_n(x) = \sup_m (\inf_{n \geq m} g_n(x)).$$

Moreover $\overline{f}(x) = \overline{f}(Tx)$, $\underline{f}(x) = \underline{f}(Tx)$ for every $x \in X$, since

$$(n + 1)^{-1} \sum_{k=0}^{n} f(T^k x) = (n + 1)^{-1} f(x) + (1 + 1/n)^{-1} n^{-1} \sum_{k=0}^{n-1} f(T^{k+1} x).$$

It is sufficient to show that

$$\int_X \overline{f} \, d\mu \leq \int_X f \, d\mu \leq \int_X \underline{f} \, d\mu.$$

For then, since $\underline{f} \leq \overline{f}$, it follows that $\overline{f}(x) = \underline{f}(x) = f^*(x)$ for μ-almost all $x \in X$ and

$$\int_X f^* \, d\mu = \int_X f \, d\mu.$$

Fix some $M > 0$ and define the 'cut-off' function $\overline{f}_M$ by

$$\overline{f}_M(x) = \min\{M, \overline{f}(x)\}.$$

Then $\overline{f}_M$ is bounded and $\overline{f}_M(Tx) = \overline{f}_M(x)$ for every $x \in X$. Fix also any $\varepsilon > 0$. By the definition of $\overline{f}(x)$, for each $x \in X$ there exists a positive integer n such that

$$\overline{f}_M(x) \leq n^{-1} \sum_{k=0}^{n-1} f(T^k x) + \varepsilon. \tag{*}$$

Thus if F_n is the set of all $x \in X$ for which (*) holds and if $E_n = \bigcup_{k=1}^{n} F_k$, then $E_1 \subseteq E_2 \subseteq \dots$ and $X = \bigcup_{n \geq 1} E_n$. Since the sets E_n are μ-measurable, we can choose N so large that $\mu(E_N) > 1 - \varepsilon/M$.

Put

$$\tilde{f}(x) = f(x) \text{ if } x \in E_N,$$
$$= \max\{f(x), M\} \text{ if } x \notin E_N.$$

Also, let $\tau(x)$ be the least positive integer $n \leq N$ for which (*) holds if $x \in E_N$, and let $\tau(x) = 1$ if $x \notin E_N$. Since $\overline{f}_M$ is T-invariant, (*) implies

$$\sum_{k=0}^{n-1} \overline{f}_M(T^k x) \leq \sum_{k=0}^{n-1} f(T^k x) + n\varepsilon$$

and hence

$$\sum_{k=0}^{\tau(x)-1} \overline{f}_M(T^k x) \leq \sum_{k=0}^{\tau(x)-1} \tilde{f}(T^k x) + \tau(x)\varepsilon.$$

To estimate the sum $\sum_{k=0}^{L-1} \overline{f}_M(T^k x)$ for any $L > N$, we partition it into blocks of the form

$$\sum_{k=0}^{\tau(y)-1} \overline{f}_M(T^k y)$$

and a remainder block. More precisely, define inductively

$$n_0(x) = 0, \ \ n_k(x) = n_{k-1}(x) + \tau(T^{n_{k-1}}x) \ \ (k = 1, 2, \dots)$$

and define h by $n_h(x) < L \leq n_{h+1}(x)$. Then

$$\sum_{k=0}^{n_1(x)-1} \overline{f}_M(T^k x) \leq \sum_{k=0}^{n_1(x)-1} \tilde{f}(T^k x) + \tau(x)\varepsilon,$$

$$\sum_{k=n_1(x)}^{n_2(x)-1} \overline{f}_M(T^k x) \leq \sum_{k=n_1(x)}^{n_2(x)-1} \tilde{f}(T^k x) + \tau(T^{n_1}x)\varepsilon,$$

$$\dots\dots$$

$$\sum_{k=n_{h-1}(x)}^{n_h(x)-1} \overline{f}_M(T^k x) \leq \sum_{k=n_{h-1}(x)}^{n_h(x)-1} \tilde{f}(T^k x) + \tau(T^{n_{h-1}}x)\varepsilon.$$

Since $n_h(x) < L$, we obtain by addition

$$\sum_{k=0}^{n_h(x)-1} \overline{f}_M(T^k x) \leq \sum_{k=0}^{n_h(x)-1} \tilde{f}(T^k x) + L\varepsilon.$$

On the other hand, since $L \leq n_{h+1}(x) \leq n_h(x) + N$, we have

$$\sum_{k=n_h(x)}^{L-1} \overline{f}_M(T^k x) \leq NM.$$

Since $\tilde{f} \geq 0$, it follows that

$$\sum_{k=0}^{L-1} \overline{f}_M(T^k x) \leq \sum_{k=0}^{L-1} \tilde{f}(T^k x) + L\varepsilon + NM.$$

Dividing by L and integrating over X, we obtain

$$\int_X \overline{f}_M \, d\mu \leq \int_X \tilde{f} \, d\mu + \varepsilon + NM/L,$$

since the measure-preserving nature of T implies that, for any $g \in L(X, \mathcal{B}, \mu)$,

$$\int_X g(Tx) \, d\mu(x) = \int_X g(x) \, d\mu(x).$$

Since

$$\int_X \tilde{f} \, d\mu \leq \int_X f \, d\mu + \int_{X \setminus E_N} M \, d\mu \leq \int_X f \, d\mu + \varepsilon,$$

it follows that

$$\int_X \overline{f}_M \, d\mu \leq \int_X f \, d\mu + 2\varepsilon + NM/L.$$

Since L may be chosen arbitrarily large and then ε arbitrarily small, we conclude that

$$\int_X \overline{f}_M \, d\mu \leq \int_X f \, d\mu.$$

Now letting $M \to \infty$, we obtain

$$\int_X \overline{f} \, d\mu \leq \int_X f \, d\mu.$$

The proof that

$$\int_X f \, d\mu \leq \int_X \underline{f} \, d\mu$$

is similar. Given $\varepsilon > 0$, there exists for each $x \in X$ a positive integer n such that

$$n^{-1} \sum_{k=0}^{n-1} f(T^k x) \leq \underline{f}(x) + \varepsilon. \tag{**}$$

If F_n is the set of all $x \in X$ for which (**) holds and if $E_n = \bigcup_{k=1}^{n} F_k$, we can choose N so large that

$$\int_{X \setminus E_N} f \, d\mu < \varepsilon.$$

Put

$$\tilde{f}(x) = f(x) \quad \text{if } x \in E_N,$$
$$= 0 \quad \text{if } x \notin E_N.$$

Let $\tau(x)$ be the least positive integer n for which (**) holds if $x \in E_N$, and $\tau(x) = 1$ otherwise. The proof now goes through in the same way as before. $\square$

It should be noticed that the preceding proof simplifies if the function f is bounded. In Birkhoff's original formulation the function f was the indicator function χ_B of an arbitrary set $B \in \mathcal{B}$. In this case the theorem says that, if $v_n(x)$ is the number of $k < n$ for which $T^k x \in B$, then $\lim_{n \to \infty} v_n(x)/n$ exists for almost all $x \in X$. That is, 'almost every point has an average sojourn time in any measurable set'.

A measure-preserving transformation T of the probability space $(X,\mathscr{B},\mu)$ is said to be *ergodic* if, for every $B \in \mathscr{B}$ with $T^{-1}B = B$, either $\mu(B) = 0$ or $\mu(B) = 1$. Part (ii) of the next proposition says that this is the case if and only if 'time means and space means are equal'.

PROPOSITION 18 *Let T be a measure-preserving transformation of the probability space $(X,\mathscr{B},\mu)$. Then T is ergodic if and only if one of the following equivalent properties holds:*

(i) *if $f \in L(X,\mathscr{B},\mu)$ satisfies $f(Tx) = f(x)$ almost everywhere, then f is constant almost everywhere;*

(ii) *if $f \in L(X,\mathscr{B},\mu)$ then, for almost all $x \in X$,*

$$n^{-1}\sum_{k=0}^{n-1} f(T^k x) \;\to\; \int_X f\,d\mu \;\; as \; n \to \infty;$$

(iii) *if $A,B \in \mathscr{B}$, then*

$$n^{-1}\sum_{k=0}^{n-1} \mu(T^{-k}A \cap B) \;\to\; \mu(A)\mu(B) \;\; as \; n \to \infty;$$

(iv) *if $C \in \mathscr{B}$ and $\mu(C) > 0$, then $\mu(\bigcup_{n\geq 1}T^{-n}C) = 1$;*

(v) *if $A,B \in \mathscr{B}$ and $\mu(A) > 0$, $\mu(B) > 0$, then $\mu(T^{-n}A \cap B) > 0$ for some $n > 0$.*

Proof Suppose first that T is ergodic and let $f \in L(X,\mathscr{B},\mu)$ satisfy $f(Tx) = f(x)$ a.e. Put

$$\overline{f}(x) \;=\; \overline{\lim}_{n\to\infty} n^{-1}\sum_{k=0}^{n-1} f(T^k x).$$

Then $\overline{f}(Tx) = \overline{f}(x)$ for every $x \in X$ and $\overline{f}(x) = f(x)$ a.e. For any $\alpha \in \mathbb{R}$, let

$$A_\alpha \;=\; \{x \in X\colon \overline{f}(x) < \alpha\}.$$

Then $\mu(A_\alpha) = 0$ or 1, since $T^{-1}A_\alpha = A_\alpha$ and T is ergodic. Since $\mu(A_\alpha)$ is a nondecreasing function of α and $\mu(A_\alpha) \to 0$ as $\alpha \to -\infty$, $\mu(A_\alpha) \to 1$ as $\alpha \to +\infty$, there exists $\beta \in \mathbb{R}$ such that $\mu(A_\alpha) = 0$ for $\alpha < \beta$ and $\mu(A_\alpha) = 1$ for $\alpha > \beta$. It follows that $\mu(A_\beta) = 0$ and $\mu(B_\beta) = 1$, where

$$B_\beta \;=\; \{x \in X\colon \overline{f}(x) \leq \beta\}.$$

Hence $f(x) = \beta$ a.e. and (i) holds.

Suppose now that (i) holds and let $f \in L(X,\mathscr{B},\mu)$. Then the function f^* in the statement of Theorem 17 must be constant a.e. Moreover, if γ is its constant value, we must have

$$\gamma \;=\; \int_X f^*\,d\mu \;=\; \int_X f\,d\mu.$$

Thus (i) implies (ii).

Suppose next that (ii) holds and let $A, B \in \mathcal{B}$. Then, for almost all $x \in X$,

$$\lim_{n \to \infty} n^{-1} \sum_{k=0}^{n-1} \chi_A(T^k x) = \int_X \chi_A \, d\mu = \mu(A).$$

Hence, for almost all $x \in X$,

$$\lim_{n \to \infty} n^{-1} \sum_{k=0}^{n-1} \chi_A(T^k x) \chi_B(x) = \mu(A) \chi_B(x)$$

and so, by the dominated convergence theorem,

$$\mu(A)\mu(B) = \int_X \lim_{n \to \infty} n^{-1} \sum_{k=0}^{n-1} \chi_A(T^k x) \chi_B(x) \, d\mu(x)$$

$$= \lim_{n \to \infty} n^{-1} \sum_{k=0}^{n-1} \int_X \chi_A(T^k x) \chi_B(x) \, d\mu(x)$$

$$= \lim_{n \to \infty} n^{-1} \sum_{k=0}^{n-1} \mu(T^{-k}A \cap B).$$

Thus (ii) implies (iii).

Suppose now that (iii) holds and choose $C \in \mathcal{B}$ with $\mu(C) > 0$. Put $A = \bigcup_{n \geq 0} T^{-n}C$ and $B = (\bigcup_{n \geq 1} T^{-n}C)^c$. Then, for every $k \geq 1$, $T^{-k}A \subseteq \bigcup_{n \geq 1} T^{-n}C$ and hence $\mu(T^{-k}A \cap B) = 0$. Thus

$$n^{-1} \sum_{k=0}^{n-1} \mu(T^{-k}A \cap B) = \mu(A \cap B)/n \to 0 \text{ as } n \to \infty.$$

Since $\mu(A) \geq \mu(C) > 0$, it follows from (iii) that $\mu(B) = 0$. Thus (iii) implies (iv).

Next choose $A, B \in \mathcal{B}$ with $\mu(A) > 0$, $\mu(B) > 0$. If (iv) holds, then $\mu(\bigcup_{n \geq 1} T^{-n}A) = 1$ and hence

$$\mu(B) = \mu(B \cap \bigcup_{n \geq 1} T^{-n}A) = \mu(\bigcup_{n \geq 1}(B \cap T^{-n}A)).$$

Since $\mu(B) > 0$, it follows that $\mu(B \cap T^{-n}A) > 0$ for some $n > 0$. Thus (iv) implies (v).

Finally choose $A \in \mathcal{B}$ with $T^{-1}A = A$ and put $B = A^c$. Then, for every $n \geq 1$, we have $\mu(T^{-n}A \cap B) = \mu(A \cap B) = 0$. If (v) holds, it follows that either $\mu(A) = 0$ or $\mu(B) = 0$. Hence (v) implies that T is ergodic. $\square$

4 Applications

We now give some examples to illustrate the general concepts and results of the previous section.

(i) Suppose $X = \mathbb{R}^d / \mathbb{Z}^d$ is a d-dimensional torus, $\mathcal{B}$ is the family of *Borel subsets* of X (i.e., the σ-algebra of subsets generated by the family of open sets), and $\mu = \lambda$ is Lebesgue measure, i.e. $\mu(B) = \int_X \chi_B(x)\, dx$ for any $B \in \mathscr{B}$, where χ_B is the indicator function of B. Every $x \in X$ is represented by a unique vector $(\xi_1,...,\xi_d)$, where $0 \le \xi_k < 1$ $(k = 1,...,d)$, and X is an abelian group with addition $z = x + y$ defined by $\zeta_k \boxminus \xi_k + \eta_k \bmod 1$ $(k = 1,...,d)$.

For any $a \in X$, the *translation* $T_a: X \to X$ defined by $T_a x = x + a$ is a measure-preserving transformation of the probability space $(X,\mathscr{B},\lambda)$.

PROPOSITION 19 *The translation* $T_a: X \to X$ *of the d-dimensional torus* $X = \mathbb{R}^d/\mathbb{Z}^d$ *is ergodic if and only if* $1,\alpha_1,...,\alpha_d$ *are linearly independent over the rational field* $\mathbb{Q}$, *where* $(\alpha_1,...,\alpha_d)$ *is the vector which represents a.*

Proof Suppose first that $1,\alpha_1,...,\alpha_d$ are not linearly independent over $\mathbb{Q}$. Then there exists a nonzero vector $n \in \mathbb{Z}^d$ such that

$$n \cdot a = v_1\alpha_1 + ... + v_d\alpha_d \in \mathbb{Z}.$$

Hence if $f(x) = e(n \cdot x)$, then $f(T_a x) = f(x)$ for all x. Since f is not constant a.e., it follows from part (i) of Proposition 18 that T_a is not ergodic.

Suppose on the other hand that $1,\alpha_1,...,\alpha_d$ are linearly independent over $\mathbb{Q}$ and let f be an integrable function such that $f(T_a x) = f(x)$ a.e. Then $f(T_a x)$ and $f(x)$ have the same Fourier coefficients:

$$\int_X f(x)e(-n \cdot x)\, dx = \int_X f(x + a)e(-n \cdot x)\, dx = e(n \cdot a) \int_X f(x)e(-n \cdot x)\, dx.$$

Since $e(n \cdot a) \ne 1$ for all $n \ne 0$, it follows that

$$\int_X f(x)e(-n \cdot x)\, dx = 0 \quad \text{for all } n \ne 0.$$

Since integrable functions with the same Fourier coefficients must agree almost everywhere, this proves that f is constant a.e. Hence, by Proposition 18 again, T_a is ergodic. ∎

If we compare Proposition 3′ and the remarks after its proof with Proposition 19, then we see from Theorems 1′-2′ and Proposition 18 that the following five statements are equivalent for $X = \mathbb{R}^d/\mathbb{Z}^d$:

(α) the sequence $(\{na\})$ is dense in X;

(β) for every $x \in X$, the sequence $(x + na)$ is uniformly distributed in X;

(γ) the translation $T_a: X \to X$ is ergodic;

(δ) for each continuous function $f: X \to \mathbb{C}$, $\lim_{n \to \infty} n^{-1} \sum_{k=0}^{n-1} f(T_a^k x) = \int_X f\, d\lambda$ for all $x \in X$;

(ε) for each function $f \in L(X,\mathscr{B},\lambda)$, $\lim_{n \to \infty} n^{-1} \sum_{k=0}^{n-1} f(T_a^k x) = \int_X f\, d\lambda$ for almost all $x \in X$.

(ii) Again suppose $X = \mathbb{R}^d/\mathbb{Z}^d$ is a d-dimensional torus, $\mathfrak{B}$ is the family of Borel subsets of X and $\mu = \lambda$ is Lebesgue measure. For any $d \times d$ matrix $A = (\alpha_{jk})$ of integers, let $R_A: X \to X$ be the map defined by $R_A x = x'$, where

$$\xi_j' \equiv \sum_{k=1}^{d} \alpha_{jk}\xi_k \mod 1 \quad (j = 1,\ldots,d).$$

If $\det A = 0$ then R_A is not measure-preserving, since the image of $\mathbb{R}^d$ under A is contained in a hyperplane of $\mathbb{R}^d$. However, if $\det A \neq 0$ then R_A is measure-preserving, since each point of X is the image under R_A of $|\det A|$ distinct points of X, and a small region B of X is the image under R_A of $|\det A|$ disjoint regions, each with volume $|\det A|^{-1}$ times that of B. (This argument is certainly valid if A is a diagonal matrix, and the general case may be reduced to this by Proposition III.41.) Thus R_A is an *endomorphism* of the torus $\mathbb{R}^d/\mathbb{Z}^d$ if and only if A is nonsingular, and an *automorphism* if and only if $\det A = \pm 1$.

PROPOSITION 20 *The endomorphism $R_A: X \to X$ of the d-dimensional torus $X = \mathbb{R}^d/\mathbb{Z}^d$ is ergodic if and only if no eigenvalue of the nonsingular matrix A is a root of unity.*

Proof For any $n \in \mathbb{Z}^d$ we have

$$e(n \cdot R_A x) = e(n \cdot Ax) = e(Dn \cdot x),$$

where $D = A^t$ is the transpose of A.

Suppose first that A, and hence also D, has an eigenvalue ω which is a root of unity: $\omega^p = 1$ for some positive integer p. Then $(D^p - I)z = 0$ for some nonzero vector z. Moreover, since D is a matrix of integers, we may assume that $z = m \in \mathbb{Z}^d$. We may also assume that $D^i m \neq D^j m$ for $0 \leq i < j < p$, by choosing p to have its least possible value. If we put

$$f(x) = e(m \cdot x) + e(m \cdot Ax) + \ldots + e(m \cdot A^{p-1}x),$$

then $f(R_A x) = f(x)$, but f is not constant a.e. Hence R_A is not ergodic, by Proposition 18.

Suppose next that R_A is not ergodic. Then, by Proposition 18 again, there exists a function $f \in L(X,\mathfrak{B},\lambda)$ such that $f(R_A x) = f(x)$ a.e., but $f(x)$ is not constant a.e. If the Fourier series of $f(x)$ is

$$\sum_{n \in \mathbb{Z}^d} c_n e(n \cdot x),$$

then the Fourier series of $f(R_A x)$ is

$$\sum_{n \in \mathbb{Z}^d} c_n e(n \cdot Ax) = \sum_{n \in \mathbb{Z}^d} c_n e(Dn \cdot x) = \sum_{n \in \mathbb{Z}^d} c_{D^{-1}n} e(n \cdot x)$$

and hence

$$c_n = c_{D^{-1}n} \quad \text{for every } n \in \mathbb{Z}^d.$$

But $c_m \neq 0$ for some nonzero $m \in \mathbb{Z}^d$, since f is not constant a.e., and $|c_n| \to 0$ as $|n| \to \infty$, since $f \in L(X,\mathcal{B},\lambda)$. Since $c_{D^{-k}m} = c_m$ for every positive integer k, it follows that the subscripts $D^{-k}m$ are not all distinct. Hence $D^p m = m$ for some positive integer p and A has an eigenvalue which is a root of unity. $\square$

(There are generalizations of Propositions 19 and 20 to translations and endomorphisms of any compact abelian group X, with Haar measure in place of Lebesgue measure.)

The preceding results have an application to the theory of 'normal numbers'. In fact, without any extra effort, we will consider also higher-dimensional generalizations. A vector $x \in \mathbb{R}^d$ is said to be *normal with respect to the matrix A*, where A is a $d \times d$ matrix of integers, if the sequence $(A^n x)$ is uniformly distributed mod 1.

PROPOSITION 21 *Let A be a d×d matrix of integers. Then (λ-)almost all vectors $x \in \mathbb{R}^d$ are normal with respect to A if and only if A is nonsingular and no eigenvalue of A is a root of unity.*

Proof If A is nonsingular and no eigenvalue of A is a root of unity then, by Proposition 20, R_A is an ergodic measure-preserving transformation of the torus $X = \mathbb{R}^d/\mathbb{Z}^d$. Hence, by Proposition 18(ii), for each nonzero $m \in \mathbb{Z}^d$,

$$n^{-1} \sum_{k=0}^{n-1} e(m \cdot A^n x) \to 0 \text{ as } n \to \infty \text{ for almost all } x \in \mathbb{R}^d.$$

Since $\mathbb{Z}^d$ is countable, and the union of a countable number of sets of measure zero is again a set of measure zero, it follows that, for almost all $x \in \mathbb{R}^d$,

$$n^{-1} \sum_{k=0}^{n-1} e(m \cdot A^n x) \to 0 \text{ as } n \to \infty \text{ for every nonzero } m \in \mathbb{Z}^d.$$

Hence, by Theorem 2', almost all $x \in \mathbb{R}^d$ are normal with respect to A.

If A is singular then, by the remark following the proof of Proposition 9, no $x \in \mathbb{R}^d$ is normal with respect to A. Suppose finally that some eigenvalue of A is a root of unity. Then there exists a positive integer p and a nonzero vector $z \in \mathbb{Z}^d$ such that $D^p z = z$, where $D = A^t$. If

$$f(x) = e(z \cdot x) + e(z \cdot Ax) + \dots + e(z \cdot A^{p-1}x),$$

then $f(Ax) = f(x)$ and hence

$$n^{-1} \sum_{k=0}^{n-1} f(A^k x) = f(x).$$

But if x is normal with respect to A then, by Theorem 1',

$$n^{-1} \sum_{k=0}^{n-1} f(A^k x) \to \int_X f \, d\lambda = 0.$$

Since f is not zero a.e., it follows that the set of all x which are normal with respect to A does not have full measure. $\square$

We consider next when normality with respect to one matrix coincides with normality with respect to another matrix.

PROPOSITION 22 *Let A be a $d{\times}d$ nonsingular matrix of integers, no eigenvalue of which is a root of unity. Then, for any positive integer q, the vector $x \in \mathbb{R}^d$ is normal with respect to A^q if and only if it is normal with respect to A.*

Proof It follows at once from Proposition 9 that if x is normal with respect to A, then it is also normal with respect to A^q.

Suppose, on the other hand, that x is normal with respect to A^q. Then, by Theorem 2', for every nonzero vector $m \in \mathbb{Z}^d$,

$$N^{-1} \sum_{n=0}^{N-1} e(m{\cdot}A^{nq}x) \to 0 \text{ as } N \to \infty.$$

Put $D = A^t$. Since D is a nonsingular matrix of integers, $D^j m$ is a nonzero vector in $\mathbb{Z}^d$ for any integer $j \geq 0$ and hence

$$N^{-1} \sum_{n=0}^{N-1} e(m{\cdot}A^{nq+j}x) = N^{-1} \sum_{n=0}^{N-1} e(D^j m{\cdot}A^{nq}x) \to 0 \text{ as } N \to \infty.$$

Adding these relations for $j = 0,1,...,q{-}1$ and dividing by q, we obtain

$$(Nq)^{-1} \sum_{n=0}^{Nq-1} e(m{\cdot}A^n x) \to 0 \text{ as } N \to \infty.$$

Since the sum of at most q terms $e(m{\cdot}A^n x)$ has absolute value at most q it follows that, also without restricting N to be a multiple of q,

$$N^{-1} \sum_{n=0}^{N-1} e(m{\cdot}A^n x) \to 0 \text{ as } N \to \infty.$$

Hence, by Theorem 2', x is normal with respect to A. $\square$

COROLLARY 23 *Let A be a $d{\times}d$ nonsingular integer matrix, no eigenvalue of which is a root of unity, and let B be a $d{\times}d$ integer matrix such that $A^p = B^q$ for some positive integers p,q. Then $x \in \mathbb{R}^d$ is normal with respect to A if and only if x is normal with respect to B.*

Proof This follows at once from Proposition 22, since the hypotheses imply that also B is nonsingular and has no eigenvalue which is a root of unity. $\square$

Brown and Moran (1993) have shown, conversely, that if A,B are *commuting* $d \times d$ nonsingular integer matrices, no eigenvalues of which are roots of unity, such that the set of all vectors normal with respect to A coincides with the set of all vectors normal with respect to B, then $A^p = B^q$ for some positive integers p,q.

These results will now be specialized to the scalar case. A real number x is said to be *normal to the base a*, where a is an integer ≥ 2, if the sequence $(a^n x)$ is uniformly distributed mod 1. It is readily shown that x is normal to the base a if and only if, in the expansion of x to the base a:

$$x = \lfloor x \rfloor + x_1/a + x_2/a^2 + \dots ,$$

where $x_i \in \{0,1,\dots,a-1\}$ for all $i \geq 1$ and $x_i = a - 1$ for at most finitely many i, every block of digits occurs with the proper frequency; i.e., for any positive integer k and any $a_1,\dots,a_k \in \{0,1,\dots,a-1\}$, the number $\nu(N)$ of i with $1 \leq i \leq N$ such that

$$x_i = a_1, x_{i+1} = a_2, \dots x_{i+k-1} = a_k,$$

satisfies $\nu(N)/N \to a^{-k}$ as $N \to \infty$. By Proposition 21, almost all real numbers x are normal to a given base a. The original proof of this by Borel (1909) was a forerunner of Birkhoff's ergodic theorem. (In fact Borel's proof was faulty, but his paper was influential. Borel used a different definition of normal number, but Wall (1949) showed that it was equivalent to the definition in terms of uniform distribution adopted here.)

The first published proof of the scalar case of Corollary 23 was given by Schmidt (1960), who also proved the scalar version of the result of Brown and Moran: the set of all numbers normal to the base a coincides with the set of all numbers normal to the base b, where a and b are integers ≥ 2, if and only if $a^p = b^q$ for some positive integers p,q.

Although almost all real numbers are normal to *every* base a, it is still not known if such familiar irrational numbers as $\sqrt{2}$, e or π are normal to some base. There are, however, various explicit constructions of normal numbers. In particular, Champernowne (1933) showed that the real number θ whose expansion to the base 10 is composed of the positive integers in their natural order, $\theta = 0.1234567891011121\dots$, is itself normal to the base 10.

(iii) Let A be a set of finite cardinality r, which for definiteness we take to be the set $\{1,\dots,r\}$, and let $p_1,\dots,p_r$ be positive real numbers with sum 1. If $\mathcal{B}_0$ is the family of all subsets of the finite set A and if, for any $B_0 \in \mathcal{B}_0$, we put $\mu_0(B_0) = \sum_{a \in B_0} p_a$, then μ_0 is a probability measure and $(A,\mathcal{B}_0,\mu_0)$ is a probability space.

Now let X be the set of all bi-infinite sequences $x = (...,x_{-2},x_{-1},x_0,x_1,x_2,...)$ with $x_i \in A$ for every $i \in \mathbb{Z}$. Thus X is the product of infinitely many copies of A. We construct a *product measure* on X in the following way.

For any finite sequence $(a_{-m},...,a_0,...,a_m)$ with $a_i \in A$ for $-m \leq i \leq m$, define the (special) *cylinder set* $[a_{-m},...,a_m]$ of *order* m to be the set of all $x \in X$ such that $x_i = a_i$ for $-m \leq i \leq m$. There are r^{2m+1} distinct cylinder sets of order m, distinct cylinder sets are disjoint and X is the union of them all.

Let $\mathcal{C}_m$ denote the collection of all unions of distinct cylinder sets of order m. Thus $X \in \mathcal{C}_m$ and, if $B \in \mathcal{C}_m$, then $B^c = X \setminus B \in \mathcal{C}_m$. Moreover $B,C \in \mathcal{C}_m$ implies $B \cup C \in \mathcal{C}_m$ and $B \cap C \in \mathcal{C}_m$. If $B \in \mathcal{C}_m$, say

$$B = [a_{-m},...,a_m] \cup ... \cup [a'_{-m},...,a'_m],$$

we define

$$\mu_m(B) = p_{a_{-m}} \cdots p_{a_m} + ... + p_{a'_{-m}} \cdots p_{a'_m}.$$

Then $\mu_m(X) = 1$, $\mu_m(B) \geq 0$ for every $B \in \mathcal{C}_m$, and

$$\mu_m(B \cup C) = \mu_m(B) + \mu_m(C) \quad \text{if } B,C \in \mathcal{C}_m \text{ and } B \cap C = \varnothing.$$

Every union of cylinder sets of order m is also a union of cylinder sets of order $m + 1$, since

$$[a_{-m},...,a_m] = \bigcup_{a,a' \in A}[a,a_{-m},...,a_m,a'].$$

Thus $\mathcal{C}_m \subsetneq \mathcal{C}_{m+1}$. Moreover μ_{m+1} continues μ_m, since

$$\mu_{m+1}([a_{-m},...,a_m]) = \sum_{j,j'=1}^{r} p_j p_{j'} p_{a_{-m}} \cdots p_{a_m}$$

$$= \mu_m([a_{-m},...,a_m])(\sum_{j=1}^{r} p_j)(\sum_{j'=1}^{r} p_{j'})$$

$$= \mu_m([a_{-m},...,a_m]).$$

Let μ denote the continuation of all μ_m to $\mathcal{C} = \mathcal{C}_0 \cup \mathcal{C}_1 \cup ...$. If $B,C \in \mathcal{C}$, then $B,C \in \mathcal{C}_m$ for some m. Hence, for given $C \in \mathcal{C}$, there are only finitely many distinct $B \in \mathcal{C}$ such that $B \subseteq C$. Consequently, if C is the union of a sequence of disjoint sets $C_n \in \mathcal{C}$ $(n = 1,2,...)$, then $C_n = \varnothing$ for all large n and $\mu(C) = \sum_{n \geq 1} \mu(C_n)$. It follows, by a construction due to Carathéodory (1914), that μ can be uniquely extended to the σ-algebra $\mathcal{B}$ of subsets of X generated by $\mathcal{C}$ so that $(X,\mathcal{B},\mu)$ is a probability space. For any $\varepsilon > 0$ there exists, for each $B \in \mathcal{B}$, some $C \in \mathcal{C}$ such that $\mu(B \Delta C) < \varepsilon$.

The *two-sided Bernoulli shift* $B_{p_1,...,p_r}$ is the map $\sigma: X \to X$ defined by $\sigma x = x'$, where $x'_i = x_{i+1}$ for every $i \in \mathbb{Z}$. It is a measure-preserving transformation of the probability space $(X, \mathcal{B}, \mu)$, since

$$\sigma^{-1}[a_{-m},...,a_m] = \bigcup_{a,a' \in A}[a,a',a_{-m},...,a_m]$$

and hence

$$\mu(\sigma^{-1}[a_{-m},...,a_m]) = \sum_{j,j'=1}^{r} p_j p_{j'} p_{a_{-m}} \cdots p_{a_m}$$

$$= \sum_{j,j'=1}^{r} p_j p_{j'} \mu([a_{-m},...,a_m]) = \mu([a_{-m},...,a_m]).$$

The Bernoulli shift $B_{1/2,1/2}$ is a model for the random process consisting of bi-infinite sequences of coin-tossings.

We may define the *general cylinder set* $C_{i_1...i_k}^{a_1...a_k}$, where $i_1,...,i_k$ are distinct integers, to be the set of all $x \in X$ such that

$$x_{i_1} = a_1, \ldots , x_{i_k} = a_k.$$

In particular, $C_i^a = \sigma^{-i}[a]$ and hence $\mu(C_i^a) = p_a$. It follows by induction on k that

$$\mu(C_{i_1...i_k}^{a_1...a_k}) = p_{a_1} \cdots p_{a_k}.$$

PROPOSITION 24 *For any given positive numbers $p_1,...,p_r$ with sum 1, the two-sided Bernoulli shift $B_{p_1,...,p_r}$ is ergodic.*

Proof Suppose $B \in \mathcal{B}$ and $\sigma^{-1}B = B$. For any $\varepsilon > 0$ there exists a set $C \in \mathcal{C}$ such that

$$\mu(B \, \Delta \, C) = \mu(B \setminus C) + \mu(C \setminus B) < \varepsilon.$$

Then

$$|\mu(B) - \mu(C)| = |\mu(C \cap B) + \mu(B \setminus C) - \mu(C \cap B) - \mu(C \setminus B)|$$

$$\leq \mu(B \setminus C) + \mu(C \setminus B) < \varepsilon$$

and hence

$$|\mu(B)^2 - \mu(C)^2| = \{\mu(C) + \mu(B)\} |\mu(B) - \mu(C)| < 2\varepsilon.$$

We may suppose that C is the union of finitely many special cylinder sets of order m. Since

$$\sigma^{-n}[a_{-m},...,a_m] = C_{-m+n,...,m+n}^{a_{-m},...,a_m},$$

for $n > 2m$ we have

$$[a'_{-m},...,a'_m] \cap \sigma^{-n}[a_{-m},...,a_m] = C_{-m,...,m,-m+n,...,m+n}^{a'_{-m},...,a'_m,a_{-m},...,a_m},$$

and hence

$$\mu([a'_{-m},...,a'_m] \cap \sigma^{-n}[a_{-m},...,a_m]) = p_{a'_{-m}} \cdots p_{a'_m} p_{a_{-m}} \cdots p_{a_m},$$

$$= \mu([a'_{-m},...,a'_m]) \, \mu([a_{-m},...,a_m]).$$

It follows that if $n > 2m$, then

$$\mu(C \cap \sigma^{-n}C) = \mu(C)^2.$$

But

$$\mu(B \setminus (C \cap \sigma^{-n}C)) \le 2\mu(B \setminus C),$$

since

$$B \setminus (C \cap \sigma^{-n}C) \subseteq (B \setminus C) \cup (B \setminus \sigma^{-n}C) \subseteq (B \setminus C) \cup \sigma^{-n}(B \setminus C),$$

and similarly

$$\mu((C \cap \sigma^{-n}C) \setminus B) \le 2\mu(C \setminus B).$$

Hence

$$|\mu(B) - \mu(C \cap \sigma^{-n}C)| \le \mu(B \setminus (C \cap \sigma^{-n}C)) + \mu((C \cap \sigma^{-n}C) \setminus B) < 2\varepsilon.$$

Thus

$$0 \le \mu(B) - \mu(B)^2 = \mu(B) - \mu(C \cap \sigma^{-n}C) + \mu(C \cap \sigma^{-n}C) - \mu(B)^2$$

$$< 2\varepsilon + \mu(C)^2 - \mu(B)^2 < 4\varepsilon.$$

Since ε is arbitrary, we conclude that $\mu(B) = \mu(B)^2$. Hence $\mu(B) = 0$ or 1, and σ is ergodic. $\square$

Similarly, if Y is the set of all infinite sequences $y = (y_1, y_2, y_3, ...)$ with $y_i \in A$ for every $i \in \mathbb{N}$, then the *one-sided Bernoulli shift* $B^+_{p_1,...,p_r}$, i.e. the map $\tau: Y \to Y$ defined by $\tau y = y'$, where $y'_i = y_{i+1}$ for every $i \in \mathbb{N}$, is a measure-preserving transformation of the analogously constructed probability space $(Y, \mathcal{B}, \mu)$. It should be noted that, although $\tau Y = Y$, τ is not invertible. In the same way as for the two-sided shift, it may be shown that the one-sided Bernoulli shift $B^+_{p_1,...,p_r}$ is always ergodic.

(iv) An example of some historical interest is the 'continued fraction' or *Gauss* map. Let $X = [0,1]$ be the unit interval and $T: X \to X$ the map defined (in the notation of §1) by

$$T\xi = \{\xi^{-1}\} \quad \text{if } \xi \in (0,1),$$
$$= 0 \quad \text{if } \xi = 0 \text{ or } 1.$$

Thus T acts as the shift operator on the continued fraction expansion of ξ: if

$$\xi = [0; a_1, a_2, ...] = \cfrac{1}{a_1 + \cfrac{1}{a_2 + ...}},$$

then $T\xi = [0; a_2, a_3, \ldots]$. (In the terminology of Chapter IV, the complete quotients of ξ are $\xi_{n+1} = 1/T^n\xi$.)

It is not difficult to show that T is a measure-preserving transformation of the probability space $(X, \mathscr{B}, \mu)$, where $\mathscr{B}$ is the family of Borel subsets of $X = [0,1]$ and μ is the 'Gauss' measure defined by

$$\mu(B) = (\log 2)^{-1}\int_B (1 + x)^{-1}\, dx.$$

It may further be shown that T is ergodic. Hence, by Birkhoff's ergodic theorem, if f is an integrable function on the interval X then, for almost all $\xi \in X$,

$$\lim_{n\to\infty} n^{-1} \sum_{k=0}^{n-1} f(T^k\xi) = (\log 2)^{-1}\int_X f(x)(1 + x)^{-1}\, dx.$$

Here it makes no difference if 'integrable' and 'almost all' refer to the invariant measure μ or to Lebesgue measure, since $1/2 \le (1 + x)^{-1} \le 1$.

Taking f to be the indicator function of the set $\{\xi \in X : a_1 = m\}$, we see that the asymptotic relative frequency of the positive integer m among the partial quotients $a_1, a_2, \ldots$ is almost always

$$(\log 2)^{-1} \int_{(m+1)^{-1}}^{m^{-1}} (1 + x)^{-1}\, dx = (\log 2)^{-1} \log ((m + 1)^2/(m(m + 2))).$$

It follows, in particular, that almost all $\xi \in X$ have unbounded partial quotients.

Again, by taking $f(\xi) = \log \xi$ it may be shown that, for almost all $\xi \in X$,

$$\lim_{n\to\infty} (1/n) \log q_n(\xi) = \pi^2/(12 \log 2),$$

where $q_n(\xi)$ is the denominator of the n-th convergent p_n/q_n of ξ. This was first proved by Lévy (1929).

In a letter to Laplace, Gauss (1812) stated that, for each $x \in (0,1)$, the proportion of $\xi \in X$ for which $T^n\xi < x$ converges as $n \to \infty$ to $\log (1 + x)/(\log 2)$ and he asked if Laplace could provide an estimate for the rapidity of convergence. If one writes

$$r_n(x) = m_n(x) - \log(1 + x)/(\log 2),$$

where $m_n(x)$ is the Lebesgue measure of the set of all $\xi \in X$ such that $T^n\xi < x$, then Gauss's statement is that $r_n(x) \to 0$ as $n \to \infty$ and his question is, how fast?

Gauss's statement was first proved by Kuz'min (1928), who also gave an estimate for the rapidity of convergence. If one regards Gauss's statement as a proposition in ergodic theory, then one needs to know that T is not only ergodic but even *mixing*, i.e. for all $A, B \in \mathscr{B}$,

$$\mu(T^{-n}A \cap B) \to \mu(A)\mu(B) \quad \text{as } n \to \infty.$$

Kuz'min's estimate $r_n(x) = O(q^{\sqrt{n}})$ for some $q \in (0,1)$ was improved by Lévy (1929) and Szüsz (1961) to $r_n(x) = O(q^n)$ with $q = 0.7$ and $q = 0.485$ respectively. A substantial advance was made by Wirsing (1974). By means of an infinite-dimensional generalization of a theorem of Perron (1907) and Frobenius (1908) on positive matrices, he showed that

$$r_n(x) = (-\lambda)^n \psi(x) + O(x(1-x)\mu^n),$$

where ψ is a twice continuously differentiable function with $\psi(0) = \psi(1) = 0$, $0 < \mu < \lambda$ and $\lambda = 0.303663...$. Wirsing's analysis has been extended by Babenko (1978) and Mayer (1990).

(v) Suppose we are given a system of ordinary differential equations

$$dx/dt = f(x), \tag{\dagger}$$

where $x \in \mathbb{R}^d$ and $f: \mathbb{R}^d \to \mathbb{R}^d$ is a continuously differentiable function. Then, for any $x \in \mathbb{R}^d$, there is a unique solution $\varphi_t(x)$ of ($\dagger$) such that $\varphi_0(x) = x$.

Suppose further that there exists an *invariant region* $X \subseteq \mathbb{R}^d$. That is, X is the closure of a bounded connected open set and $x \in X$ implies $\varphi_t(x) \in X$. Then the map $T_t: X \to X$ given by $T_t x = \varphi_t(x)$ is defined for every $t \in \mathbb{R}$ and satisfies $T_{t+s}x = T_t(T_s x)$.

Suppose finally that div $f = 0$ for every $x \in \mathbb{R}^d$, where $x = (x_1,...,x_d)$, $f = (f_1,...,f_d)$ and

$$\operatorname{div} f: = \sum_{k=1}^{d} \partial f_k/\partial x_k.$$

Then, by a theorem due to Liouville, the map T_t sends an arbitrary region into a region of the same volume. (For the statement and proof of Liouville's theorem see, for example, V.I. Arnold, *Mathematical methods of classical mechanics*, Springer-Verlag, New York, 1978.) It follows that if $\mathcal{B}$ is the family of Borel subsets of X and μ Lebesgue measure, normalized so that $\mu(X) = 1$, then T_t is a measure-preserving transformation of the probability space $(X, \mathcal{B}, \mu)$.

An important special case is the Hamiltonian system of ordinary differential equations

$$dp_i/dt = -\partial H/\partial q_i, \quad dq_i/dt = \partial H/\partial p_i \quad (i = 1,...,n),$$

where $H(p_1,...,p_n,q_1,...,q_n)$ is a twice continuously differentiable real-valued function. The divergence does indeed vanish identically in this case, since

$$-\sum_{i=1}^{n} \partial^2 H/\partial p_i \partial q_i + \sum_{i=1}^{n} \partial^2 H/\partial q_i \partial p_i = 0.$$

Furthermore, for any $h \in \mathbb{R}$, the energy surface $X: H(p,q) = h$ is invariant, since

$$dH[p(t),q(t)]/dt = \sum_{i=1}^{n} \partial H/\partial p_i(-\partial H/\partial q_i) + \sum_{i=1}^{n} \partial H/\partial q_i \partial H/\partial p_i = 0.$$

It is not difficult to show that if σ is the volume element on X induced by the Euclidean metric $\| \|$ on $\mathbb{R}^{2n}$, and if

$$\nabla H = (\partial H/\partial p_1,...,\partial H/\partial p_n,\partial H/\partial q_1,...,\partial H/\partial q_n)$$

is the gradient of H, then the maps T_t preserve the measure μ on X defined by

$$\mu(B) = \int_B d\sigma/\|\nabla H\|.$$

If X is compact, this measure can be normalized and we obtain a family of measure-preserving transformations T_t ($t \in \mathbb{R}$) of the corresponding probability space.

(vi) Many problems arising in mechanics may be reduced by a change of variables to the geometric problem of *geodesic flow*. If M is a smooth Riemannian manifold then the set of all pairs (x,v), where $x \in M$ and v is a unit vector in the tangent space to M at x, can be given the structure of a Riemannian manifold, the *unit tangent bundle* T_1M. Evidently T_1M is a $(2n{-}1)$-dimensional manifold if M is n-dimensional. There is a natural measure μ on T_1M such that $d\mu = dv_q d\omega_q$, where dv_q is the volume element at q of the Riemannian manifold M and ω_q is Lebesgue measure on the unit sphere S^{n-1} in the tangent space to M at x. If M is compact, then the measure μ can be normalized so that $\mu(T_1M) = 1$.

A *geodesic* on M is a curve $\gamma \subseteq M$ such that the length of every curve in M joining a point $x \in \gamma$ to any sufficiently close point $y \in \gamma$ is not less than the length of the arc of γ which joins x and y. Given any point $(x,v) \in T_1M$, there is a unique geodesic passing through x in the direction of v. The geodesic flow on T_1M is the flow $\varphi_t: T_1M \to T_1M$ defined by $\varphi_t(x,v) = (x_t,v_t)$, where x_t is the point of M reached from x after time t by travelling with unit speed along the geodesic determined by (x,v) and v_t is the unit tangent vector to this geodesic at x_t. If M is compact then, for every real t, φ_t is defined and is a measure-preserving transformation of the corresponding probability space $(T_1M,\mathcal{B},\mu)$.

The geodesics on a compact 2-dimensional manifold M whose curvature at each point is negative were profoundly studied by Hadamard (1898). It was first shown by E. Hopf (1939) that in this case φ_t is ergodic for every $t > 0$. (We must exclude $t = 0$, since φ_0 is the identity map.) This result has been considerably generalized by Anosov (1967) and others. In particular, the geodesic flow on a compact n-dimensional Riemannian manifold is ergodic if at each point the curvature of every 2-dimensional section is negative.

Although the preceding examples look quite different, some of them are not 'really' different, i.e. apart from sets of measure zero. More precisely, if $(X_1,\mathcal{B}_1,\mu_1)$ and $(X_2,\mathcal{B}_2,\mu_2)$ are probability spaces with measure-preserving transformations $T_1: X_1 \to X_1$ and $T_2: X_2 \to X_2$, we say that T_1 is *isomorphic* to T_2 if there exist sets $X_1' \in \mathcal{B}_1$, $X_2' \in \mathcal{B}_2$ with $\mu_1(X_1') = 1$, $\mu_2(X_2') = 1$ and $T_1 X_1' \subseteq X_1'$, $T_2 X_2' \subseteq X_2'$, and a bijective map φ of X_1' onto X_2' such that

(i) for any $B_1 \subseteq X_1'$, $B_1 \in \mathcal{B}_1$ if and only if $\varphi(B_1) \in \mathcal{B}_2$ and then $\mu_1(B_1) = \mu_2(\varphi(B_1))$;
(ii) $\varphi(T_1 x) = T_2 \varphi(x)$ for every $x \in X_1'$.

For example, it is easily shown that the Bernoulli shift $B_{p_1,\dots,p_r}$ is isomorphic to the following transformation of the unit square, equipped with Lebesgue measure. Divide the square into r vertical strips of width $p_1,\dots,p_r$; then contract the height of the i-th strip and expand its width so that it has height p_i and width 1; finally combine these rectangles to form the unit square again by regarding them as horizontal strips of height $p_1,\dots,p_r$. (For $r = 2$ and $p_1 = p_2 = 1/2$, this transformation of the unit square is allegedly used by bakers when kneading dough.)

It is easily shown also that isomorphism is an equivalence relation and that it preserves ergodicity. However, it is usually quite difficult to show that two measure-preserving transformations are indeed isomorphic. A period of rapid growth was initiated with the definition by Kolmogorov (1958), and its practical implementation by Sinai (1959), of a new numerical isomorphism invariant, the *entropy* of a measure-preserving transformation. For the formal definition of entropy we refer to the texts on ergodic theory cited at the end of the chapter. Here we merely state its value for some of the preceding examples.

Any translation T_a of the torus $\mathbb{R}^d/\mathbb{Z}^d$ has entropy zero, whereas the endomorphism R_A of $\mathbb{R}^d/\mathbb{Z}^d$ has entropy

$$\sum_{i:\, |\lambda_i| > 1} \log |\lambda_i|,$$

where $\lambda_1,\dots,\lambda_d$ are the eigenvalues of the matrix A and the summation is over those of them which lie outside the unit circle.

The two-sided Bernoulli shift $B_{p_1,\dots,p_r}$ has entropy

$$- \sum_{j=1}^{r} p_j \log p_j,$$

and the entropy of the one-sided Bernoulli shift $B^+_{p_1,\dots,p_r}$ is given by the same formula. It follows that $B_{1/2,1/2}$ is not isomorphic to $B_{1/3,1/3,1/3}$, since the first has entropy $\log 2$ and the second has entropy $\log 3$. Ornstein (1970) established the remarkable result that two-sided Bernoulli shifts are completely classified by their entropy: $B_{p_1,\dots,p_r}$ is isomorphic to $B_{q_1,\dots,q_s}$ if and only if

$$- \sum_{j=1}^{r} p_j \log p_j \; = \; - \sum_{k=1}^{s} q_k \log q_k.$$

This is no longer true for one-sided Bernoulli shifts. Walters (1973) has shown that $B^+_{p_1,...,p_r}$ is isomorphic to $B^+_{q_1,...,q_s}$ if and only if $r = s$ and $q_1,...,q_s$ is a permutation of $p_1,...,p_r$.

The Gauss map $Tx = \{x^{-1}\}$ has entropy $\pi^2/6 \log 2$. Although it is mixing, it is not isomorphic to a Bernoulli shift.

Katznelson (1971) showed that any ergodic automorphism of the torus $\mathbb{R}^d/\mathbb{Z}^d$ is isomorphic to a two-sided Bernoulli shift, and Lind (1977) has extended this result to ergodic automorphisms of any compact abelian group.

Ornstein and Weiss (1973) showed that, if φ_t is the geodesic flow on a smooth (of class C^3) compact two-dimensional Riemannian manifold whose curvature at each point is negative, then φ_t is isomorphic to a two-sided Bernoulli shift for every $t > 0$. Although, as Hilbert showed, a compact surface of negative curvature cannot be imbedded in $\mathbb{R}^3$, the geodesic flow on a surface of negative curvature can be realized as the motion of a particle constrained to move on a surface in $\mathbb{R}^3$ subject to centres of attraction and repulsion in the ambient space. The isomorphism with a Bernoulli shift shows that a deterministic mechanical system can generate a random process. Thus philosophical objections to 'Laplacian determinism' or to 'God playing dice' do not seem to have much point.

5 Recurrence

It was shown by Poincaré (1890) that the paths of a Hamiltonian system of differential equations almost always return to any neighbourhood, however small, of their initial points. Poincaré's proof was inevitably incomplete, since at the time measure theory did not exist. However, Carathéodory (1919) showed that his argument could be made rigorous with the aid of Lebesgue measure:

PROPOSITION 25   *Let $T: X \to X$ be a measure-preserving transformation of the probability space $(X, \mathcal{B}, \mu)$. Then almost all points of any $B \in \mathcal{B}$ return to B infinitely often, i.e. for each $x \in B$, apart from a set of μ-measure zero, there exists an increasing sequence (n_k) of positive integers such that $T^{n_k}x \in B$ ($k = 1,2,...$).*

Furthermore, if $\mu(B) > 0$, then $\mu(B \cap T^{-n}B) > 0$ for infinitely many $n \geq 1$.

Proof For any $N \geq 0$, put $B_N = \bigcup_{n \geq N} T^{-n}B$. Then

$$A: = \bigcap_{N \geq 0} B_N$$

is the set of all points $x \in X$ such that $T^n x \in B$ for infinitely many positive integers n. Since $B_{N+1} = T^{-1}B_N$, we have $\mu(B_{N+1}) = \mu(B_N)$ and hence $\mu(B_N) = \mu(B_0)$ for all $N \geq 1$. Since $B_{N+1} \subseteq B_N$, it follows that

$$\mu(A) = \lim_{N \to \infty} \mu(B_N) = \mu(B_0).$$

Since $A \subseteq B_0$, this implies

$$\mu(B_0 \setminus A) = \mu(B_0) - \mu(A) = 0$$

and hence, since $B \subseteq B_0$, $\mu(B \setminus A) = 0$.

This proves the first statement of the proposition. If $\mu(B \cap T^{-n}B) = 0$ for all $n \geq m$, then $\mu(B \cap B_N) = 0$ for all $N \geq m$ and hence

$$\mu(B \cap A) = \lim_{N \to \infty} \mu(B \cap B_N) = 0.$$

Consequently

$$\mu(B) = \mu(B \setminus A) + \mu(B \cap A) = 0,$$

which proves the second statement of the proposition. $\square$

Furstenberg (1977) extended Proposition 25 in the following way:

Let T be a measure-preserving transformation of the probability space $(X,\mathfrak{B},\mu)$. If $B \in \mathfrak{B}$ with $\mu(B) > 0$ and if $p \geq 2$, then $\mu(B \cap T^{-n}B \cap ... \cap T^{-(p-1)n}B) > 0$ for some $n \geq 1$.

Furstenberg's proof of his theorem made heavy use of ergodic theory and, in particular, of a new structure theory for measure-preserving transformations. From his theorem Furstenberg was able to deduce quite easily a result for which Szemeredi (1975) had given a complicated combinatorial proof:

Let S be a subset of the set $\mathbb{N}$ of positive integers which has positive upper density; i.e., for some $\alpha \in (0,1)$, there exist arbitrarily long intervals $I \subseteq \mathbb{N}$ containing at least $\alpha|I|$ elements of S. Then S contains arithmetic progressions of arbitrary finite length.

Furstenberg's approach to this result is not really shorter than Szemeredi's, but it is much more systematic. In fact the following generalization of Furstenberg's theorem was given soon afterwards by Furstenberg and Katznelson (1978):

If $T_1,...,T_p$ are commuting measure-preserving transformations of the probability space $(X,\mathfrak{B},\mu)$ and if $B \in \mathfrak{B}$ with $\mu(B) > 0$, then $\mu(B \cap T_1^{-n}B \cap ... \cap T_p^{-n}B) > 0$ for infinitely many $n \geq 1$.

Furstenberg and Katznelson could then deduce quite easily a multi-dimensional extension of Szemeredi's theorem which is still beyond the reach of combinatorial methods. Szemeredi's theorem was itself a far-reaching generalization of a famous theorem of van der Waerden (1927):

If $\mathbb{N} = S_1 \cup \ldots \cup S_r$ *is a partition of the set of all positive integers into finitely many subsets, then one of the subsets* S_j *contains arithmetic progressions of arbitrary finite length.*

Besides being more general, Szemeredi's result also shows how the subset S_j may be chosen.

Poincaré's measure-theoretic recurrence theorem has a topological counterpart due to Birkhoff (1912):

If X *is a compact metric space and* $T: X \rightarrow X$ *a continuous map, then there exists a point* $z \in X$ *and an increasing sequence* (n_k) *of positive integers such that* $T^{n_k}z \rightarrow z$ *as* $k \rightarrow \infty$.

Before Furstenberg and Katznelson proved their measure-theoretic theorem, Furstenberg and Weiss (1978) had already proved its topological counterpart:

If X *is a compact metric space and* $T_1,\ldots,T_p$ *commuting continuous maps of* X *into itself, then there exists a point* $z \in X$ *and an increasing sequence* (n_k) *of positive integers such that* $T_i^{n_k}z \rightarrow z$ *as* $k \rightarrow \infty$ $(i = 1,\ldots,p)$.

From their theorem Furstenberg and Weiss were able to deduce quite easily both van der Waerden's theorem and a known multi-dimensional generalization of it, due to Grünwald. It would take too long to prove here Szemeredi's theorem by the method of Furstenberg and Katznelson, but we will prove van der Waerden's theorem by the method of Furstenberg and Weiss. The proof illustrates how results in one area of mathematics can find application in another area which is apparently unrelated.

PROPOSITION 26 *Let* (X,d) *be a compact metric space and* $T: X \rightarrow X$ *a continuous map. Then, for any real* $\varepsilon > 0$ *and any* $p \in \mathbb{N}$, *there exists some* $z \in X$ *and* $n \in \mathbb{N}$ *such that*

$$\text{d}(T^n z, z) < \varepsilon, \quad \text{d}(T^{2n} z, z) < \varepsilon, \quad \ldots, \quad \text{d}(T^{pn} z, z) < \varepsilon.$$

Proof (i) A subset A of X is said to be *invariant* under T if $TA \subseteq A$. The closure $\overline{A}$ of an invariant set A is again invariant since, by the continuity of T, $T\overline{A} \subseteq \overline{TA}$. Let $\mathscr{F}$ be the collection of all nonempty closed invariant subsets of X. Clearly $\mathscr{F}$ is not empty, since $X \in \mathscr{F}$.

If we regard $\mathscr{F}$ as partially ordered by inclusion then, by *Hausdorff's maximality theorem*, $\mathscr{F}$ contains a maximal totally ordered subcollection $\mathscr{T}$. The intersection Z of all the subsets in $\mathscr{T}$ is both closed and invariant. It is also nonempty, since X is compact. Hence $Z \in \mathscr{T}$ and, by construction, no nonempty proper closed subset of Z is invariant.

By replacing X by its compact subset Z we may now assume that the only closed invariant subsets of X itself are X and $\varnothing$.

(ii) For any given $z \in X$, the closure of the set $(T^n z)_{n \geq 1}$ is a nonempty closed invariant subset of X and therefore coincides with X. Thus for every $\varepsilon > 0$ there exists $n = n(\varepsilon) \geq 1$ such that $d(T^n z, z) < \varepsilon$. This proves the proposition for $p = 1$.

We suppose now that $p > 1$ and the proposition holds with p replaced by $p - 1$.

(iii) We show next that, for any $\varepsilon > 0$, there exists a finite set K of positive integers such that, for all $x, x' \in X$,

$$d(T^k x', x) < \varepsilon/2 \quad \text{for some } k \in K.$$

If B is a nonempty open subset of X, then for every $z \in X$ there exists some $n \geq 1$ such that $T^n z \in B$. Hence $X = \bigcup_{n \geq 1} T^{-n} B$. Since X is compact and the sets $T^{-n} B$ are open, there is a finite set $K(B)$ of positive integers such that

$$X = \bigcup_{k \in K(B)} T^{-k} B.$$

Since X is compact again, there exist finitely many open balls $B_1, ..., B_r$ with radius $\varepsilon/4$ such that $X = B_1 \cup ... \cup B_r$. If $x, x' \in X$, then $x \in B_i$ for some $i \in \{1, ..., r\}$ and $x' \in T^{-k} B_i$ for some $k \in K(B_i)$. Thus we can take $K = K(B_1) \cup ... \cup K(B_r)$.

(iv) We now show that, for any $\varepsilon > 0$ and any $x \in X$, there exists $y \in X$ and $n \geq 1$ such that

$$d(T^n y, x) < \varepsilon, \quad d(T^{2n} y, x) < \varepsilon, \quad ... , \quad d(T^{pn} y, x) < \varepsilon.$$

In fact, since each T^k $(k \in K)$ is uniformly continuous on X, we can choose $\rho > 0$ so that $d(x_1, x_2) < \rho$ implies $d(T^k x_1, T^k x_2) < \varepsilon/2$ for all $x_1, x_2 \in X$ and all $k \in K$. By the induction hypothesis, there exist $x' \in X$ and $n \geq 1$ such that

$$d(T^n x', x') < \rho, \quad ... , \quad d(T^{(p-1)n} x', x') < \rho.$$

But the invariant set TX is closed, since X is compact, and so $TX = X$. Hence $T^n X = X$ and we can choose $y' \in X$ so that $T^n y' = x'$. Thus

$$d(T^n y', x') = 0, \quad d(T^{2n} y', x') < \rho, \quad ... , \quad d(T^{pn} y', x') < \rho.$$

It follows that, for all $k \in K$,

$$d(T^{n+k}y', T^k x') < \varepsilon/2, \; \ldots \; , \; d(T^{pn+k}y', T^k x') < \varepsilon/2.$$

For each $x \in X$ there is a $k \in K$ such that $d(T^k x', x) < \varepsilon/2$. Thus if $y = T^k y'$, then

$$d(T^n y, x) < \varepsilon, \; \ldots \; , \; d(T^{pn} y, x) < \varepsilon.$$

(v) Let $\varepsilon_0 > 0$ and $x_0 \in X$ be given. By (iv) there exist $x_1 \in X$ and $n_1 \geq 1$ such that

$$d(T^{n_1} x_1, x_0) < \varepsilon_0, \; \ldots \; , \; d(T^{pn_1} x_1, x_0) < \varepsilon_0.$$

We can now choose $\varepsilon_1 \in (0, \varepsilon_0)$ so that $d(x, x_1) < \varepsilon_1$ implies

$$d(T^{n_1} x, x_0) < \varepsilon_0, \; \ldots \; , \; d(T^{pn_1} x, x_0) < \varepsilon_0.$$

Suppose we have defined points $x_1, \ldots, x_k$, positive integers $n_1, \ldots, n_k$, and $\varepsilon_1, \ldots, \varepsilon_k \in (0, \varepsilon_0)$ such that, for $i = 1, \ldots, k$,

$$d(T^{n_i} x_i, x_{i-1}) < \varepsilon_{i-1}, \; \ldots \; , \; d(T^{pn_i} x_i, x_{i-1}) < \varepsilon_{i-1},$$

and $d(x, x_i) < \varepsilon_i$ implies

$$d(T^{n_i} x, x_{i-1}) < \varepsilon_{i-1}, \; \ldots \; , \; d(T^{pn_i} x, x_{i-1}) < \varepsilon_{i-1}.$$

By (iv) there exist $x_{k+1} \in X$ and $n_{k+1} \geq 1$ such that

$$d(T^{n_{k+1}} x_{k+1}, x_k) < \varepsilon_k, \; \ldots \; , \; d(T^{pn_{k+1}} x_{k+1}, x_k) < \varepsilon_k,$$

and we can then choose $\varepsilon_{k+1} \in (0, \varepsilon_0)$ so that $d(x, x_{k+1}) < \varepsilon_{k+1}$ implies

$$d(T^{n_{k+1}} x, x_k) < \varepsilon_k, \; \ldots \; , \; d(T^{pn_{k+1}} x, x_k) < \varepsilon_k.$$

Thus the process can be continued indefinitely.

By taking successively $i = j - 1, j - 2, \ldots$ we see that, if $i < j$, then

$$d(T^{n_{i+1} + \ldots + n_{j-1} + n_j} x_j, x_i) < \varepsilon_i, \; \ldots \; , \; d(T^{p(n_{i+1} + \ldots + n_{j-1} + n_j)} x_j, x_i) < \varepsilon_i.$$

Since X is compact, it is covered by a finite number r of open balls with radius $\varepsilon_0/2$. Hence there exist i, j with $0 \leq i < j \leq r$ such that $d(x_i, x_j) < \varepsilon_0$. If we put $n = n_{i+1} + \ldots + n_{j-1} + n_j$ then, since $\varepsilon_i < \varepsilon_0$, we obtain from the triangle inequality

$$d(T^n x_j, x_j) < 2\varepsilon_0, \; \ldots \; , \; d(T^{pn} x_j, x_j) < 2\varepsilon_0.$$

But $\varepsilon_0 > 0$ was arbitrary. $\square$

It may be deduced from Proposition 26, by means of *Baire's category theorem*, that under the same hypotheses there exists a point $z \in X$ and an increasing sequence (n_k) of positive integers such that $T^{in_k}z \to z$ as $k \to \infty$ $(i = 1,...,p)$. However, as we now show, Proposition 26 already suffices to proves van der Waerden's theorem.

The set X^* of all infinite sequences $x = (x_1,x_2,...)$, where $x_i \in \{1,2,...,r\}$ for every $i \geq 1$, can be given the structure of a compact metric space by defining $d(x,x) = 0$ and $d(x,y) = 2^{-k}$ if $x \neq y$ and k is the least positive integer such that $x_k \neq y_k$. The shift map $\tau: X^* \to X^*$, defined by $\tau((x_1,x_2,...)) = (x_2,x_3,...)$, is continuous, since

$$d(\tau(x),\tau(y)) \leq 2\, d(x,y).$$

With the partition $\mathbb{N} = S_1 \cup ... \cup S_r$ in the statement of van der Waerden's theorem we associate the infinite sequence $x \in X^*$ defined by $x_i = j$ if $i \in S_j$.

Let X denote the closure of the set $(\tau^n x)_{n \geq 1}$. Then X is a closed subset of X^* which is invariant under τ. By Proposition 26, there exists a point $z \in X$ and a positive integer n such that

$$d(\tau^n z,z) < 1/2, \;\; d(\tau^{2n}z,z) < 1/2, \; ... \; , \;\; d(\tau^{pn}z,z) < 1/2;$$

i.e. $z_1 = z_{n+1} = z_{2n+1} = ... = z_{pn+1}$. Since $z \in X$, there is a positive integer m such that $d(\tau^m x,z) < 2^{-pn-1}$, i.e. $x_{m+i} = z_i$ for $1 \leq i \leq pn + 1$. It follows that

$$x_{m+1} = x_{m+n+1} = ... = x_{m+pn+1}.$$

Thus for every positive integer p there is a set $S_{j(p)}$ which contains an arithmetic progression of length p. Since there are only r possible values for $j(p)$, one of the sets S_j must contain arithmetic progressions of arbitrary finite length.

A far-reaching generalization of van der Waerden's theorem has been given by Hales and Jewett (1963). Let $A = \{a_1,...,a_q\}$ be a finite set and let A^n be the set of all n-tuples with elements from A. A set $W = \{w^1,...,w^q\} \subseteq A^n$ of q n-tuples $w^k = (w_1^k,...,w_n^k)$ is said to be a *combinatorial line* if there exists a partition

$$\{1,...,n\} = I \cup J, \;\; I \cap J = \varnothing,$$

such that

$$w_i^k = a_k \;\; (k = 1,...,q) \text{ for } i \in I; \;\; w_j^1 = ... = w_j^q \;\; \text{for } j \in J.$$

The Hales–Jewett theorem says that, for any positive integer r, there exists a positive integer $N = N(q,r)$ such that, if A^N is partitioned into r classes, then at least one of these classes contains a combinatorial line.

If one takes $A = \{0,1,...,q-1\}$ and interprets A^n as the set of expansions to base q of all non-negative integers less than q^n, then a combinatorial line is an arithmetic progression. On the other hand, if one takes $A = \mathbb{F}_q$ to be a finite field with q elements and interprets A^n as the n-dimensional vector space $\mathbb{F}_q^n$, then a combinatorial line is an affine line. The interesting feature of the Hales–Jewett theorem is that it is purely combinatorial and does not involve any notion of addition.

6 Further remarks

Uniform distribution and discrepancy are thoroughly discussed in Kuipers and Niederreiter [30]. For later results, see Drmota and Tichy [13]. Since these two books have extensive bibliographies, we will be sparing with references. However, it would be remiss not to recommend the great paper of Weyl [52], which remains as fresh as when it was written.

Lemma 0 is often attributed to Polya (1920), but it was already proved by Buchanan and Hildebrandt [9].

Fejér's proof that continuous periodic functions can be uniformly approximated by trigonometric polynomials is given in Dym and McKean [15]. The theorem also follows directly from the the theorem of Weierstrass (1885) on the uniform approximation of continuous functions by ordinary polynomials. A remarkable generalization of both results was given by Stone (1937); see Stone [49]. The 'Stone–Weierstrass theorem' is also proved in Rudin [44], for example.

Chen [11] gives a quantitative version of Kronecker's theorem of a different type from Proposition 3'.

The converse of Proposition 10 is proved by Kemperman [27]. For the history of the problem of mean motion, and generalizations to almost periodic functions, see Jessen and Tornehave [24]. Methods for estimating exponential sums were developed in connection with the theory of uniform distribution, but then found other applications. See Chandrasekharan [10] and Graham and Kolesnik [21].

For applications of discrepancy to numerical integration, see Niederreiter [36],[37]. For the basic properties of functions of bounded variation and the definition of total variation see, for example, Riesz and Sz.-Nagy [42].

Sharper versions of the original Erdös–Turan inequality are proved by Niederreiter and Philipp [38] and in Montgomery [35]. The discrepancy of the sequence $(\{n\alpha\})$, where α is an irrational number whose continued fraction expansion has bounded partial quotients (i.e., is *badly approximable*), is discussed by Dupain and Sós [14]. The discrepancy of the sequence ($\{n\alpha\}$), where $\alpha \in \mathbb{R}^d$, has been deeply studied by Beck [3]. The work of Roth, Schmidt and others is treated in Beck and Chen [4].

For accounts of measure theory, see Billingsley [6], Halmos [22], Loève [32] and Saks [46]. More detailed treatments of ergodic theory are given in the books of Petersen [39], Walters [51] and Cornfeld *et al.* [12]. The prehistory of ergodic theory is described by the Ehrenfests [16]. However, they do not refer to the paper of Poincaré (1894), which is reproduced in [41].

The proof of Birkhoff's ergodic theorem given here follows Katznelson and Weiss [26]. A different proof is given in the book of Walters.

Many other ergodic theorems besides Birkhoff's are discussed in Krengel [29]. We mention only the *subadditive ergodic theorem* of Kingman (1968): if T is a measure-preserving transformation of the probability space $(X,\mathcal{B},\mu)$ and if (g_n) is a sequence of functions in $L(X,\mathcal{B},\mu)$ such that $\inf_n n^{-1} \int_X g_n \, d\mu > -\infty$ and, for all $m,n \geq 1$,

$$g_{n+m}(x) \leq g_n(x) + g_m(T^n x) \quad \text{a.e.,}$$

then $n^{-1} g_n(x) \to g^*(x)$ a.e., where $g^*(Tx) = g^*(x)$ a.e., $g^* \in L(X,\mathcal{B},\mu)$ and

$$\int_X g^* \, d\mu = \lim_{n\to\infty} n^{-1} \int_X g_n \, d\mu = \inf_n n^{-1} \int_X g_n \, d\mu.$$

Birkhoff's ergodic theorem may be regarded as a special case by taking $g_n(x) = \sum_{k=0}^{n-1} f(T^k x)$. A simple proof of Kingman's theorem is given by Steele [48]. For applications of Kingman's theorem to percolation processes and products of random matrices, see Kingman [28]. The multiplicative ergodic theorem of Oseledets is derived from Kingman's theorem by Ruelle [45].

The book of Kuipers and Niederreiter cited above has an extensive discussion of normal numbers. For normality with respect to a matrix, see also Brown and Moran [8].

Proofs of Gauss's statement on the continued fraction map are contained in the books by Billingsley [7] and Rockett and Szusz [43]. For more recent work, see Wirsing [53], Babenko [2] and Mayer [33]. For the deviation of $(1/n) \log q_n(\xi)$ from its (a.e.) limiting value $\pi^2/(12 \log 2)$ there are analogues of the central limit theorem and the law of the iterated logarithm; see Philipp and Stackelberg [40]. For higher-dimensional generalizations of Gauss's invariant measure, see Hardcastle and Khanin [23].

Applications of ergodic theory to classical mechanics are discussed in the books of Arnold and Avez [1] and Katok and Hasselblatt [25]. For connections between ergodic theory and the '$3x + 1$ problem', see Lagarias [31].

Ergodic theory has been used to generalize considerably some of the results on lattices in Chapter VIII. A *lattice* in a locally compact group G is a discrete subgroup Γ such that the G-invariant measure of the quotient space G/Γ is finite. (In Chapter VIII, $G = \mathbb{R}^n$ and $\Gamma = \mathbb{Z}^n$.) Zimmer [54] gives a good introduction to the results which have been obtained in this area.

An attractive account of the work of Furstenberg and his collaborators is given in Furstenberg [17]. See also Graham *et al.* [20] and the book of Petersen cited above. The discovery of van der Waerden's theorem is described in van der Waerden [50]. For a recent direct proof, see Mills [34].

The direct proofs reduce the theorem to an equivalent finite form: *for any positive integer p, there exists a positive integer N such that, whenever the set* $\{1,2,...,N\}$ *is partitioned into two subsets, at least one subset contains an arithmetic progression of length p.* The original proofs provided an upper bound for the least possible value $N(p)$ of N, but it was unreasonably large. Some progress towards obtaining reasonable upper bounds has recently been made by Shelah [47] and Gowers [19].

The Hales–Jewett theorem is proved, and then extensively generalized, in Bergelson and Leibman [5]. Furstenberg and Katznelson [18] prove a density version of the Hales–Jewett theorem, analogous to Szemeredi's density version of van der Waerden's theorem.

7 Selected references

[1] V.I. Arnold and A. Avez, *Ergodic problems of classical mechanics*, Benjamin, New York, 1968.

[2] K.I. Babenko, On a problem of Gauss, *Soviet Math. Dokl.* **19** (1978), 136-140.

[3] J. Beck, Probabilistic diophantine approximation, I. Kronecker sequences, *Ann. of Math.* **140** (1994), 451-502.

[4] J. Beck and W.W.L. Chen, *Irregularities of distribution*, Cambridge University Press, 1987.

[5] V. Bergelson and A. Leibman, Set polynomials and polynomial extension of the Hales–Jewett theorem, *Ann. of Math.* **150** (1999), 33-75.

[6] P. Billingsley, *Probability and measure*, 3rd ed., Wiley, New York, 1995.

[7]   P. Billingsley, *Ergodic theory and information*, reprinted, Krieger, Huntington, N.Y., 1978.

[8]   G. Brown and W. Moran, Schmidt's conjecture on normality for commuting matrices, *Invent. Math.* **111** (1993), 449-463.

[9]   H.E. Buchanan and H.T. Hildebrandt, Note on the convergence of a sequence of functions of a certain type, *Ann. of Math.* **9** (1908), 123-126.

[10] K. Chandrasekharan, Exponential sums in the development of number theory, *Proc. Steklov Inst. Math.* **132** (1973), 3-24.

[11] Y.-G. Chen, The best quantitative Kronecker's theorem, *J. London Math. Soc.* (2) **61** (2000), 691-705.

[12] I.P. Cornfeld, S.V. Fomin and Ya. G. Sinai, *Ergodic theory*, Springer-Verlag, New York, 1982.

[13] M. Drmota and R.F. Tichy, *Sequences, discrepancies and applications*, Lecture Notes in Mathematics **1651**, Springer, Berlin, 1997.

[14] I. Dupain and V.T. Sós, On the discrepancy of $(n\alpha)$ sequences, *Topics in classical number theory* (ed. G. Halász), Vol. I, pp. 355-387, North-Holland, Amsterdam, 1984.

[15] H. Dym and H.P. McKean, *Fourier series and integrals*, Academic Press, Orlando, FL, 1972.

[16] P. and T. Ehrenfest, *The conceptual foundations of the statistical approach in mechanics*, English translation by M.J. Moravcsik, Cornell University Press, Ithaca, 1959. [German original, 1912]

[17] H. Furstenberg, *Recurrence in ergodic theory and combinatorial number theory*, Princeton University Press, 1981.

[18] H. Furstenberg and Y. Katznelson, A density version of the Hales–Jewett theorem, *J. Analyse Math.* **57** (1991), 64-119.

[19] W.T. Gowers, A new proof of Szemeredi's theorem, *Geom. Funct. Anal.* **11** (2001), 465-588.

[20] R.L. Graham, B.L. Rothschild and J.H. Spencer, *Ramsey theory*, 2nd ed., Wiley, New York, 1990.

[21] S.W. Graham and G. Kolesnik, *Van der Corput's method of exponential sums*, London Math. Soc. Lecture Notes **126**, Cambridge University Press, 1991.

[22] P.R. Halmos, *Measure theory*, 2nd printing, Springer-Verlag, New York, 1974.

[23] D.M. Hardcastle and K. Khanin, Continued fractions and the d-dimensional Gauss transformation, *Comm. Math. Phys.* **215** (2001), 487-515.

[24] B. Jessen and H. Tornehave, Mean motion and zeros of almost periodic functions, *Acta Math.* **77** (1945), 137-279.

[25] A. Katok and B. Hasselblatt, *Introduction to the modern theory of dynamical systems*, Cambridge University Press, 1995.

[26] Y. Katznelson and B. Weiss, A simple proof of some ergodic theorems, *Israel J. Math.* **42** (1982), 291-296.

[27] J.H.B. Kemperman, Distributions modulo 1 of slowly changing sequences, *Nieuw Arch. Wisk.* (3) **21** (1973), 138-163.

[28] J.F.C. Kingman, Subadditive processes, *Ecole d'Eté de Probabilités de Saint-Flour* V-1975 (ed. A. Badrikian), pp. 167-223, Lecture Notes in Mathematics **539**, Springer-Verlag, 1976.

[29] U. Krengel, *Ergodic theorems*, de Gruyter, Berlin, 1985.

[30] L. Kuipers and H. Niederreiter, *Uniform distribution of sequences*, Wiley, New York, 1974.

[31] J.C. Lagarias, The $3x + 1$ problem and its generalizations, *Amer. Math. Monthly* **92** (1985), 3-23.

[32] M. Loève, *Probability theory*, 4th ed. in 2 vols., Springer-Verlag, New York, 1978.

[33] D.H. Mayer, On the thermodynamic formalism for the Gauss map, *Comm. Math. Phys.* **130** (1990), 311-333.

[34] G. Mills, A quintessential proof of van der Waerden's theorem on arithmetic progressions, *Discrete Math.* **47** (1983), 117-120.

[35] H.L. Montgomery, *Ten lectures on the interface between analytic number theory and harmonic analysis*, CBMS Regional Conference Series in Mathematics **84**, American Mathematical Society, Providence, R.I., 1994.

[36] H. Niederreiter, Quasi-Monte Carlo methods and pseudo-random numbers, *Bull. Amer. Math. Soc.* **84** (1978), 957-1041.

[37] H. Niederreiter, *Random number generation and quasi-Monte Carlo methods*, CBMS–NSF Regional Conference Series in Applied Mathematics **63**, SIAM, Philadelphia, 1992.

[38] H. Niederreiter and W. Philipp, Berry–Esseen bounds and a theorem of Erdös and Turán on uniform distribution mod 1, *Duke Math. J.* **40** (1973), 633-649.

[39] K. Petersen, *Ergodic theory*, Cambridge University Press, 1983.

[40] W. Philipp and O.P. Stackelberg, Zwei Grenzwertsätze für Kettenbrüche, *Math. Ann.* **181** (1969), 152-156.

[41] H. Poincaré, Sur la théorie cinétique des gaz, *Oeuvres*, t. X, pp. 246-263, Gauthier-Villars, Paris, 1954.

[42] F. Riesz and B. Sz.-Nagy, *Functional analysis*, English transl. by L.F. Boron, Ungar, New York, 1955.

[43] A. Rockett and P. Szusz, *Continued fractions*, World Scientific, Singapore, 1992.

[44] W. Rudin, *Principles of mathematical analysis*, 3rd ed., McGraw-Hill, New York, 1976.

[45] D. Ruelle, Ergodic theory of differentiable dynamical systems, *Inst. Hautes Études Sci. Publ. Math.* **50** (1979), 27-58.

[46] S. Saks, *Theory of the integral*, 2nd revised ed., English transl. by L.C. Young, reprinted, Dover, New York, 1964.

[47] S. Shelah, Primitive recursion bounds for van der Waerden numbers, *J. Amer. Math. Soc.* **1** (1988), 683-697.

[48] J.M. Steele, Kingman's subadditive ergodic theorem, *Ann. Inst. H. Poincaré Sect. B* **25** (1989), 93-98.

[49] M.H. Stone, A generalized Weierstrass approximation theorem, *Studies in modern analysis* (ed. R.C. Buck), pp. 30-87, Mathematical Association of America, 1962.

[50] B.L. van der Waerden, How the proof of Baudet's conjecture was found, *Studies in Pure Mathematics* (ed. L. Mirsky), pp. 251-260, Academic Press, London, 1971.

[51] P. Walters, *An introduction to ergodic theory*, Springer-Verlag, New York, 1982.

[52] H. Weyl, Über die Gleichverteilung von Zahlen mod Eins, *Math. Ann.* **77** (1916), 313-352. [Reprinted in *Selecta Hermann Weyl*, pp. 111-147, Birkhäuser, Basel, 1956 and in *Hermann Weyl, Gesammelte Abhandlungen* (ed. K. Chandrasekharan), *Band I*, pp. 563-599, Springer-Verlag, Berlin, 1968]

[53] E. Wirsing, On the theorem of Gauss–Kusmin–Lévy and a Frobenius type theorem for function spaces, *Acta Arith.* **24** (1974), 507-528.

[54] R.J. Zimmer, *Ergodic theory and semi-simple groups*, Birkhäuser, Boston, 1984.

XII
Elliptic functions

Our discussion of elliptic functions may be regarded as an essay in revisionism, since we do not use Liouville's theorem, Riemann surfaces or the Weierstrassian functions. We wish to show that the methods used by the founding fathers of the subject provide a natural and rigorous approach, which is very well suited for applications.

The work is arranged so that the initial sections are mutually independent, although motivation for each section is provided by those which precede it. To some extent we have also separated the discussion for real and for complex parameters, so that those interested only in the real case may skip the complex one.

1 Elliptic integrals

After the development of the integral calculus in the second half of the 17th century, it was natural to apply it to the determination of the arc length of an ellipse since, by Kepler's first law, the planets move in elliptical orbits with the sun at one focus.

An ellipse is described in rectangular coordinates by an equation

$$x^2/a^2 + y^2/b^2 = 1,$$

where a and b are the *semi-axes* of the ellipse $(a > b > 0)$. It is also given parametrically by

$$x = a \sin \theta, \ y = b \cos \theta \ (0 \le \theta \le 2\pi).$$

The arc length $s(\Theta)$ from $\theta = 0$ to $\theta = \Theta$ is given by

$$s(\Theta) = \int_0^\Theta [(dx/d\theta)^2 + (dy/d\theta)^2]^{1/2} \, d\theta$$

$$= \int_0^\Theta (a^2 \cos^2\theta + b^2 \sin^2\theta)^{1/2} \, d\theta$$

$$= \int_0^\Theta [a^2 - (a^2 - b^2) \sin^2\theta]^{1/2} \, d\theta.$$

If we put $b^2 = a^2(1 - k^2)$, where k $(0 < k < 1)$ is the *eccentricity* of the ellipse, this takes the form

$$s(\Theta) = a \int_0^\Theta (1 - k^2 \sin^2\theta)^{1/2} \, d\theta.$$

If we further put $z = \sin \theta = x/a$ and restrict attention to the first quadrant, this assumes the algebraic form

$$a \int_0^Z [(1 - k^2 z^2)/(1 - z^2)]^{1/2} \, dz.$$

Since the arc length of the whole quadrant is obtained by taking $Z = 1$, the arc length of the whole ellipse is

$$4a \int_0^1 [(1 - k^2 z^2)/(1 - z^2)]^{1/2} \, dz.$$

Consider next Galileo's problem of the simple pendulum. If θ is the angle of deflection from the downward vertical, the equation of motion of the pendulum is

$$d^2\theta/dt^2 + (g/l) \sin \theta = 0,$$

where l is the length of the pendulum and g is the gravitational constant. This differential equation has the first integral

$$(d\theta/dt)^2 = (2g/l)(\cos \theta - a),$$

where a is a constant. In fact $a < 1$ for a real motion, and for oscillatory motion we must also have $a > -1$. We can then put $a = \cos \alpha$ $(0 < \alpha < \pi)$, where α is the maximum value of θ, and integrate again to obtain

$$t = (l/2g)^{1/2} \int_0^\Theta (\cos \theta - \cos \alpha)^{-1/2} \, d\theta$$

$$= (l/4g)^{1/2} \int_0^\Theta (\sin^2 \alpha/2 - \sin^2 \theta/2)^{-1/2} \, d\theta.$$

Putting $k = \sin \alpha/2$ and $kx = \sin \theta/2$, we can rewrite this in the form

$$t = (l/g)^{1/2} \int_0^X [(1 - k^2 x^2)(1 - x^2)]^{-1/2} \, dx.$$

The angle of deflection θ attains its maximum value α when $X = 1$, and the motion is periodic with period

$$T = 4(l/g)^{1/2} \int_0^1 [(1 - k^2 x^2)(1 - x^2)]^{-1/2} \, dx.$$

Attempts to evaluate the integrals in both these problems in terms of algebraic and elementary transcendental functions proved fruitless. Thus the idea arose of treating them as fundamental entities in terms of which other integrals could be expressed.

An example is the determination of the arc length of a *lemniscate*. This curve, which was studied by Jacob Bernoulli (1694), has the form of a figure of eight and is the locus of all points $z \in \mathbb{C}$ such that $|2z^2 - 1| = 1$ or, in polar coordinates,

$$r^2 = \cos 2\theta \qquad (-\pi/4 \leq \theta \leq \pi/4 \cup 3\pi/4 \leq \theta \leq 5\pi/4).$$

If $-\pi/4 \leq \Theta \leq 0$, the arc length $s(\Theta)$ from $\theta = -\pi/4$ to $\theta = \Theta$ is given by

$$
\begin{aligned}
s(\Theta) &= \int_{-\pi/4}^{\Theta} [r^2 + (dr/d\theta)^2]^{1/2} \, d\theta \\
&= \int_{-\pi/4}^{\Theta} [r^2 + (1 - r^4)/r^2]^{1/2} \, d\theta \\
&= \int_0^R (1 - r^4)^{-1/2} \, dr.
\end{aligned}
$$

If we make the change of variables $x = \sqrt{2}r/(1 + r^2)^{1/2}$, then $dx/dr = \sqrt{2}/(1 + r^2)^{3/2}$ and

$$s(\Theta) = 2^{-1/2} \int_0^X [(1 - x^2/2)(1 - x^2)]^{-1/2} \, dx.$$

Another example is the determination of the surface area of an ellipsoid. Suppose the ellipsoid is described in rectangular coordinates by the equation

$$x^2/a^2 + y^2/b^2 + z^2/c^2 = 1,$$

where $a > b > c > 0$. The total surface area is $8S$, where S is the surface area of the part contained in the positive octant. In this octant we have

$$z = c[1 - (x/a)^2 - (y/b)^2]^{1/2}$$

and hence

$$1 + (\partial z/\partial x)^2 + (\partial z/\partial y)^2 = [1 - (\alpha x/a)^2 - (\beta y/b)^2]/[1 - (x/a)^2 - (y/b)^2],$$

where

$$\alpha = (a^2 - c^2)^{1/2}/a, \quad \beta = (b^2 - c^2)^{1/2}/b.$$

Consequently

$$S = \int_0^a \int_0^{b(1-(x/a)^2)^{1/2}} [1 - (\alpha x/a)^2 - (\beta y/b)^2]^{1/2} [1 - (x/a)^2 - (y/b)^2]^{-1/2} \, dy \, dx.$$

If we make the change of variables

$$x = ar \cos \theta, \quad y = br \sin \theta,$$

with Jacobian $J = abr$, we obtain

$$S = ab \int_0^{\pi/2} d\theta \int_0^1 (1 - \sigma r^2)^{1/2} (1 - r^2)^{-1/2} r \, dr,$$

where
$$\sigma = \alpha^2 \cos^2 \theta + \beta^2 \sin^2 \theta.$$

If we now put
$$u^2 = (1 - r^2)/(1 - \sigma r^2),$$

then $r^2 = (1 - u^2)/(1 - \sigma u^2)$ and

$$r \, dr/du = -(1 - \sigma)u/(1 - \sigma u^2)^2.$$

Hence
$$S = ab \int_0^{\pi/2} d\theta \int_0^1 (1 - \sigma)(1 - \sigma u^2)^{-2} \, du.$$

Inverting the order of integration and giving σ its value, we obtain

$$S = ab \int_0^1 du \int_0^{\pi/2} [(1-\alpha^2) \cos^2\theta + (1-\beta^2) \sin^2 \theta][(1 - \alpha^2 u^2)\cos^2\theta + (1 - \beta^2 u^2)\sin^2 \theta]^{-2} \, d\theta.$$

It is readily verified that

$$\int_0^{\pi/2} \cos^2\theta \, (m \cos^2\theta + n \sin^2 \theta)^{-2} \, d\theta = \pi/4m(mn)^{1/2},$$

$$\int_0^{\pi/2} \sin^2\theta \, (m \cos^2\theta + n \sin^2 \theta)^{-2} \, d\theta = \pi/4n(mn)^{1/2}.$$

Thus we obtain finally

$$S = (\pi ab/4) \int_0^1 [(1-\alpha^2)/(1 - \alpha^2 u^2) + (1-\beta^2)/(1 - \beta^2 u^2)][(1 - \alpha^2 u^2)(1 - \beta^2 u^2)]^{-1/2} \, du.$$

By an *elliptic integral* one understands today any integral of the form

$$\int R(x,w) \, dx,$$

where $R(x,w)$ is a rational function of x and w, and where $w^2 = g(x)$ is a polynomial in x of degree 3 or 4 without repeated roots. The elliptic integral is said to be *complete* if it is a definite integral in which the limits of integration are distinct roots of $g(x)$.

The case of a quartic is easily reduced to that of a cubic. In the preceding examples we can simply put $y = x^2$. Thus, for the lemniscate,

$$s(\Theta) = 2^{-1/2} \int_0^Y [4y(1- y)(1 - y/2)]^{-1/2} dy.$$

In general, suppose $g(x) = (x - \alpha)h(x)$, where h is a cubic. If

$$h(x) = h_0(x - \alpha)^3 + h_1(x - \alpha)^2 + h_2(x - \alpha) + h_3$$

and we make the change of variables $x = \alpha + 1/y$, then $g(x) = g^*(y)/y^4$, where

$$g^*(y) = h_0 + h_1 y + h_2 y^2 + h_3 y^3,$$

and

$$\int R(x,w)\, dx \;=\; \int R^*(y,v)\, dy,$$

where $R^*(y,v)$ is a rational function of y and v, and $v^2 = g^*(y)$.

Since any even power of w is a polynomial in x, the integrand can be written in the form $R(x,w) = (A + Bw)/(C + Dw)$, where A,B,C,D are polynomials in x. Multiplying numerator and denominator by $(C - Dw)w$, we obtain

$$R(x,w) \;=\; N/L + M/Lw,$$

where L,M,N are polynomials in x. By decomposing the rational function N/L into partial fractions its integral can be evaluated in terms of rational functions and (real or complex) logarithms. By similarly decomposing the rational function M/L into partial fractions, we are reduced to evaluating the integrals

$$I_0 = \int dx/w, \quad I_n = \int x^n\, dx/w, \quad J_n(\gamma) = \int (x - \gamma)^{-n}\, dx/w,$$

where $n \in \mathbb{N}$ and $\gamma \in \mathbb{C}$.

The argument of the preceding paragraph is actually valid if $w^2 = g$ is any polynomial. Suppose now that g is a cubic without repeated roots, say

$$g(x) \;=\; a_0 x^3 + a_1 x^2 + a_2 x + a_3.$$

By differentiation we obtain, for any integer $m \geq 0$,

$$(x^m w)' \;=\; mx^{m-1}w + x^m g'/2w \;=\; (2mx^{m-1}g + x^m g')/2w.$$

Since the numerator on the right is the polynomial

$$(2m + 3)a_0 x^{m+2} + (2m + 2)a_1 x^{m+1} + (2m + 1)a_2 x^m + 2ma_3 x^{m-1},$$

it follows on integration that

$$2x^m w \;=\; (2m + 3)a_0 I_{m+2} + (2m + 2)a_1 I_{m+1} + (2m + 1)a_2 I_m + 2ma_3 I_{m-1}.$$

It follows by induction that, for each integer $n > 1$,

$$I_n \;=\; p_n(x)w + c_n I_0 + c_n' I_1,$$

where $p_n(x)$ is a polynomial of degree $n - 2$ and c_n, c_n' are constants. Thus the evaluation of I_n for $n > 1$ reduces to the evaluation of I_0 and I_1.

Consider now the integral $J_n(\gamma)$. In the same way as before, for any integer $m \geq 1$,

$$d\{(x - \gamma)^{-m}w\}/dx \;=\; -m(x - \gamma)^{-m-1}w + (x - \gamma)^{-m}g'/2w$$

$$=\; \{-2mg + (x - \gamma)g'\}/2w(x - \gamma)^{m+1}.$$

We can write

$$g(x) \;=\; b_0 + b_1(x - \gamma) + b_2(x - \gamma)^2 + b_3(x - \gamma)^3$$

and the numerator on the right of the previous equation is then

$$-2mb_0 + (1 - 2m)b_1(x - \gamma) + (2 - 2m)b_2(x - \gamma)^2 + (3 - 2m)b_3(x - \gamma)^3.$$

It follows on integration that

$$2(x - \gamma)^{-m}w \;=\; -2mb_0J_{m+1}(\gamma) + (1 - 2m)b_1J_m(\gamma) + (2 - 2m)b_2J_{m-1}(\gamma) + (3 - 2m)b_3J_{m-2}(\gamma),$$

where $J_{-1}(\gamma) = \int (x - \gamma)\,dx/w$ is a constant linear combination of I_0 and I_1. Since g does not have repeated roots, $b_1 \neq 0$ if $b_0 = 0$.

It follows by induction that if $g(\gamma) = b_0 \neq 0$ then, for any $n > 1$,

$$J_n(\gamma) \;=\; q_n((x - \gamma)^{-1})w + d_nJ_1(\gamma) + d_n'I_0 + d_n''I_1,$$

where $q_n(t)$ is a polynomial of degree $n - 1$ and d_n, d_n', d_n'' are constants. On the other hand, if $g(\gamma) = 0$ then $g'(\gamma) = b_1 \neq 0$ and, for any $n \geq 1$,

$$J_n(\gamma) \;=\; r_n((x - \gamma)^{-1})w + e_nI_0 + e_n'I_1,$$

where $r_n(t)$ is a polynomial of degree n and e_n, e_n' are constants.

Thus the evaluation of an arbitrary elliptic integral can be reduced to the evaluation of

$$I_0 = \int dx/w, \;\; I_1 = \int x\,dx/w, \;\; J_1(\gamma) = \int (x - \gamma)^{-1}\,dx/w,$$

where $w^2 = g$ is a cubic without repeated roots, $\gamma \in \mathbb{C}$ and $g(\gamma) \neq 0$. Following Legendre (1793), to whom this reduction is due, integrals of these types are called respectively *elliptic integrals of the first, second and third kinds*.

The cubic g can itself be simplified. If α is a root of g then, by replacing x by $x - \alpha$, we may assume that $g(0) = 0$. If β is now another root of g then, by replacing x by x/β, we may further assume that $g(1) = 0$. Thus the evaluation of an arbitrary elliptic integral may be reduced to one for which g has the form

$$g_\lambda(x) \colon \;=\; 4x(1 - x)(1 - \lambda x),$$

where $\lambda \in \mathbb{C}$ and $\lambda \neq 0,1$. This normal form, which was used by Riemann (1858) in lectures, is obtained from the normal form of Legendre by the change of variables $x = \sin^2 \theta$. To draw attention to the difference, it is convenient to call it *Riemann's normal form*.

The range of λ can be further restricted by linear changes of variables. The transformation $y = (1 - \lambda x)/(1 - \lambda)$ replaces Riemann's normal form by one of the same type with λ replaced by $U\lambda = 1 - \lambda$. Similarly, the transformation $y = 1 - \lambda x$ replaces Riemann's normal form by one of the same type with λ replaced by $V\lambda = 1/(1 - \lambda)$. The transformations U and V together generate a group $\mathcal{G}$ of order 6 (isomorphic to the symmetric group $\mathcal{S}_3$ of all permutations of three letters), since

$$U^2 = V^3 = (UV)^2 = I.$$

The values of λ corresponding to the elements I, V, V^2, U, UV, UV^2 of $\mathcal{G}$ are

$$\lambda, \ 1/(1 - \lambda), \ (\lambda - 1)/\lambda, \ 1 - \lambda, \ \lambda/(\lambda - 1), \ 1/\lambda.$$

The region $\mathcal{F}$ of the complex plane $\mathbb{C}$ defined by the inequalities

$$|\lambda - 1| < 1, \ \ 0 < \Re\lambda < 1/2,$$

is a *fundamental domain* for the group $\mathcal{G}$; i.e., no point of $\mathcal{F}$ is mapped to a different point of $\mathcal{F}$ by an element of $\mathcal{G}$ and each point of $\mathbb{C}$ is mapped to a point of $\mathcal{F}$ or its boundary $\partial\mathcal{F}$ by some element of $\mathcal{G}$. Consequently the sets $\{G(\mathcal{F}): G \in \mathcal{G}\}$ form a *tiling* of $\mathbb{C}$; i.e.,

$$\mathbb{C} = \bigcup_{G \in \mathcal{G}} G(\mathcal{F} \cup \partial\mathcal{F}), \ \ G(\mathcal{F}) \cap G'(\mathcal{F}) = \varnothing \text{ if } G, G' \in \mathcal{G} \text{ and } G \neq G'.$$

This is illustrated in Figure 1, where the set $G(\mathcal{F})$ is represented simply by the group element G and, in particular, $\mathcal{F}$ is represented by I. It follows that in Riemann's normal form we may

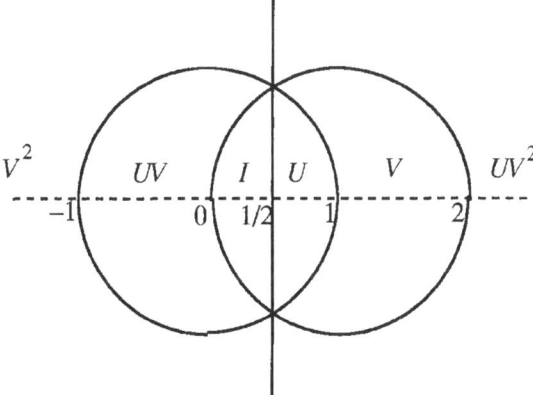

Figure 1: Fundamental domain for λ

suppose $\lambda \in \mathscr{F} \cup \partial\mathscr{F}$.

The changes of variable in the preceding reduction to Riemann's normal form may be complex, even though the original integrand was real. It will now be shown that any real elliptic integral can be reduced by a real change of variables to one in Riemann's normal form, where $0 < \lambda < 1$ and the independent variable is restricted to the interval $0 \le x \le 1$.

If g is a cubic or quartic with only real roots, this can be achieved by a linear fractional transformation, mapping roots of g to roots of g_λ. Appropriate transformations are listed in Tables 1 and 2. It should be noted that λ is always a *cross-ratio* of the four roots of g in Table 2, and that λ is always a cross-ratio of the three roots of g and the point '∞' in Table 1.

Suppose now that g is a real cubic or quartic with a pair of conjugate complex roots. Then we can write

$$g(x) = Q_1 Q_2 = (a_1 x^2 + 2b_1 x + c_1)(a_2 x^2 + 2b_2 x + c_2),$$

where the coefficients are real, $a_1 c_1 - b_1^2 > 0$ and $a_2 c_2 - b_2^2 \ne 0$, but a_2 may be zero.

Consider first the case where $a_2 \ne 0$ and $b_1 = b_2 a_1/a_2$. Then

$$Q_1 = a_1(x + b_1/a_1)^2 + b_1', \quad Q_2 = a_2(x + b_1/a_1)^2 + b_2',$$

where

$$b_1' = (a_1 c_1 - b_1^2)/a_1, \quad b_2' = (a_2 c_2 - b_2^2)/a_2.$$

If we put $y = (x + b_1/a_1)^2$, then

$$R(x) = R_1(y) + R_2(y)y^{1/2},$$

where the rational functions R_1, R_2 are determined by the rational function R, and

$$dx/g(x)^{1/2} = \pm\, dy/2[y(a_1 y + b_1')(a_2 y + b_2')]^{1/2}.$$

Thus we are reduced to the case of a cubic with 3 distinct real roots.

In the remaining cases there exist distinct real values s_1, s_2 of s such that the polynomial $Q_1 + sQ_2$ is proportional to a perfect square. For $Q_1 + sQ_2$ is proportional to a perfect square if

$$D(s) := (a_1 + sa_2)(c_1 + sc_2) - (b_1 + sb_2)^2 = 0.$$

We have $D(0) = a_1 c_1 - b_1^2 > 0$. If $a_2 = 0$, then $b_2 \ne 0$ and $D(\pm\infty) = -\infty$. On the other hand, if $a_2 \ne 0$, then $D(-a_1/a_2) < 0$, since $b_1 \ne b_2 a_1/a_2$, and $D(s)$ has the sign of $a_2 c_2 - b_2^2$ for both large positive and large negative s. Thus the quadratic $D(s)$ has distinct real roots s_1, s_2. Hence

$$Q_1 + s_1 Q_2 = (a_1 + s_1 a_2)(x + d_1)^2, \quad Q_1 + s_2 Q_2 = (a_1 + s_2 a_2)(x + d_2)^2,$$

where $a_1 + s_j a_2 \ne 0$ $(j = 1,2)$ and

$$dx/g(x)^{1/2} \;=\; dy/\mu g_\lambda(y)^{1/2}$$

$$g(x) \;=\; A(x-\alpha_1)(x-\alpha_2)(x-\alpha_3), \text{ where } \alpha_1 > \alpha_2 > \alpha_3;\; \alpha_{jk} = \alpha_j - \alpha_k$$
$$g_\lambda(y) \;=\; 4y(1-y)(1-\lambda y), \text{ where } 0 < \lambda < 1,\, y \in (0,1)$$
$$\mu = (\alpha_{13})^{1/2}/2,\;\; \lambda_0 = \alpha_{23}/\alpha_{13},\;\; 1-\lambda_0 = \alpha_{12}/\alpha_{13}.$$

A	λ	*Range*	*Transformation*	*Corresponding values*	
$+1$	λ_0	$x \ge \alpha_1$	$y = (x-\alpha_1)/(x-\alpha_2)$	$x = \infty$ α_1	$y = 1$ 0
-1	$1-\lambda_0$	$\alpha_2 \le x \le \alpha_1$	$= \alpha_{13}(x-\alpha_2)/\alpha_{12}(x-\alpha_3)$	α_1 α_2	1 0
$+1$	λ_0	$\alpha_3 \le x \le \alpha_2$	$= (x-\alpha_3)/\alpha_{23}$	α_2 α_3	1 0
-1	$1-\lambda_0$	$x \le \alpha_3$	$= \alpha_{13}/(\alpha_1-x)$	α_3 $-\infty$	1 0

Table 1: *Reduction to Riemann's normal form, g a cubic with all roots real*

$$dx/g(x)^{1/2} \;=\; dy/\mu g_\lambda(y)^{1/2}$$

$$g(x) \;=\; A(x-\alpha_1)(x-\alpha_2)(x-\alpha_3)(x-\alpha_4), \text{ where } \alpha_1 > \alpha_2 > \alpha_3 > \alpha_4;\; \alpha_{jk} = \alpha_j - \alpha_k$$
$$g_\lambda(y) \;=\; 4y(1-y)(1-\lambda y), \text{ where } 0 < \lambda < 1,\, y \in (0,1)$$
$$\mu = (\alpha_{13}\alpha_{24})^{1/2}/2,\;\; \lambda_0 = \alpha_{23}\alpha_{14}/\alpha_{13}\alpha_{24},\;\; 1-\lambda_0 = \alpha_{12}\alpha_{34}/\alpha_{13}\alpha_{24}.$$

A	λ	*Range*	*Transformation*	*Corresponding values*	
$+1$	λ_0	$x \ge \alpha_1$	$y = \alpha_{24}(x-\alpha_1)/\alpha_{14}(x-\alpha_2)$	$x = \infty$ α_1	$y = \alpha_{24}/\alpha_{14}$ 0
-1	$1-\lambda_0$	$\alpha_2 \le x \le \alpha_1$	$= \alpha_{13}(x-\alpha_2)/\alpha_{12}(x-\alpha_3)$	α_1 α_2	1 0
$+1$	λ_0	$\alpha_3 \le x \le \alpha_2$	$= \alpha_{24}(x-\alpha_3)/\alpha_{23}(x-\alpha_4)$	α_2 α_3	1 0
-1	$1-\lambda_0$	$\alpha_4 \le x \le \alpha_3$	$= \alpha_{13}(x-\alpha_4)/\alpha_{34}(\alpha_1-x)$	α_3 α_4	1 0
$+1$	λ_0	$x \le \alpha_4$	$= \alpha_{24}(x-\alpha_1)/\alpha_{14}(x-\alpha_2)$	α_4 $-\infty$	1 α_{24}/α_{14}

Table 2: *Reduction to Riemann's normal form, g a quartic with all roots real*

$$d_1 = (b_1 + s_1 b_2)/(a_1 + s_1 a_2), \quad d_2 = (b_1 + s_2 b_2)/(a_1 + s_2 a_2).$$

Consequently

$$Q_1 = A_1(x + d_1)^2 + B_1(x + d_2)^2, \quad Q_2 = A_2(x + d_1)^2 + B_2(x + d_2)^2,$$

where

$$A_1 = -s_2(a_1 + s_1 a_2)/(s_1 - s_2), \quad B_1 = s_1(a_1 + s_2 a_2)/(s_1 - s_2),$$
$$A_2 = (a_1 + s_1 a_2)/(s_1 - s_2), \quad B_2 = -(a_1 + s_2 a_2)/(s_1 - s_2).$$

If we put $y = \{(x + d_1)/(x + d_2)\}^2$, then

$$R(x) = R_1(y) + R_2(y)y^{1/2},$$

where again the rational functions R_1, R_2 are determined by the rational function R, and

$$dx/g(x)^{1/2} = \pm dy/2|d_2 - d_1|[y(A_1 y + B_1)(A_2 y + B_2)]^{1/2}.$$

Thus we are again reduced to the case of a cubic with 3 distinct real roots.

The preceding argument may be applied also when g has only real roots, provided the factors Q_1 and Q_2 are chosen so that their zeros do not interlace. Suppose (without loss of generality) that $g = g_\lambda$ is in Riemann's normal form and take

$$Q_1 = (1 - x)(1 - \lambda x), \quad Q_2 = 4x.$$

In this case we can write

$$Q_1 = \{(1 + \sqrt{\lambda})^2(x - 1/\sqrt{\lambda})^2 - (1 - \sqrt{\lambda})^2(x + 1/\sqrt{\lambda})^2\}\sqrt{\lambda}/4,$$
$$Q_2 = -\sqrt{\lambda}\{(x - 1/\sqrt{\lambda})^2 - (x + 1/\sqrt{\lambda})^2\}.$$

If we put

$$1 - 4\sqrt{\lambda}y/(1 + \sqrt{\lambda})^2 = \{(x - 1/\sqrt{\lambda})/(x + 1/\sqrt{\lambda})\}^2,$$

we obtain

$$dx/g_\lambda(x)^{1/2} = dy/\mu g_\rho(y)^{1/2},$$

where

$$\mu = 1 + \sqrt{\lambda}, \quad \rho = 4\sqrt{\lambda}/(1 + \sqrt{\lambda})^2.$$

The usefulness of this change of variables will be seen in the next section.

2 The arithmetic-geometric mean

Let a and b be positive real numbers, with $a > b$, and let

$$a_1 = (a + b)/2, \quad b_1 = (ab)^{1/2}$$

be respectively their arithmetic and geometric means. Then

$$a_1 < (a + a)/2 = a, \quad b_1 > (bb)^{1/2} = b,$$

and

$$a_1 - b_1 = (a^{1/2} - b^{1/2})^2/2 > 0.$$

Thus a_1, b_1 satisfy the same hypotheses as a, b and the procedure can be repeated. If we define sequences $\{a_n\}, \{b_n\}$ inductively by

$$a_0 = a, \quad b_0 = b,$$
$$a_{n+1} = (a_n + b_n)/2, \quad b_{n+1} = (a_n b_n)^{1/2} \qquad\qquad (n = 0,1,\ldots),$$

then

$$0 < b_0 < b_1 < b_2 < \ldots < a_2 < a_1 < a_0.$$

It follows that $a_n \to \lambda$ and $b_n \to \mu$ as $n \to \infty$, where $\lambda \geq \mu > 0$. In fact $\lambda = \mu$, as one sees by letting $n \to \infty$ in the relation $a_{n+1} = (a_n + b_n)/2$. The convergence of the sequences $\{a_n\}$ and $\{b_n\}$ to their common limit is extremely rapid, since

$$a_n - b_n = (a_{n-1} - b_{n-1})^2/8a_{n+1}.$$

(As an example, if $a = \sqrt{2}$ and $b = 1$, calculation shows that a_4 and b_4 differ by only one unit in the 20th decimal place.)

The common limit of the sequences $\{a_n\}$ and $\{b_n\}$ will be denoted by $M(a,b)$. The definition can be extended to arbitrary positive real numbers a, b by putting

$$M(a,a) = a, \quad M(b,a) = M(a,b).$$

Following Gauss (1818), $M(a,b)$ is known as the *arithmetic-geometric mean* of a and b. However, the preceding algorithm, which we will call the *AGM* algorithm, was first introduced by Lagrange (1784/5), who showed that it had a remarkable application to the numerical calculation of arbitrary elliptic integrals. The first tables of elliptic integrals, which made them as accessible as logarithms, were constructed in this way under the supervision of Legendre (1826). Today the algorithm can be used directly by electronic computers.

By putting $1 - \lambda x = t^2/a^2$ in Riemann's normal form, it may be seen that any real elliptic integral may be brought to the form

$$\int \varphi(t)[(a^2 - t^2)(t^2 - b^2)]^{-1/2} \, dt,$$

where $\varphi(t)$ is a rational function of t^2 with real coefficients, $a > b > 0$ and $t \in [b,a]$. We will restrict attention here to the *complete* elliptic integral

$$J = \int_b^a \varphi(t)[(a^2 - t^2)(t^2 - b^2)]^{-1/2}\, dt,$$

but at the cost of some complication the discussion may be extended to *incomplete* elliptic integrals (where the interval of integration is a proper subinterval of $[b,a]$).

If we make the change of variables

$$t^2 = a^2 \sin^2 \theta + b^2 \cos^2 \theta \quad (0 \le \theta \le \pi/2),$$

then

$$t\, dt/d\theta = (a^2 - b^2) \sin \theta \cos \theta = [(a^2 - t^2)(t^2 - b^2)]^{1/2}$$

and

$$J = \int_0^{\pi/2} \varphi((a^2 \sin^2 \theta + b^2 \cos^2 \theta)^{1/2})\, d\theta/(a^2 \sin^2 \theta + b^2 \cos^2 \theta)^{1/2}.$$

Now put

$$t_1 = (1/2)(t + ab/t)$$

and, as before,

$$a_1 = (a + b)/2, \quad b_1 = (ab)^{1/2}.$$

Then

$$a_1{}^2 - t_1{}^2 = (a^2 - t^2)(t^2 - b^2)/4t^2,$$
$$t_1{}^2 - b_1{}^2 = (t^2 - ab)^2/4t^2,$$
$$dt_1/dt = (t^2 - ab)/2t^2.$$

As t increases from b to b_1, t_1 decreases from a_1 to b_1, and as t further increases from b_1 to a, t_1 increases from b_1 back to a_1. Since

$$t = t_1 \pm (t_1{}^2 - b_1{}^2)^{1/2},$$

it follows from these observations that

$$\int_b^a \varphi(t)[(a^2 - t^2)(t^2 - b^2)]^{-1/2}\, dt \;=\; \int_{b_1}^{a_1} \psi(t_1)[(a_1{}^2 - t_1{}^2)(t_1{}^2 - b_1{}^2)]^{-1/2}\, dt_1,$$

where

$$\psi(t_1) \;=\; (1/2)\{\varphi[(t_1 + (t_1{}^2 - b_1{}^2)^{1/2}] + \varphi[(t_1 - (t_1{}^2 - b_1{}^2)^{1/2}]\}.$$

In particular, if we take $\varphi(t) = 1$ and put

$$\mathcal{H}(a,b) := \int_b^a [(a^2 - t^2)(t^2 - b^2)]^{-1/2}\, dt,$$

we obtain

$$\mathcal{K}(a,b) \;=\; \mathcal{K}(a_1,b_1).$$

Hence, by repeating the process, $\mathcal{K}(a,b) = \mathcal{K}(a_n,b_n)$. But

$$\mathcal{K}(a_n,b_n) \;=\; \int_0^{\pi/2} (a_n^2 \sin^2\theta + b_n^2 \cos^2\theta)^{-1/2}\, d\theta$$

and

$$b_n \le (a_n^2 \sin^2\theta + b_n^2 \cos^2\theta)^{1/2} \le a_n.$$

Consequently, by letting $n \to \infty$ we obtain

$$\mathcal{K}(a,b) \;=\; \pi/2M(a,b). \tag{1}$$

Now take $\varphi(t) = a^2 - t^2$ and put

$$\mathcal{E}(a,b)\colon \;=\; \int_b^a [(a^2 - t^2)/(t^2 - b^2)]^{1/2}\, dt.$$

In this case

$$\psi(t_1) \;=\; (a^2 - b^2)/2 + 2(a_1^2 - t_1^2)$$

and hence

$$\mathcal{E}(a,b) \;=\; (a^2 - b^2)\mathcal{K}(a,b)/2 + 2\mathcal{E}(a_1,b_1).$$

If we write

$$e_n \;=\; 2^n(a_n^2 - b_n^2)$$

then, since $\mathcal{K}(a,b) = \mathcal{K}(a_n,b_n)$, by repeating the process we obtain

$$\mathcal{E}(a,b)/\mathcal{K}(a,b) \;=\; (e_0 + e_1 + \ldots + e_{n-1})/2 + 2^n\mathcal{E}(a_n,b_n)/\mathcal{K}(a_n,b_n).$$

But

$$2^n\mathcal{E}(a_n,b_n) \;=\; e_n \int_0^{\pi/2} \cos^2\theta\, (a_n^2 \sin^2\theta + b_n^2 \cos^2\theta)^{-1/2}\, d\theta$$

and $e_n \to 0$ (rapidly) as $n \to \infty$, since

$$e_n \;=\; 2^n(a_{n-1} - b_{n-1})^2/4 \;=\; e_{n-1}(a_{n-1} - b_{n-1})/4a_n.$$

Hence

$$\mathcal{E}(a,b)/\mathcal{K}(a,b) \;=\; (e_0 + e_1 + e_2 + \ldots)/2. \tag{2}$$

To avoid taking differences of nearly equal quantities, the constants e_n may be calculated by means of the recurrence relations

$$e_n \;=\; e_{n-1}^2/2^{n+2}a_n^2 \quad (n = 1,2,\ldots).$$

Next take

$$\varphi(t) \;=\; p[(p^2 - a^2)(p^2 - b^2)]^{1/2}/(p^2 - t^2),$$

where either $p > a$ or $0 < p < b$, and put

$$\mathscr{P}(a,b,p) := \int_b^a p[(p^2 - a^2)(p^2 - b^2)]^{1/2} \, dt/(p^2 - t^2)[(a^2 - t^2)(t^2 - b^2)]^{1/2}.$$

In this case

$$\psi(t_1) = q_1 \pm p_1[(p_1^2 - a_1^2)(p_1^2 - b_1^2)]^{1/2}/(p_1^2 - t_1^2),$$

where

$$p_1 = (1/2)(p + ab/p),$$
$$q_1 = (p_1^2 - a_1^2)^{1/2} = [(p^2 - a^2)(p^2 - b^2)]^{1/2}/2p,$$

and the $+$ or $-$ sign is taken according as $p > a$ or $0 < p < b$. Since $p_1 > a_1$ in either event, without loss of generality *we now assume that* $p > a$. Then also $p_1 < p$ and

$$\mathscr{P}(a,b,p) = q_1\mathscr{K}(a,b) + \mathscr{P}(a_1,b_1,p_1).$$

Define the sequence $\{p_n\}$ inductively by

$$p_0 = p, \quad p_{n+1} = (1/2)(p_n + a_nb_n/p_n) \quad (n = 0,1,\dots),$$

and put

$$q_{n+1} = (p_{n+1}^2 - a_{n+1}^2)^{1/2} = [(p_n^2 - a_n^2)(p_n^2 - b_n^2)]^{1/2}/2p_n.$$

Then $p_n \to v \geq M(a,b)$ as $n \to \infty$, since $a_n < p_n < p_{n-1}$. In fact $v = M(a,b)$, as one sees by letting $n \to \infty$ in the recurrence relation defining the sequence $\{p_n\}$. Moreover

$$\delta_n := (a_n^2 - b_n^2)/(p_n^2 - a_n^2) \to 0 \quad \text{as } n \to \infty,$$

since

$$\delta_{n+1} = \delta_n \left(\frac{p_n^2}{4a_{n+1}^2}\right)\left(\frac{a_n^2 - b_n^2}{p_n^2 - b_n^2}\right) < \delta_n \, p_n^2/4a_{n+1}^2.$$

Hence

$$(p_n^2 - b_n^2)/(p_n^2 - a_n^2) = 1 + \delta_n \to 1.$$

Since

$$\mathscr{P}(a_n,b_n,p_n) = p_n \, [(p_n^2 - a_n^2)(p_n^2 - b_n^2)]^{1/2}\int_0^{\pi/2} \frac{(a_n^2\sin^2\theta + b_n^2\cos^2\theta)^{-1/2} \, d\theta}{(p_n^2 - a_n^2)\sin^2\theta + (p_n^2 - b_n^2)\cos^2\theta},$$

it follows that $\mathscr{P}(a_n,b_n,p_n) \to \pi/2$ as $n \to \infty$. Hence

$$\mathscr{P}(a,b,p) = (q_1 + q_2 + \dots)\mathscr{K}(a,b) + \pi/2. \tag{3}$$

To avoid repeated root extractions, the constants q_n may be calculated by means of the recurrence relations

$$q_{n+1} = (p_{n-1} - p_n)q_n/2p_n \quad (n = 1,2,\dots).$$

Using (1)-(3), complete elliptic integrals of all three kinds can be calculated by the *AGM* algorithm. We now consider another application, the utility of which will be seen in §6.

By putting $t_1 = (1/2)(t + ab/t)$ again, one sees that

$$\int_a^\infty [(t^2 - a^2)(t^2 - b^2)]^{-1/2} \, dt = (1/2)\int_{a_1}^\infty [(t_1^2 - a_1^2)(t_1^2 - b_1^2)]^{-1/2} \, dt_1.$$

But the change of variables $u = a(1 - b^2/t^2)^{1/2}$ shows that

$$\int_a^\infty [(t^2 - a^2)(t^2 - b^2)]^{-1/2} \, dt = \mathcal{K}(a,c),$$

where $c = (a^2 - b^2)^{1/2}$. It follows that

$$\mathcal{K}(a,c) = \mathcal{K}(a_1,c_1)/2 = \dots = \mathcal{K}(a_n,c_n)/2^n,$$

where $c_n = (a_n^2 - b_n^2)^{1/2}$. The asymptotic behaviour of $\mathcal{K}(a_n,c_n)$ may be determined in the following way.

If we put $s = ac/t$, then s decreases from a to c as t increases from c to a, and

$$ds/dt = -[(a^2 - s^2)(s^2 - c^2)]^{1/2}/[(a^2 - t^2)(t^2 - c^2)]^{1/2}.$$

Since $s = t$ when $t = h: = (ac)^{1/2}$, it follows that

$$\mathcal{K}(a,c) = 2\int_c^h [(a^2 - t^2)(t^2 - c^2)]^{-1/2} \, dt.$$

But, for $c \le t \le h$,

$$b^{-1} = (a^2 - c^2)^{-1/2} \le (a^2 - t^2)^{-1/2} \le (a^2 - h^2)^{-1/2} = a^{-1}(1 - c/a)^{-1/2}.$$

Hence

$$2b^{-1}L \le \mathcal{K}(a,c) \le 2a^{-1}(1 - c/a)^{-1/2}L,$$

where

$$L: = \int_c^h (t^2 - c^2)^{-1/2} \, dt = \log \{(a/c)^{1/2} + (a/c - 1)^{1/2}\}.$$

If we now replace a,b,c by a_n,b_n,c_n then, since $a_n/c_n \to \infty$ and $a_n,b_n \to M(a,b)$, we deduce that

$$2^n \, \mathcal{K}(a,c)/\log (4a_n/c_n) \to 1/M(a,b) = 2\mathcal{K}(a,b)/\pi.$$

But $4a_n/c_n = (4a_n/c_{n-1})^2$, since $c_n = (a_{n-1} - b_{n-1})/2$, and hence

$$2^{-n} \log (4a_n/c_n) = 2^{1-n} \log (4a_{n-1}/c_{n-1}) - 2^{1-n} \log (a_{n-1}/a_n)$$

$$= \dots$$

$$= \log (4a_0/c_0) - \log (a_0/a_1) - 2^{-1} \log (a_1/a_2) - \dots - 2^{1-n} \log (a_{n-1}/a_n).$$

It follows that

$$\pi \mathcal{H}(a,c)/2\mathcal{H}(a,b) = \log (4a_1/c_0) - \sum_{n=1}^{\infty} 2^{-n} \log (a_n/a_{n+1}). \tag{4}$$

Finally, to determine $\mathcal{E}(a,c)$ we can use the relation

$$\mathcal{H}(a,b)\mathcal{E}(a,c) + \mathcal{H}(a,c)\mathcal{E}(a,b) - a^2\mathcal{H}(a,b)\mathcal{H}(a,c) = \pi/2.$$

By homogeneity we need only establish this relation for $a = 1$. Since

$$\mathcal{H}(1,(1 - \lambda)^{1/2}) = \int_0^1 [4x(1 - x)(1 - \lambda x)]^{-1/2} \, dx,$$
$$\mathcal{E}(1,(1 - \lambda)^{1/2}) = \int_0^1 [(1 - \lambda x)/4x(1 - x)]^{1/2} \, dx,$$

it is in fact equivalent to the following relation, due to Legendre, between the complete elliptic integrals of the first and second kinds:

PROPOSITION 1 *If*

$$K(\lambda) = \int_0^1 [4x(1 - x)(1 - \lambda x)]^{-1/2} \, dx, \quad E(\lambda) = \int_0^1 [(1 - \lambda x)/4x(1 - x)]^{1/2} \, dx,$$

then

$$K(\lambda)E(1 - \lambda) + K(1 - \lambda)E(\lambda) - K(\lambda)K(1 - \lambda) = \pi/2 \quad \text{for } 0 < \lambda < 1. \tag{5}$$

Proof We show first that the derivative of the left side of (5) is zero. Evidently

$$dE/d\lambda = -(1/2)\int_0^1 x[4x(1 - x)(1 - \lambda x)]^{-1/2} \, dx = [E(\lambda) - K(\lambda)]/2\lambda.$$

Similarly,

$$dK/d\lambda = (1/2)\int_0^1 x(1 - \lambda x)^{-1}[4x(1 - x)(1 - \lambda x)]^{-1/2} \, dx.$$

Substituting $x = (1 - u)/(1 - \lambda u)$ and writing $\lambda' = 1 - \lambda$, we obtain

$$dK/d\lambda = (1/2\lambda')\int_0^1 [(1 - u)/4u(1 - \lambda u)]^{1/2} \, du$$

$$= [E(\lambda) - \lambda'K(\lambda)]/2\lambda\lambda'.$$

It follows that

$$d(\lambda\lambda' \, dK/d\lambda)/d\lambda = K/4.$$

Thus $y_1(\lambda) = K(\lambda)$ is a solution of the second order linear differential equation

$$d(\lambda\lambda' \, dy/d\lambda)/d\lambda - y/4 = 0. \tag{6}$$

By symmetry, $y_2(\lambda) = K(\lambda')$ is also a solution. It follows that the 'Wronskian'

$$W = \lambda\lambda' (y_2 \, dy_1/d\lambda - y_1 \, dy_2/d\lambda)$$

has derivative zero and so is constant. But, writing

$$K'(\lambda) = K(1 - \lambda), \quad E'(\lambda) = E(1 - \lambda),$$

we have

$$2W \;=\; K'(E - \lambda'K) + K(E' - \lambda K') \;=\; KE' + K'(E - K).$$

To evaluate this constant we let $\lambda \to 0$. Putting $x = \sin^2 \theta$, we obtain

$$K(\lambda) = \int_0^{\pi/2} (1 - \lambda \sin^2 \theta)^{-1/2} \, d\theta, \quad E(\lambda) = \int_0^{\pi/2} (1 - \lambda \sin^2 \theta)^{1/2} \, d\theta$$

and hence, as $\lambda \to 0$,

$$K(\lambda) \to \pi/2, \; E(\lambda) \to \pi/2, \; E(\lambda') \to 1.$$

Moreover

$$K(\lambda')[E(\lambda) - K(\lambda)] \to 0,$$

since

$$K(\lambda) - E(\lambda) \;=\; \lambda \int_0^1 x[4x(1 - x)(1 - \lambda x)]^{-1/2} \, dx \;=\; O(\lambda)$$

and

$$0 \le K(\lambda') \le \int_0^{\pi/2} [1 - (1 - \lambda)]^{-1/2} \, d\theta \;=\; O(\lambda^{-1/2}).$$

It follows that $2W = \pi/2$. $\square$

If $\lambda = 1/2$, then $\lambda' = \lambda$ and (5) takes the simple form

$$K(1/2)[2E(1/2) - K(1/2)] = \pi/2.$$

By the remarks preceding the statement of Proposition 1, the left side can be evaluated by the *AGM* algorithm. In this way π has recently been calculated to millions of decimal places. (It will be recalled that the value $\lambda = 1/2$ occurred in the rectification of the lemniscate.)

3 Elliptic functions

According to Jacobi, the theory of elliptic functions was conceived on 23 December 1751, the day on which the Berlin Academy asked Euler to report on the *Produzioni Matematiche* of Count Fagnano, a copy of which had been sent them by the author. The papers which aroused Euler's interest had in fact already appeared in an obscure Italian journal between 1715 and 1720. Fagnano had shown first how a quadrant of a lemniscate could be halved, then how it could be divided algebraically into 2^m, $3 \cdot 2^m$ or $5 \cdot 2^m$ equal parts. He had also established an algebraic relation between the length of an elliptic arc, the length of another suitably chosen arc and the length of a quadrant. By analysing and extending his arguments, Euler was led ultimately (1761) to a general addition theorem for elliptic integrals. An elegant proof of Euler's theorem was given by Lagrange (1768/9), using differential equations. We follow this approach here.

Let

$$g_\lambda(x) = 4x(1-x)(1-\lambda x) = 4\lambda x^3 - 4(1+\lambda)x^2 + 4x$$

be Riemann's normal form and let $2f_\lambda(x)$ be its derivative:

$$f_\lambda(x) = 6\lambda x^2 - 4(1+\lambda)x + 2.$$

By the fundamental existence and uniqueness theorem for ordinary differential equations, the second order differential equation

$$x'' = f_\lambda(x) \tag{7}$$

has a unique solution $S(t) = S(t,\lambda)$, defined (and holomorphic) for $|t|$ sufficiently small, which satisfies the initial conditions

$$S(0) = S'(0) = 0. \tag{8}$$

The solution $S(t,\lambda)$ is an elementary function if $\lambda = 0$ or 1:

$$S(t,0) = \sin^2 t, \quad S(t,1) = \tanh^2 t.$$

(For other values of λ, $S(t)$ coincides with the Jacobian elliptic function $\mathrm{sn}^2 t$.)

Evidently $S(t)$ is an even function of t, since $S(-t)$ is also a solution of (7) and satisfies the same initial conditions (8).

For any solution $x(t)$ of (7), the function $x'(t)^2 - g_\lambda[x(t)]$ is a constant, since its derivative is zero. In particular,

$$S'(t)^2 = g_\lambda[S(t)], \tag{9}$$

since both sides vanish for $t = 0$.

If $|\tau|$ is sufficiently small, then $x_1(t) = S(t+\tau)$ and $x_2(t) = S(t-\tau)$ are solutions of (7) near $t = 0$. Moreover,

$$x_j'(t)^2 = g_\lambda[x_j(t)] \quad (j = 1,2),$$

since these relations hold for $t = 0$. From

$$(x_1x_2' + x_1'x_2)' = x_1 f_\lambda(x_2) + x_2 f_\lambda(x_1) + 2x_1'x_2'$$

and

$$(x_1x_2' + x_1'x_2)^2 = x_1^2 g_\lambda(x_2) + x_2^2 g_\lambda(x_1) + 2x_1x_2x_1'x_2'$$

we obtain

$$2x_1x_2 (x_1x_2' + x_1'x_2)' - (x_1x_2' + x_1'x_2)^2 - 2x_1x_2x_1'x_2'$$

$$= x_1^2\{2x_2 f_\lambda(x_2) - g_\lambda(x_2)\} + x_2^2\{2x_1 f_\lambda(x_1) - g_\lambda(x_1)\}.$$

But if $g_\lambda(x) = \alpha x^3 + \beta x^2 + \gamma x$ and $f_\lambda(x) = g_\lambda'(x)/2$, then

$$2x f_\lambda(x) - g_\lambda(x) = x^2 (2\alpha x + \beta).$$

Hence

$$2x_1x_2(x_1x_2' + x_1'x_2)' - (x_1x_2' + x_1'x_2)^2 = 2x_1^2x_2^2\{\alpha(x_1 + x_2) + \beta\} + 2x_1x_2x_1'x_2'.$$

On the other hand,

$$(x_1' - x_2')(x_1x_2' + x_1'x_2) = x_2 g_\lambda(x_1) - x_1 g_\lambda(x_2) + (x_1 - x_2)x_1'x_2'$$

$$= x_1x_2(x_1 - x_2)\{\alpha(x_1 + x_2) + \beta\} + (x_1 - x_2)x_1'x_2'.$$

Comparing these two relations, we obtain

$$\{2x_1x_2 (x_1x_2' + x_1'x_2)' - (x_1x_2' + x_1'x_2)^2\}(x_1 - x_2) = 2x_1x_2(x_1' - x_2')(x_1x_2' + x_1'x_2).$$

If we divide by $2x_1x_2(x_1 - x_2)(x_1x_2' + x_1'x_2)$, this takes the form

$$\frac{(x_1x_2'+x_1'x_2)'}{x_1x_2'+x_1'x_2} - \frac{x_1x_2'+x_1'x_2}{2x_1x_2} = \frac{x_1'-x_2'}{x_1 - x_2},$$

which can be integrated to give

$$(x_1x_2' + x_1'x_2)^2 = C(\tau)x_1x_2(x_1 - x_2)^2,$$

where the constant $C(\tau)$ depends on τ. Equivalently,

$$[S(u)S'(v) - S'(u)S(v)]^2 = C((u+v)/2)S(u)S(v)[S(u) - S(v)]^2.$$

To evaluate the constant, we divide throughout by $S(v)$ and let $v \to 0$. By (9), this yields $C(u/2) = \gamma/S(u)$. Since $\gamma = 4$, we obtain finally

$$S(u + v) = 4S(u)S(v)[S(u) - S(v)]^2/[S(u)S'(v) - S'(u)S(v)]^2 . \tag{10}$$

Thus $S(u + v)$ is a rational function of $S(u), S(v), S'(u), S'(v)$. Moreover, since $(S')^2 = g_\lambda(S)$, there exists a polynomial $p(x,y,z)$, not identically zero and with coefficients independent of u and v, such that $p[S(u + v), S(u), S(v)] = 0$. In other words, the function $S(u)$ has an *algebraic addition theorem*.

The relation (10) can also be written in the form

$$S(u + v) = [S(u)S'(v) + S'(u)S(v)]^2/4S(u)S(v)[1 - \lambda S(u)S(v)]^2, \tag{11}$$

since

$$S(u)^2S'(v)^2 - S'(u)^2S(v)^2 = S(u)^2g_\lambda[S(v)] - S(v)^2g_\lambda[S(u)]$$

$$= 4S(u)S(v)[S(u) - S(v)][1 - \lambda S(u)S(v)].$$

Replacing v by $-v$ in (11) and subtracting the result from (11), we obtain

$$S(u + v) - S(u - v) = S'(u)S'(v)/[1 - \lambda S(u)S(v)]^2. \tag{12}$$

In particular, for $v = u$,

$$S(2u) = g_\lambda[S(u)]/[1 - \lambda S^2(u)]^2. \tag{13}$$

We recall that a function is *meromorphic* in a connected open set D if it is holomorphic throughout D, except for isolated singularities which are poles. Since, by (13), $S(2t)$ is a rational function of $S(t)$, it follows that if $S(t)$ is meromorphic and a solution (wherever it is finite) of the differential equation (7) in an open disc $|t| < R$, then its definition can be extended so that it is meromorphic and a solution (wherever it is finite) of the differential equation (7) throughout the disc $|t| < 2R$. But the fundamental existence and uniqueness theorem guarantees that $S(t)$ is holomorphic in a neighbourhood of the origin. Consequently we can extend its definition so that it is meromorphic and a solution of (7) in the whole complex plane $\mathbb{C}$.

Further properties of the function $S(t)$ may be derived from the differential equation (7). For any constants α, β, if $y(t) = \alpha S(\beta t)$, then $y(0) = y'(0) = 0$. It is readily seen that $y(t)$ satisfies a differential equation of the form (7) if and only if either $\alpha = 1$, $\beta = \pm 1$ or $\alpha = \lambda$, $\lambda\beta^2 = 1$, and in the latter case with λ replaced by $1/\lambda$ in (7). It follows that, for any $\lambda \neq 0$,

$$S(t, 1/\lambda) = \lambda S(\lambda^{-1/2}t, \lambda). \tag{14}$$

By differentiation it may be shown also that $S(it, \lambda)/[S(it, \lambda) - 1]$, where $i^2 = -1$, is a solution of the differential equation (7) with λ replaced by $1 - \lambda$. It follows that

$$S(t, 1 - \lambda) = S(it, \lambda)/[S(it, \lambda) - 1]. \tag{15}$$

By combining (14) and (15) we obtain, for any $\lambda \neq 0, 1$, three more relations:

$$S(t, 1/(1 - \lambda)) = (1 - \lambda)S(i(1 - \lambda)^{-1/2}t, \lambda)/[S(i(1 - \lambda)^{-1/2}t, \lambda) - 1], \tag{16}$$

$$S(t, (\lambda - 1)/\lambda) = \lambda S(i\lambda^{-1/2}t, \lambda)/[\lambda S(i\lambda^{-1/2}t, \lambda) - 1], \tag{17}$$

$$S(t, \lambda/(\lambda - 1)) = (1 - \lambda)S((1 - \lambda)^{-1/2}t, \lambda)\big/[1 - \lambda S((1 - \lambda)^{-1/2}t, \lambda)]. \tag{18}$$

As in §1, it follows from (14)-(18) that the evaluation of $S(t, \lambda)$ for all $t, \lambda \in \mathbb{C}$ reduces to its evaluation for λ in the region $|\lambda - 1| \leq 1$, $0 \leq \mathfrak{R}\lambda \leq 1/2$. Similarly it follows from (14) and (18) that the evaluation of $S(t, \lambda)$ for all $t, \lambda \in \mathbb{R}$ reduces to its evaluation for λ in the interval $0 < \lambda < 1$. We now show that $S(t, \lambda)$ can then be calculated by the *AGM* algorithm.

It is easily verified that if

$$z(t) = (1 + \sqrt{\lambda})^2 S(t, \lambda)\big/[1 + \sqrt{\lambda}S(t, \lambda)]^2,$$

then

$$(dz/dt)^2 = (1 + \sqrt{\lambda})^2\{4\lambda_0 z^3 - 4(1 + \lambda_0)z^2 + 4z\},$$

where

$$\lambda_0 = 4\sqrt{\lambda}/(1 + \sqrt{\lambda})^2. \tag{19}$$

Since $z(0) = z'(0) = 0$ and $z''(0) \neq 0$, it follows that $z(t) = S((1 + \sqrt{\lambda})t, \lambda_0)$. Thus

$$S((1 + \sqrt{\lambda})t, \lambda_0) = (1 + \sqrt{\lambda})^2 S(t, \lambda)\big/[1 + \sqrt{\lambda}S(t, \lambda)]^2. \tag{20}$$

The inequality $0 < \lambda < 1$ implies $\lambda < \lambda_0 < 1$. Hence, by regarding (19) as a quadratic equation for $\sqrt{\lambda}$, we obtain

$$\sqrt{\lambda} = [1 - (1 - \lambda_0)^{1/2}]^2/\lambda_0. \tag{21}$$

If we write $\sqrt{\lambda_0} = c_0/a_0$, where $c_0 = (a_0^2 - b_0^2)^{1/2}$ and $0 < b_0 < a_0$, then

$$\sqrt{\lambda} = (a_0 - b_0)/(a_0 + b_0) = c_1/a_1,$$

where

$$a_1 = (a_0 + b_0)/2, \quad b_1 = (a_0 b_0)^{1/2}, \quad c_1 = (a_1^2 - b_1^2)^{1/2}.$$

Since $1 + \sqrt{\lambda} = a_0/a_1$, we can rewrite (20) in the form

$$S(a_0 t, \lambda_0) = (1 + c_1/a_1)^2 S(a_1 t, \lambda_1)\big/[1 + (c_1/a_1)S(a_1 t, \lambda_1)]^2,$$

where $\lambda_1 = \lambda = (c_1/a_1)^2$. Repeating the process, we obtain

$$S(a_{n-1} t, \lambda_{n-1}) = (1 + c_n/a_n)^2 S(a_n t, \lambda_n)\big/[1 + (c_n/a_n)S(a_n t, \lambda_n)]^2,$$

where $\lambda_n = (c_n/a_n)^2$. As $n \to \infty$,

$$a_n \to \mu := M(a, b), \quad c_n \to 0, \quad \lambda_n \to 0.$$

Since $S(t,0) = \sin^2 t$, for some (not very large) $n = N$ we have $S(a_N t, \lambda_N) \approx \sin^2 \mu t$, which may be considered as known. Then, by taking successively $n = N, N-1, ..., 1$ we can calculate $S(a_0 t, \lambda_0)$. Moreover, we can start the process by taking $a_0 = 1$, $b_0 = (1 - \lambda_0)^{1/2}$.

We now consider periodicity properties. If $\lambda \neq 1$ and $S(h) = 1$ for some nonzero $h \in \mathbb{C}$ then, by (13), $S(2h) = 0$. Furthermore $S'(2h) = 0$, by (9). It follows that $S(t)$ has period $2h$, since $S(t + 2h)$ is a solution of the differential equation (7) which satisfies the same initial conditions (8) as $S(t)$. It remains to show that there exists such an h.

Suppose first that $\lambda \in \mathbb{R}$ and $0 < \lambda < 1$. Since $S''(0) = 2$, we have $S'(t) > 0$ for small $t > 0$. If $S'(t) > 0$ for $0 < t < T$, then $S(t)$ is a positive increasing function for $0 < t < T$. Since $g_\lambda[S(t)] > 0$, we must also have $S(t) < 1$ for $0 < t < T$. From the relation

$$t = \int_0^{S(t)} dx/g_\lambda(x)^{1/2},$$

it follows that $T \leq K(\lambda)$, where

$$K(\lambda) := \int_0^1 dx/g_\lambda(x)^{1/2}.$$

Hence $S'(t)$ vanishes for some t such that $0 < t \leq K(\lambda)$ and we can now take T to be the least $t > 0$ for which $S'(t) = 0$. Then $S'(T) = 0$, $S(T) = 1$ and by letting $t \to T$ we obtain $T = K(\lambda)$.

This shows that $S(u)$ maps the interval $[0, K(\lambda)]$ bijectively onto $[0,1]$, and if

$$u(\xi) = \int_0^\xi dx/g_\lambda(x)^{1/2} \quad (0 \leq \xi \leq 1),$$

then $S[u(\xi)] = \xi$. Thus, in the real domain, the elliptic integral of the first kind is *inverted* by the function $S(u)$.

Since $\lambda \neq 1$, it follows that $S(t) = S(t,\lambda)$ has period $2K(\lambda)$. Since $\lambda \neq 0$, it follows from (15) that $S(t,\lambda)$ also has period $2iK(1 - \lambda)$. Thus $S(t,\lambda)$ is a *doubly-periodic* function, with a real period and a pure imaginary period. We will show that all periods are given by

$$2mK(\lambda) + 2niK(1 - \lambda) \quad (m,n \in \mathbb{Z}).$$

The periods of a nonconstant meromorphic function f form a discrete additive subgroup of $\mathbb{C}$. If f has two periods whose ratio is not real then, by the simple case $n = 2$ of Proposition VIII.7, it has periods ω_1, ω_2 such that all periods are given by

$$m\omega_1 + n\omega_2 \quad (m,n \in \mathbb{Z}).$$

In the present case we can take $\omega_1 = 2K(\lambda)$, $\omega_2 = 2iK(1 - \lambda)$ since, by construction, $2K(\lambda)$ is the least positive period.

Suppose next that $\lambda \in \mathbb{R}$ and either $\lambda > 1$ or $\lambda < 0$. Then, by (14) and (15), $S(t,\lambda)$ is again a doubly-periodic function with a real period and a pure imaginary period.

Suppose finally that $\lambda \in \mathbb{C} \setminus \mathbb{R}$. Without loss of generality, *we assume $\Im\lambda > 0$*. Then $g_\lambda(z)$ does not vanish in the upper half-plane $\mathfrak{H}$. It follows that there exists a unique function $h_\lambda(z)$, holomorphic for $z \in \mathfrak{H}$ with $\Re h_\lambda(z) > 0$ for z near 0, such that

$$h_\lambda(z)^2 = g_\lambda(z). \tag{22}$$

Moreover, we may extend the definition so that $h_\lambda(z)$ is continuous and (22) continues to hold for $z \in \mathfrak{H} \cup \mathbb{R}$.

We can write $S(t) = \psi(t^2)$, where

$$\psi(w) = w + a_2 w^2 + \ldots$$

is holomorphic at the origin. By inversion of series, there exists a function

$$\phi(z) = z + b_2 z^2 + \ldots \,,$$

which is holomorphic at the origin, such that $\psi[\phi(z)] = z$. For $z \in \mathfrak{H}$ near 0, put

$$u(z) = \phi(z)^{1/2},$$

where the square root is chosen so that $\Re u(z) > 0$. Then $S[u(z)] = z$. Differentiating and then squaring, we obtain

$$S'[u(z)]u'(z) = 1, \quad u'(z)^2 = 1/g_\lambda(z).$$

But $u'(z)$ also has positive real part, since $S'[u(z)] \sim 2u(z)$ for $z \to 0$. Consequently $u'(z) = 1/h_\lambda(z)$. Since $u(z) \to 0$ as $z \to 0$, we conclude that

$$u(z) = \int_0^z d\zeta/h_\lambda(\zeta), \tag{23}$$

where the path of integration is (say) a straight line segment. However, the function on the right is holomorphic for all $z \in \mathfrak{H}$. Consequently, if we define $u(z)$ by (23) then, by analytic continuation, the relation $S[u(z)] = z$ continues to hold for all $z \in \mathfrak{H}$. Letting $z \to 1$, we now obtain $S(h) = 1$ for $h = K(\lambda)$, where

$$K(\lambda) := \int_0^1 dx/g_\lambda(x)^{1/2}$$

and the square root is chosen so that $g_\lambda(x)^{1/2}$ is continuous and has positive real part for small $x > 0$ and actually, as we will see in a moment, for $0 < x < 1$. Hence $S(t)$ has period $2K(\lambda)$. Furthermore, by (15), $S(t)$ also has period $2iK(1-\lambda)$.

For $0 < x < 1$ we have

$$1/g_\lambda(x)^{1/2} = (1 - \bar{\lambda}x)^{1/2}/[4x(1-x)]^{1/2}|1 - \lambda x|.$$

If $\lambda = \mu + i\nu$, where $\nu > 0$, then $1 - \bar{\lambda}x = \gamma + i\delta$, where $\gamma = 1 - \mu x$ and $\delta = \nu x > 0$ for $0 < x < 1$. Hence

$$(1 - \bar{\lambda}x)^{1/2} = \alpha + i\beta,$$

where

$$\alpha = \{\gamma + (\gamma^2 + \delta^2)^{1/2}\}^{1/2}/\sqrt{2}, \quad 2\alpha\beta = \delta,$$

first for small $x > 0$ and then, by continuity, for $0 < x < 1$. Thus α and β are positive for $0 < x < 1$. Consequently $\Re g_\lambda(x)^{1/2} > 0$ for $0 < x < 1$ and

$$K(\lambda) = A + iB,$$

where $A > 0, B > 0$.

Similarly, for $0 < y < 1$ we have

$$1/g_{1-\lambda}(y)^{1/2} = (1 - (1 - \bar{\lambda})y)^{1/2}/[4y(1 - y)]^{1/2}|1 - (1 - \lambda)y|$$

and $1 - (1 - \bar{\lambda})y = \gamma' - i\delta'$, where $\gamma' = 1 - (1 - \mu)y$ and $\delta' = \nu y > 0$ for $0 < y < 1$. Hence

$$(1 - (1 - \bar{\lambda})y)^{1/2} = \alpha' - i\beta',$$

where

$$\alpha' = \{\gamma' + (\gamma'^2 + \delta'^2)^{1/2}\}^{1/2}/\sqrt{2}, \quad 2\alpha'\beta' = \delta'.$$

Thus α' and β' are positive for $0 < y < 1$, and

$$K(1 - \lambda) = A' - iB',$$

where $A' > 0, B' > 0$.

We will now show that the period ratio $iK(1 - \lambda)/K(\lambda)$ is not real by showing that the quotient $K(1 - \lambda)/K(\lambda)$ has positive real part. Since this is equivalent to showing that

$$AA' - BB' > 0,$$

it is sufficient to show that $\alpha\alpha' - \beta\beta' > 0$ for all $x, y \in (0,1)$. The inequality is certainly satisfied for all x, y near 0, since $\alpha \to 1, \beta \to 0$ as $x \to 0$ and $\alpha' \to 1, \beta' \to 0$ as $y \to 0$. Thus we need only show that we never have $\alpha\alpha' = \beta\beta'$. But

$$2\alpha^2 = (\gamma^2 + \delta^2)^{1/2} + \gamma, \quad 2\beta^2 = (\gamma^2 + \delta^2)^{1/2} - \gamma,$$

with analogous expressions for $2\alpha'^2, 2\beta'^2$. Hence, if $\alpha\alpha' = \beta\beta'$, then by squaring we obtain

$$[(\gamma^2 + \delta^2)^{1/2} + \gamma][(\gamma'^2 + \delta'^2)^{1/2} + \gamma'] = [(\gamma^2 + \delta^2)^{1/2} - \gamma][(\gamma'^2 + \delta'^2)^{1/2} - \gamma'],$$

which reduces to

$$\gamma(\gamma'^2 + \delta'^2)^{1/2} = -\gamma'(\gamma^2 + \delta^2)^{1/2}.$$

Squaring again, we obtain $\gamma^2\delta'^2 = \gamma'^2\delta^2$. Since the previous equation shows that γ and γ' do not have the same sign, it follows that

$$\gamma\delta' + \gamma'\delta = 0.$$

Substituting for $\gamma,\delta,\gamma',\delta'$ their explicit expressions, this takes the form $v(x + y - xy) = 0$. Hence $x(1 - y) + y = 0$, which is impossible if $0 < y < 1$ and $x > 0$.

The relation $S[u(z)] = z$, where $u(z)$ is defined by (23), shows that the elliptic integral of the first kind is *inverted* by the elliptic function $S(u)$. We may use this to simplify other elliptic integrals. The change of variables $x = S(u)$ replaces the integral

$$\int R(x)\, dx/g_\lambda(x)^{1/2}$$

by $\int R[S(u)]\, du$. Following Jacobi, we take

$$E(u): = \int_0^u [1 - \lambda S(v)]\, dv \tag{24}$$

as the standard elliptic integral of the second kind, and

$$\Pi(u,a): = (\lambda/2)\int_0^u S'(a)S(v)dv/[1 - \lambda S(a)S(v)] \tag{25}$$

as the standard elliptic integral of the third kind.

Many properties of these functions may be obtained by integration from corresponding properties of the function $S(u)$. By way of example, we show that

$$E(u + a) - E(u - a) - 2E(a) = -\lambda S'(a)S(u)/[1 - \lambda S(a)S(u)]. \tag{26}$$

Indeed it is evident that both sides vanish when $u = 0$, and it follows from (12) that they have the same derivative with respect to u. Integrating (26) with respect to u, we further obtain

$$\Pi(u,a) = uE(a) - (1/2)\int_{u-a}^{u+a} E(v)dv. \tag{27}$$

Thus the function $\Pi(u,a)$, which depends on two variables (as well as the parameter λ) can be expressed in terms of functions of only one variable. Furthermore, we have the *interchange property* (due, in other notation, to Legendre)

$$\Pi(u,a) - uE(a) = \Pi(a,u) - aE(u). \tag{28}$$

If we take $u = 2K = 2K(\lambda)$, then $S'(u) = 0$ and hence $\Pi(a,u) = 0$. Thus

$$\Pi(2K,a) = 2KE(a) - aE(2K), \tag{29}$$

which shows that the complete elliptic integral of the third kind can be expressed in terms of complete and incomplete elliptic integrals of the first and second kinds.

In order to justify taking $\Pi(u,a)$ as the standard elliptic integral of the third kind, we show finally that $S(a)$ takes all complex values. Otherwise, if $S(u) \neq c$ for all $u \in \mathbb{C}$, then $c \neq 0$ and

$$f(u) = S(u)/[S(u) - c]$$

is holomorphic in the whole complex plane. Furthermore, it is doubly-periodic with two periods ω_1, ω_2 whose ratio is not real. Since it is bounded in the parallelogram with vertices $0, \omega_1, \omega_2, \omega_1 + \omega_2$, it follows that it is bounded in $\mathbb{C}$. Hence, by Liouville's theorem, f is a constant. Since S is not constant and $c \neq 0$, this is a contradiction.

4 Theta functions

Theta functions arise not only in connection with elliptic functions (as we will see), but also in problems of heat conduction, statistical mechanics and number theory.

Consider the bi-infinite series

$$\sum_{n=-\infty}^{n=\infty} q^{n^2} z^n = 1 + \sum_{n=1}^{n=\infty} q^{n^2} z^n + \sum_{n=1}^{n=\infty} q^{n^2} z^{-n},$$

where $q, z \in \mathbb{C}$ and $z \neq 0$. Both series on the right converge if $|q| < 1$, both diverge if $|q| > 1$, and at most one converges if $|q| = 1$. Thus we now assume $|q| < 1$.

A remarkable representation for the series on the left was given by Jacobi (1829), in §64 of his *Fundamenta Nova*, and is now generally known as *Jacobi's triple product formula*:

PROPOSITION 2 *If $|q| < 1$ and $z \neq 0$, then*

$$\sum_{n=-\infty}^{n=\infty} q^{n^2} z^n = \prod_{n=1}^{\infty} (1 + q^{2n-1}z)(1 + q^{2n-1}z^{-1})(1 - q^{2n}). \tag{30}$$

Proof Put

$$f_N(z) = \prod_{n=1}^{N} (1 + q^{2n-1}z)(1 + q^{2n-1}z^{-1}).$$

Then we can write

$$f_N(z) = c_0^N + c_1^N (z + z^{-1}) + \ldots + c_N^N (z^N + z^{-N}). \tag{31}$$

To determine the coefficients c_n^N we use the functional relation

$$f_N(q^2 z) = (1 + q^{2N+1}z)(1 + q^{-1}z^{-1})f_N(z)/(1 + qz)(1 + q^{2N-1}z^{-1})$$
$$= (1 + q^{2N+1}z)f_N(z)/(qz + q^{2N}).$$

Multiplying both sides by $qz + q^{2N}$ and equating coefficients of z^{n+1} we get, for $n = 0,1,...,$ $N - 1$,

$$q^{2n+1} c_n^N + q^{2N+2n+2} c_{n+1}^N = c_{n+1}^N + q^{2N+1} c_n^N,$$

i.e.,

$$q^{2n+1}(1 - q^{2N-2n}) c_n^N = (1 - q^{2N+2n+2}) c_{n+1}^N.$$

But, since $\sum_{n=1}^{N} (2n - 1) = N^2$, it follows from the definition of $f_N(z)$ that $c_N^N = q^{N^2}$. Hence

$$c_n^N = (1 - q^{2N+2n+2})(1 - q^{2N+2n+4})..(1 - q^{4N})q^{n^2}/(1 - q^2)(1 - q^4)..(1 - q^{2N-2n}) \quad (0 \leq n < N).$$

If $|q| < 1$ and $z \neq 0$, then the infinite products

$$\prod_{n=1}^{\infty} (1 + q^{2n-1}z), \quad \prod_{n=1}^{\infty} (1 + q^{2n-1}z^{-1}), \quad \prod_{n=1}^{\infty} (1 - q^{2n})$$

are all convergent. From the convergence of the last it follows that, for each fixed n,

$$\lim_{N \to \infty} c_n^N = q^{n^2}/\prod_{k=1}^{\infty} (1 - q^{2k}).$$

Moreover, there exists a constant $A > 0$, depending on q but not on n or N, such that

$$|c_n^N| \leq A|q|^{n^2}.$$

For we can choose $B > 0$ so that $|\prod_{k=1}^{m} (1 - q^{2k})| \geq B$ for all m, we can choose $C > 0$ so that $|\prod_{k=1}^{m} (1 - q^{2k})| \leq C$ for all m, and we can then take $A = C/B^2$. Since the series $\sum_{n=-\infty}^{n=\infty} q^{n^2}z^n$ is absolutely convergent, it follows that we can proceed to the limit term by term in (31) to obtain (30). $\square$

In the series $\sum_{n=-\infty}^{n=\infty} q^{n^2}z^n$ we now put

$$q = e^{\pi i \tau}, \quad z = e^{2\pi i v},$$

so that $|q| < 1$ corresponds to $\Im\tau > 0$, and we define the *theta function*

$$\theta(v;\tau) = \sum_{n=-\infty}^{n=\infty} e^{\pi i \tau n^2} e^{2\pi i v n}.$$

The function $\theta(v;\tau)$ is holomorphic in v and τ for all $v \in \mathbb{C}$ and $\tau \in \mathfrak{H}$ (the upper half-plane). Since initially we will be more interested in the dependence on v, with τ just a parameter, we will often write $\theta(v)$ in place of $\theta(v;\tau)$. Furthermore, we will still use q as an abbreviation for $e^{\pi i \tau}$.

Evidently

$$\theta(v + 1) = \theta(v) = \theta(-v).$$

Moreover,

$$\theta(v + \tau) = \sum_{n=-\infty}^{n=\infty} q^{n^2+2n} e^{2\pi i v n}$$

$$= q^{-1} e^{-2\pi i v} \sum_{n=-\infty}^{n=\infty} q^{(n+1)^2} e^{2\pi i v(n+1)}$$

$$= e^{-\pi i(2v+\tau)} \theta(v).$$

It may be immediately verified that

$$\partial^2\theta/\partial v^2 = -4\pi^2 q \, \partial\theta/\partial q = 4\pi i \, \partial\theta/\partial\tau,$$

which becomes the partial differential equation of heat conduction in one dimension on putting $\tau = 4\pi i t$.

By Proposition 2, we have also the product representation

$$\theta(v) = \prod_{n=1}^{\infty} (1 + q^{2n-1}e^{2\pi i v})(1 + q^{2n-1}e^{-2\pi i v})(1 - q^{2n}).$$

It follows that the points

$$v = 1/2 + \tau/2 + m + n\tau \quad (m,n \in \mathbb{Z})$$

are simple zeros of $\theta(v)$, and that these are the only zeros.

One important property of the theta function is almost already known to us:

PROPOSITION 3 *For all $v \in \mathbb{C}$ and $\tau \in \mathfrak{H}$,*

$$\theta(v; -1/\tau) = (\tau/i)^{1/2} e^{\pi i \tau v^2} \theta(\tau v; \tau), \tag{32}$$

where the square root is chosen to have positive real part.

Proof Suppose first that $\tau = iy$, where $y > 0$. We wish to show that

$$\sum_{n=-\infty}^{n=\infty} e^{-n^2\pi/y} e^{2n\pi i v} = y^{1/2} \sum_{n=-\infty}^{n=\infty} e^{-(v+n)^2\pi y}.$$

But this was already proved in Proposition IX.10.

Thus (32) holds when τ is pure imaginary. Since, with the stated choice of square root, both sides of (32) are holomorphic functions for $v \in \mathbb{C}$ and $\tau \in \mathfrak{H}$, the relation continues to hold throughout this extended domain, by analytic continuation. $\square$

Following Hermite (1858), for any integers α,β we now put

$$\theta_{\alpha,\beta}(v) = \theta_{\alpha,\beta}(v;\tau) = \sum_{n=-\infty}^{n=\infty} (-1)^{\beta n} e^{\pi i \tau(n+\alpha/2)^2} e^{2\pi i v(n+\alpha/2)}.$$

(The factor $(-1)^{\beta n}$ may be made less conspicuous by writing it as $e^{\pi i \beta n}$.) Since

$$\theta_{\alpha+2,\beta}(v) = (-1)^{\beta} \theta_{\alpha,\beta}(v), \quad \theta_{\alpha,\beta+2}(v) = \theta_{\alpha,\beta}(v),$$

there are only four essentially distinct functions, namely

$$\theta_{00}(v) = \sum_{n=-\infty}^{n=\infty} e^{\pi i \tau n^2} e^{2\pi i v n},$$

$$\theta_{01}(v) = \sum_{n=-\infty}^{n=\infty} (-1)^n e^{\pi i \tau n^2} e^{2\pi i v n},$$

$$\theta_{10}(v) = \sum_{n=-\infty}^{n=\infty} e^{\pi i \tau (n+1/2)^2} e^{\pi i v (2n+1)},$$

$$\theta_{11}(v) = \sum_{n=-\infty}^{n=\infty} (-1)^n e^{\pi i \tau (n+1/2)^2} e^{\pi i v (2n+1)}.$$

(33)

Moreover,

$$\theta_{00}(v;\tau) = \theta(v;\tau), \quad \theta_{01}(v;\tau) = \theta(v+1/2;\tau),$$

$$\theta_{10}(v;\tau) = e^{\pi i (v+\tau/4)} \theta(v+\tau/2;\tau), \quad \theta_{11}(v;\tau) = e^{\pi i (v+\tau/4)} \theta(v+1/2+\tau/2;\tau).$$

In fact, for all integers m,n,

$$\theta_{\alpha,\beta}(v+m\tau/2+n/2) = \theta_{\alpha+m,\beta+n}(v) e^{-\pi i (mv+m^2\tau/4-\alpha n/2)}.$$

(34)

Since the zeros of $\theta(v;\tau)$ are the points $v = 1/2 + \tau/2 + m\tau + n$, the zeros of $\theta_{\alpha,\beta}(v)$ are the points

$$v = (\beta+1)/2 + (\alpha+1)\tau/2 + m\tau + n \quad (m,n \in \mathbb{Z}).$$

The notation for theta functions is by no means standardized. Hermite's notation reflects the underlying symmetry, but for purposes of comparison we indicate its connection with the more commonly used notation in Whittaker and Watson [29]:

$$\theta_{00}(v;\tau) = \vartheta_3(\pi v, q), \quad \theta_{01}(v;\tau) = \vartheta_4(\pi v, q),$$

$$\theta_{10}(v;\tau) = \vartheta_2(\pi v, q), \quad \theta_{11}(v;\tau) = i\, \vartheta_1(\pi v, q).$$

It follows from the definitions that $\theta_{00}(v;\tau)$, $\theta_{01}(v;\tau)$ and $\theta_{10}(v;\tau)$ are even functions of v, whereas $\theta_{11}(v;\tau)$ is an odd function of v. Moreover $\theta_{00}(v;\tau)$ and $\theta_{01}(v;\tau)$ are periodic with period 1 in v, but $\theta_{10}(v;\tau)$ and $\theta_{11}(v;\tau)$ change sign when v is increased by 1.

All four theta functions satisfy the same partial differential equation as $\theta(v;\tau)$. From the product expansion of $\theta(v;\tau)$ we obtain the product expansions

$$\theta_{00}(v) = Q_0 \prod_{n=1}^{\infty} (1 + q^{2n-1}e^{2\pi i v})(1 + q^{2n-1}e^{-2\pi i v}),$$

$$\theta_{01}(v) = Q_0 \prod_{n=1}^{\infty} (1 - q^{2n-1}e^{2\pi i v})(1 - q^{2n-1}e^{-2\pi i v}),$$

$$\theta_{10}(v) = 2Q_0 e^{\pi i \tau/4} \cos \pi v \prod_{n=1}^{\infty} (1 + q^{2n}e^{2\pi i v})(1 + q^{2n}e^{-2\pi i v}),$$

$$\theta_{11}(v) = 2iQ_0 e^{\pi i \tau/4} \sin \pi v \prod_{n=1}^{\infty} (1 - q^{2n}e^{2\pi i v})(1 - q^{2n}e^{-2\pi i v}),$$

(35)

where $q = e^{\pi i \tau}$ and

$$Q_0 = \prod_{n=1}^{\infty} (1 - q^{2n}).$$

In particular,

$$\theta_{00}(0) = Q_0 \prod_{n=1}^{\infty} (1 + q^{2n-1})^2,$$

$$\theta_{01}(0) = Q_0 \prod_{n=1}^{\infty} (1 - q^{2n-1})^2,$$

$$\theta_{10}(0) = 2q^{1/4} Q_0 \prod_{n=1}^{\infty} (1 + q^{2n})^2.$$

By differentiating with respect to v and then putting $v = 0$, we obtain also $\theta_{11}'(0) = 2\pi i q^{1/4}Q_0^3$.
But

$$Q_0 = \prod_{n=1}^{\infty} (1 - q^n)(1 + q^n)$$

$$= \prod_{n=1}^{\infty} (1 - q^{2n})(1 - q^{2n-1})(1 + q^{2n})(1 + q^{2n-1}),$$

which implies

$$\prod_{n=1}^{\infty} (1 - q^{2n-1})(1 + q^{2n})(1 + q^{2n-1}) = 1.$$

It follows that

$$\theta_{00}(0)\theta_{01}(0)\theta_{10}(0) = 2q^{1/4} Q_0^3$$

and hence

$$\theta_{11}'(0) = \pi i \theta_{00}(0)\theta_{01}(0)\theta_{10}(0). \tag{36}$$

It is evident from their series definitions that, when q is replaced by $-q$, the functions θ_{00} and θ_{01} are interchanged, whereas the functions $q^{-1/4}\theta_{10}$ and $q^{-1/4}\theta_{11}$ are unaltered. Hence

$$\theta_{00}(v;\tau + 1) = \theta_{01}(v;\tau), \quad \theta_{10}(v;\tau + 1) = e^{\pi i/4} \theta_{10}(v;\tau),$$

$$\theta_{01}(v;\tau + 1) = \theta_{00}(v;\tau), \quad \theta_{11}(v;\tau + 1) = e^{\pi i/4} \theta_{11}(v;\tau). \tag{37}$$

From Proposition 3 we obtain also the transformation formulas

$$\theta_{00}(v; -1/\tau) = (\tau/i)^{1/2} e^{\pi i \tau v^2} \theta_{00}(\tau v;\tau), \quad \theta_{10}(v; -1/\tau) = (\tau/i)^{1/2} e^{\pi i \tau v^2} \theta_{01}(\tau v;\tau),$$

$$\theta_{01}(v; -1/\tau) = (\tau/i)^{1/2} e^{\pi i \tau v^2} \theta_{10}(\tau v;\tau), \quad \theta_{11}(v; -1/\tau) = -i(\tau/i)^{1/2} e^{\pi i \tau v^2} \theta_{11}(\tau v;\tau). \tag{38}$$

Up to this point we have used Hermite's notation just to dress up old results in new clothes. The next result breaks fresh ground.

PROPOSITION 4 *For all* $v,w \in \mathbb{C}$ *and* $\tau \in \mathfrak{H}$,

$$\theta_{00}(v;\tau)\theta_{00}(w;\tau) = \theta_{00}(v+w;2\tau)\theta_{00}(v-w;2\tau) + \theta_{10}(v+w;2\tau)\theta_{10}(v-w;2\tau),$$

$$\theta_{10}(v;\tau)\theta_{10}(w;\tau) = \theta_{10}(v+w;2\tau)\theta_{00}(v-w;2\tau) + \theta_{00}(v+w;2\tau)\theta_{10}(v-w;2\tau),$$

$$\theta_{00}(v;\tau)\theta_{01}(w;\tau) = \theta_{01}(v+w;2\tau)\theta_{01}(v-w;2\tau) + \theta_{11}(v+w;2\tau)\theta_{11}(v-w;2\tau),$$

$$\theta_{01}(v;\tau)\theta_{01}(w;\tau) = \theta_{00}(v+w;2\tau)\theta_{00}(v-w;2\tau) - \theta_{10}(v+w;2\tau)\theta_{10}(v-w;2\tau),$$

$$\theta_{10}(v;\tau)\theta_{11}(w;\tau) = \theta_{11}(v+w;2\tau)\theta_{01}(v-w;2\tau) - \theta_{01}(v+w;2\tau)\theta_{11}(v-w;2\tau),$$

$$\theta_{11}(v;\tau)\theta_{11}(w;\tau) = \theta_{10}(v+w;2\tau)\theta_{00}(v-w;2\tau) - \theta_{00}(v+w;2\tau)\theta_{10}(v-w;2\tau).$$

Proof From the definition of θ_{00},

$$\theta_{00}(v;\tau)\theta_{00}(w;\tau) = \sum_{j,k} e^{\pi i \tau(j^2+k^2)} e^{2\pi i v j} e^{2\pi i w k} = \sum_{j+k \text{ even}} + \sum_{j+k \text{ odd}} \cdot$$

In the first sum on the right we can write $j + k = 2m, j - k = 2n$. Then $j = m + n, k = m - n$ and

$$\sum_{j+k \text{ even}} = \sum_{m,n \in \mathbb{Z}} e^{2\pi i \tau(m^2+n^2)} e^{2\pi i(v+w)m} e^{2\pi i(v-w)n} = \theta_{00}(v+w;2\tau)\theta_{00}(v-w;2\tau).$$

In the second sum we can write $j + k = 2m + 1, j - k = 2n + 1$. Then $j = m + n + 1, k = m - n$ and

$$\sum_{j+k \text{ odd}} = \sum_{m,n \in \mathbb{Z}} e^{2\pi i \tau\{(m+1/2)^2+(n+1/2)^2\}} e^{2\pi i v(m+n+1)} e^{2\pi i w(m-n)} = \theta_{10}(v+w;2\tau)\theta_{10}(v-w;2\tau).$$

Adding, we obtain the first relation of the proposition.

We obtain the second relation from the first by replacing v by $v + \tau/2$ and w by $w + \tau/2$. The remaining relations are obtained from the first two by increasing v and/or w by $1/2$. $\square$

By taking $w = v$ in Proposition 4, and adding or subtracting pairs of equations whose right sides differ only in one sign, we obtain the *duplication formulas*:

PROPOSITION 5 *For all* $v \in \mathbb{C}$ *and* $\tau \in \mathfrak{H}$,

$$\theta_{00}(2v;2\tau) = [\theta_{00}{}^2(v;\tau) + \theta_{01}{}^2(v;\tau)]/2\theta_{00}(0;2\tau) = [\theta_{10}{}^2(v;\tau) - \theta_{11}{}^2(v;\tau)]/2\theta_{10}(0;2\tau),$$

$$\theta_{10}(2v;2\tau) = [\theta_{00}{}^2(v;\tau) - \theta_{01}{}^2(v;\tau)]/2\theta_{10}(0;2\tau) = [\theta_{10}{}^2(v;\tau) + \theta_{11}{}^2(v;\tau)]/2\theta_{00}(0;2\tau),$$

$$\theta_{01}(2v;2\tau) = \theta_{00}(v;\tau)\theta_{01}(v;\tau)/\theta_{01}(0;2\tau),$$

$$\theta_{11}(2v;2\tau) = \theta_{10}(v;\tau)\theta_{11}(v;\tau)/\theta_{01}(0;2\tau). \quad \square$$

From Proposition 4 we can also derive the following *addition formulas*:

PROPOSITION 6 *For all* $v, w \in \mathbb{C}$ *and* $\tau \in \mathfrak{H}$,

$$\theta_{01}{}^2(0)\theta_{01}(v+w)\theta_{01}(v-w) = \theta_{01}{}^2(v)\theta_{01}{}^2(w) - \theta_{11}{}^2(v)\theta_{11}{}^2(w)$$
$$= \theta_{00}{}^2(v)\theta_{00}{}^2(w) - \theta_{10}{}^2(v)\theta_{10}{}^2(w),$$

$$\theta_{00}(0)\theta_{01}(0)\theta_{00}(v+w)\theta_{01}(v-w) = \theta_{00}(v)\theta_{01}(v)\theta_{00}(w)\theta_{01}(w) + \theta_{10}(v)\theta_{11}(v)\theta_{10}(w)\theta_{11}(w),$$

$$\theta_{01}(0)\theta_{10}(0)\theta_{10}(v+w)\theta_{01}(v-w) = \theta_{01}(v)\theta_{10}(v)\theta_{01}(w)\theta_{10}(w) + \theta_{00}(v)\theta_{11}(v)\theta_{00}(w)\theta_{11}(w),$$

$$\theta_{00}(0)\theta_{10}(0)\theta_{11}(v+w)\theta_{01}(v-w) = \theta_{01}(v)\theta_{11}(v)\theta_{00}(w)\theta_{10}(w) + \theta_{00}(v)\theta_{10}(v)\theta_{01}(w)\theta_{11}(w),$$

where all theta functions have the same second argument τ.

Proof Consider the second relation. If we use the first and fourth relations of Proposition 4 to evaluate the products $\theta_{00}(v)\theta_{00}(w)$ and $\theta_{01}(v)\theta_{01}(w)$, we obtain

$$\theta_{00}(v)\theta_{01}(v)\theta_{00}(w)\theta_{01}(w) = \theta_{00}{}^2(v+w;2\tau)\theta_{00}{}^2(v-w;2\tau) - \theta_{10}{}^2(v+w;2\tau)\theta_{10}{}^2(v-w;2\tau).$$

Similarly, if we use the second and sixth relations of Proposition 4 to evaluate the products $\theta_{10}(v)\theta_{10}(w)$ and $\theta_{11}(v)\theta_{11}(w)$, we obtain

$$\theta_{10}(v)\theta_{11}(v)\theta_{10}(w)\theta_{11}(w) = \theta_{10}{}^2(v+w;2\tau)\theta_{00}{}^2(v-w;2\tau) - \theta_{00}{}^2(v+w;2\tau)\theta_{10}{}^2(v-w;2\tau).$$

Hence, in the second relation of the present proposition the right side is equal to

$$[\theta_{00}{}^2(v+w;2\tau) + \theta_{10}{}^2(v+w;2\tau)][\theta_{00}{}^2(v-w;2\tau) - \theta_{10}{}^2(v-w;2\tau)].$$

On the other hand, if we use the first and fourth relations of Proposition 4 to evaluate the products $\theta_{00}(0)\theta_{00}(v+w)$ and $\theta_{01}(0)\theta_{01}(v-w)$, we see that the left side is likewise equal to

$$[\theta_{00}{}^2(v+w;2\tau) + \theta_{10}{}^2(v+w;2\tau)][\theta_{00}{}^2(v-w;2\tau) - \theta_{10}{}^2(v-w;2\tau)].$$

This proves the second relation of the proposition, and the others may be proved similarly. ∎

COROLLARY 7 *For all* $v \in \mathbb{C}$ *and* $\tau \in \mathfrak{H}$,

$$\theta_{00}{}^2(0)\theta_{01}{}^2(v) + \theta_{10}{}^2(0)\theta_{11}{}^2(v) = \theta_{01}{}^2(0)\theta_{00}{}^2(v), \qquad (39)$$

$$\theta_{10}{}^2(0)\theta_{01}{}^2(v) + \theta_{00}{}^2(0)\theta_{11}{}^2(v) = \theta_{01}{}^2(0)\theta_{10}{}^2(v). \qquad (40)$$

Moreover, for all $\tau \in \mathfrak{H}$,

$$\theta_{00}{}^4(0) = \theta_{01}{}^4(0) + \theta_{10}{}^4(0). \qquad (41)$$

Proof We obtain (39) and (40) from the first relation of Proposition 6 by taking $w = 1/2$ and $w = (1 + \tau)/2$ respectively. We obtain (41) from (39) by taking $v = 1/2$. $\square$

If we regard (39) and (40) as a system of simultaneous linear equations for the unknowns $\theta_{01}^2(v), \theta_{11}^2(v)$, then the determinant of this system is $\theta_{00}^4(0) - \theta_{10}^4(0) = \theta_{01}^4(0) \neq 0$. It follows that the square of any theta function may be expressed as a linear combination of the squares of any other two theta functions.

By substituting for the theta functions their expansions as infinite products, the formula (41) may be given the following remarkable form:

$$\Pi_{n=1}^\infty (1 + q^{2n-1})^8 = \Pi_{n=1}^\infty (1 - q^{2n-1})^8 + 16q \, \Pi_{n=1}^\infty (1 + q^{2n})^8.$$

PROPOSITION 8 *For all* $v \in \mathbb{C}$ *and* $\tau \in \mathfrak{H}$,

$$\{\theta_{00}(v)/\theta_{01}(v)\}' = \pi i \, \theta_{10}^2(0)\theta_{10}(v)\theta_{11}(v)/\theta_{01}^2(v), \tag{42}$$

$$\{\theta_{10}(v)/\theta_{01}(v)\}' = \pi i \, \theta_{00}^2(0)\theta_{00}(v)\theta_{11}(v)/\theta_{01}^2(v), \tag{43}$$

$$\{\theta_{11}(v)/\theta_{01}(v)\}' = \pi i \, \theta_{01}^2(0)\theta_{00}(v)\theta_{10}(v)/\theta_{01}^2(v), \tag{44}$$

$$\{\theta_{01}'(v)/\theta_{01}(v)\}' = \theta_{01}''(0)/\theta_{01}(0) + \pi^2 \, \theta_{00}^2(0)\theta_{10}^2(0)\theta_{11}^2(v)/\theta_{01}^2(v). \tag{45}$$

Proof By differentiating the second relation of Proposition 6 with respect to w and then putting $w = 0$, we obtain

$$\theta_{00}(0)\theta_{01}(0)[\theta_{00}'(v)\theta_{01}(v) - \theta_{00}(v)\theta_{01}'(v)] = \theta_{10}(0)\theta_{11}'(0)\theta_{10}(v)\theta_{11}(v),$$

since not only $\theta_{11}(0) = 0$ but also $\theta_{00}'(0) = \theta_{01}'(0) = \theta_{10}'(0) = 0$. Dividing by $\theta_{01}^2(v)$ and recalling the expression (36) for $\theta_{11}'(0)$, we obtain (42). Similarly, from the third and fourth relations of Proposition 6 we obtain (43) and (44).

In the same way, if we differentiate the first relation of Proposition 6 twice with respect to w and then put $w = 0$, we obtain

$$\theta_{01}^2(0)[\theta_{01}''(v)\theta_{01}(v) - \theta_{01}'(v)^2] = \theta_{01}(0)\theta_{01}''(0)\theta_{01}^2(v) - \theta_{11}'(0)^2\theta_{11}^2(v).$$

Hence, using (36) again, we obtain (45). $\square$

We are now in a position to make the connection between theta functions and elliptic functions.

5   Jacobian elliptic functions

The behaviour of the theta functions when their argument is increased by 1 or τ makes it clear that doubly-periodic functions may be constructed from their quotients. We put

$$\operatorname{sn} u = \operatorname{sn}(u;\tau) := -i\,\theta_{00}(0)\theta_{11}(v)/\theta_{10}(0)\theta_{01}(v),$$

$$\operatorname{cn} u = \operatorname{cn}(u;\tau) := \theta_{01}(0)\theta_{10}(v)/\theta_{10}(0)\theta_{01}(v), \tag{46}$$

$$\operatorname{dn} u = \operatorname{dn}(u;\tau) := \theta_{01}(0)\theta_{00}(v)/\theta_{00}(0)\theta_{01}(v),$$

where $u = \pi\,\theta_{00}^2(0)v$.

The constant multiples are chosen so that, in addition to $\operatorname{sn} 0 = 0$, we have $\operatorname{cn} 0 = \operatorname{dn} 0 = 1$. The independent variable is scaled so that, by (42)-(44),

$$d(\operatorname{sn} u)/du = \operatorname{cn} u\,\operatorname{dn} u,$$

$$d(\operatorname{cn} u)/du = -\operatorname{sn} u\,\operatorname{dn} u, \tag{47}$$

$$d(\operatorname{dn} u)/du = -\lambda\,\operatorname{sn} u\,\operatorname{cn} u,$$

where

$$\lambda = \lambda(\tau) := \theta_{10}^4(0;\tau)/\theta_{00}^4(0;\tau). \tag{48}$$

It follows at once from the definitions that $\operatorname{sn} u$ is an odd function of u, whereas $\operatorname{cn} u$ and $\operatorname{dn} u$ are even functions of u. It follows from (41) that

$$1 - \lambda(\tau) = \theta_{01}^4(0;\tau)/\theta_{00}^4(0;\tau), \tag{49}$$

and from (39)-(40) that

$$\operatorname{cn}^2 u = 1 - \operatorname{sn}^2 u, \quad \operatorname{dn}^2 u = 1 - \lambda\,\operatorname{sn}^2 u. \tag{50}$$

Evidently (47) implies

$$d(\operatorname{sn}^2 u)/du = 2\operatorname{sn} u\,\operatorname{cn} u\,\operatorname{dn} u,$$

$$d^2(\operatorname{sn}^2 u)/du^2 = 2(\operatorname{cn}^2 u\,\operatorname{dn}^2 u - \operatorname{sn}^2 u\,\operatorname{dn}^2 u - \lambda\,\operatorname{sn}^2 u\,\operatorname{cn}^2 u).$$

If we write $S(u) = S(u;\tau) := \operatorname{sn}^2 u$ and use (50), we can rewrite this in the form

$$d^2 S/du^2 = 2[(1 - S)(1 - \lambda S) - S(1 - \lambda S) - \lambda S(1 - S)]$$

$$= 6\lambda S^2 - 4(1 + \lambda)S + 2.$$

Since $S(0) = S'(0) = 0$, we conclude that $S(u)$ coincides with the function denoted by the same symbol in §3. However, it should be noted that now λ is not given, but is determined by τ. Thus the question arises: can we choose $\tau \in \mathfrak{H}$ (the upper half-plane) so that $\lambda(\tau)$ is any prescribed complex number other than 0 or 1?

For many applications it is sufficient to know that we can choose $\tau \in \mathfrak{H}$ so that $\lambda(\tau)$ is any prescribed real number between 0 and 1. Since this case is much simpler, we will deal with it now and defer treatment of the general case until the next section. We have

$$\lambda(\tau) = 1 - \theta_{01}^4(0;\tau)/\theta_{00}^4(0;\tau) = 1 - \prod_{n=1}^{\infty} \{(1 - q^{2n-1})/(1 + q^{2n-1})\}^8,$$

where $q = e^{\pi i \tau}$. If $\tau = iy$, where $y > 0$, then $0 < q < 1$. Moreover, as y increases from 0 to ∞, q decreases from 1 to 0 and the infinite product increases from 0 to 1. Thus $\lambda(\tau)$ decreases continuously from 1 to 0 and, for each $w \in (0,1)$, there is a unique pure imaginary $\tau \in \mathfrak{H}$ such that $\lambda(\tau) = w$.

It should be mentioned that, also with our previous approach, $S(u)$ could have been recognized as the square of a meromorphic function by defining sn u, cn u, dn u to be the solution, for *given* $\lambda \in \mathbb{C}$, of the system of differential equations (47) which satisfies the initial condition sn $0 = 0$, cn $0 = $ dn $0 = 1$.

Elliptic functions were first defined by Abel (1827) as the inverses of elliptic integrals. His definitions were modified by Jacobi (1829) to accord with Legendre's normal form for elliptic integrals, and the functions sn u, cn u, dn u are generally known as the *Jacobian elliptic functions*. The actual notation is due to Gudermann (1838). The definition by means of theta functions was given later by Jacobi (1838) in lectures.

Several properties of the Jacobian elliptic functions are easy consequences of the later definition. In the first place, all three are meromorphic in the whole u-plane, since the theta functions are everywhere holomorphic. Their poles are determined by the zeros of $\theta_{01}(v)$ and are all simple. Similarly, the zeros of sn u, cn u and dn u are determined by the zeros of $\theta_{11}(v), \theta_{10}(v)$ and $\theta_{00}(v)$ respectively and are all simple. If we put

$$\mathbf{K} = \mathbf{K}(\tau) := \pi \theta_{00}^2(0;\tau)/2, \quad \mathbf{K'} = \mathbf{K'}(\tau) := \tau \mathbf{K}(\tau)/i, \tag{51}$$

then we have

Poles of sn u, cn u, dn u: $u = 2m\mathbf{K} + (2n + 1)i\mathbf{K'}$ $(m,n \in \mathbb{Z})$.   (52)

Zeros of sn u: $u = 2m\mathbf{K} + 2ni\mathbf{K'}$,

cn u: $u = (2m + 1)\mathbf{K} + 2ni\mathbf{K'}$, $(m,n \in \mathbb{Z})$ (53)

dn u: $u = (2m + 1)\mathbf{K} + (2n + 1)i\mathbf{K'}$.

From the definitions (46) of the Jacobian elliptic functions and the behaviour of the theta functions when v is increased by 1 or τ we further obtain

$$\text{sn } u = -\text{ sn } (u + 2\mathbf{K}) = \text{sn } (u + 2i\mathbf{K}'),$$

$$\text{cn } u = -\text{ cn } (u + 2\mathbf{K}) = -\text{ cn } (u + 2i\mathbf{K}'), \tag{54}$$

$$\text{dn } u = \text{ dn } (u + 2\mathbf{K}) = -\text{ dn } (u + 2i\mathbf{K}').$$

It follows that all three functions are *doubly-periodic*. In fact sn u has periods $4\mathbf{K}$ and $2i\mathbf{K}'$, cn u has periods $4\mathbf{K}$ and $2\mathbf{K}+ 2i\mathbf{K}'$, and dn u has periods $2\mathbf{K}$ and $4i\mathbf{K}'$. In each case the ratio of the two periods is not real, since $\tau \in \mathfrak{H}$.

Since any period must equal a difference between two poles, it must have the form $2m\mathbf{K} + 2ni\mathbf{K}'$ for some $m,n \in \mathbb{Z}$. Since $4\mathbf{K}$ and $2i\mathbf{K}'$ are periods of sn u, but $2\mathbf{K}$ is not, and since any integral linear combination of periods is again a period, it follows that the periods of sn u are precisely the integral linear combinations of $4\mathbf{K}$ and $2i\mathbf{K}'$. Similarly the periods of cn u are the integral linear combinations of $4\mathbf{K}$ and $2\mathbf{K} + 2i\mathbf{K}'$, and the periods of dn u are the integral linear combinations of $2\mathbf{K}$ and $4i\mathbf{K}'$.

It was shown in §3 that, if $0 < \lambda < 1$, then $S(t,\lambda)$ has least positive period $2K(\lambda)$, where

$$K(\lambda) = \int_0^1 dx/g_\lambda(x)^{1/2}.$$

But, as we have seen, there is a unique pure imaginary $\tau \in \mathfrak{H}$ such that $\lambda = \lambda(\tau)$, and $2K[\lambda(\tau)]$ is then the least positive period of $\text{sn}^2 (u;\tau)$. Since the periods of $\text{sn}^2 (u;\tau)$ are $2m\mathbf{K} + 2ni\mathbf{K}'$ $(m,n \in \mathbb{Z})$, and since $\mathbf{K},\mathbf{K}'$ are real and positive when τ is pure imaginary, it follows that

$$K[\lambda(\tau)] = \mathbf{K}(\tau).$$

The domain of validity of this relation may be extended by appealing to results which will be established in §6. In fact it holds, by analytic continuation, for all τ in the region $\mathfrak{D}$ illustrated in Figure 3, since $\lambda(\tau) \in \mathfrak{H}$ for $\tau \in \mathfrak{D}$.

From the definitions (46) of the Jacobian elliptic functions, the addition formulas for the theta functions (Proposition 6) and the expression (48) for λ, we obtain *addition formulas* for the Jacobian functions:

$$\text{sn } (u_1 + u_2) = (\text{sn } u_1 \text{ cn } u_2 \text{ dn } u_2 + \text{sn } u_2 \text{ cn } u_1 \text{ dn } u_1)/(1 - \lambda \text{ sn}^2 u_1 \text{ sn}^2 u_2),$$

$$\text{cn } (u_1 + u_2) = (\text{cn } u_1 \text{ cn } u_2 - \text{sn } u_1 \text{ sn } u_2 \text{ dn } u_1 \text{ dn } u_2)/(1 - \lambda \text{ sn}^2 u_1 \text{ sn}^2 u_2), \tag{55}$$

$$\text{dn } (u_1 + u_2) = (\text{dn } u_1 \text{ dn } u_2 - \lambda \text{ sn } u_1 \text{ sn } u_2 \text{ cn } u_1 \text{ cn } u_2)/(1 - \lambda \text{ sn}^2 u_1 \text{ sn}^2 u_2).$$

The addition formulas show that the evaluation of the Jacobian elliptic functions for arbitrary complex argument may be reduced to their evaluation for real and pure imaginary arguments.

The usual addition formulas for the sine and cosine functions may be regarded as limiting cases of (55). For if $\tau = iy$ and $y \to \infty$, the product expansions (35) show that

$$\theta_{00}(v) \to 1, \quad \theta_{01}(v) \to 1,$$

$$\theta_{10}(v) \sim 2e^{\pi i\tau/4} \cos \pi v, \quad \theta_{11}(v) \sim 2i\, e^{\pi i\tau/4} \sin \pi v,$$

and hence

$$\lambda \to 0, \quad u \to \pi v,$$
$$\operatorname{sn} u \to \sin u, \quad \operatorname{cn} u \to \cos u, \quad \operatorname{dn} u \to 1.$$

The definitions (46) of the Jacobian elliptic functions and the transformation formulas (37)-(38) for the theta functions imply also *transformation formulas* for the Jacobian functions:

PROPOSITION 9 *For all* $u \in \mathbb{C}$ *and* $\tau \in \mathfrak{H}$,

$$\operatorname{sn}(u;\tau+1) = (1-\lambda(\tau))^{1/2} \operatorname{sn}(u';\tau)/\operatorname{dn}(u';\tau),$$
$$\operatorname{cn}(u;\tau+1) = \operatorname{cn}(u';\tau)/\operatorname{dn}(u';\tau),$$
$$\operatorname{dn}(u;\tau+1) = 1/\operatorname{dn}(u';\tau),$$

where

$$u' = u/(1-\lambda(\tau))^{1/2}$$

and

$$(1-\lambda(\tau))^{1/2} = \theta_{01}^2(0;\tau)/\theta_{00}^2(0;\tau).$$

Furthermore,

$$\lambda(\tau+1) = \lambda(\tau)/[\lambda(\tau)-1],$$
$$\mathbf{K}(\tau+1) = (1-\lambda(\tau))^{1/2}\, \mathbf{K}(\tau).$$

Proof With $v = u/\pi\theta_{00}^2(0;\tau+1)$ we have, by (37),

$$\operatorname{dn}(u;\tau+1) = \theta_{00}(0;\tau)\theta_{01}(v;\tau)/\theta_{01}(0;\tau)\theta_{00}(v;\tau) = 1/\operatorname{dn}(u';\tau),$$

where

$$u' = \pi\theta_{00}^2(0;\tau)v = \theta_{00}^2(0;\tau)u/\theta_{01}^2(0;\tau) = u/(1-\lambda(\tau))^{1/2}.$$

Similarly, from (37) and (48)-(49), we obtain

$$\lambda(\tau+1) = -\theta_{10}^4(0;\tau)/\theta_{01}^4(0;\tau) = \lambda(\tau)/[\lambda(\tau)-1].$$

The other relations are established in the same way. $\quad\square$

PROPOSITION 10 *For all $u \in \mathbb{C}$ and $\tau \in \mathfrak{H}$,*

$$\text{sn } (u; -1/\tau) = -i \text{ sn } (iu;\tau)/\text{cn } (iu;\tau),$$
$$\text{cn } (u; -1/\tau) = 1/\text{cn } (iu;\tau),$$
$$\text{dn } (u; -1/\tau) = \text{dn } (iu;\tau)/\text{cn } (iu;\tau),$$

Furthermore,

$$\lambda(-1/\tau) = 1 - \lambda(\tau),$$
$$K(-1/\tau) = K'(\tau).$$

Proof With $v = u/\pi\theta_{00}^2(0;-1/\tau)$ we have, by (38),

$$\text{sn } (u; -1/\tau) = -i\,\theta_{00}(0;-1/\tau)\theta_{11}(v;-1/\tau)/\theta_{10}(0;-1/\tau)\theta_{01}(v;-1/\tau)$$
$$= -\theta_{00}(0;\tau)\theta_{11}(\tau v;\tau)/\theta_{01}(0;\tau)\theta_{10}(\tau v;\tau).$$

On the other hand, with $v' = iu/\pi\theta_{00}^2(0;\tau)$ we have

$$\text{sn } (iu; \tau)/\text{cn } (iu;\tau) = -i\,\theta_{00}(0;\tau)\theta_{11}(v';\tau)/\theta_{01}(0;\tau)\theta_{10}(v';\tau).$$

Since $\tau v = v'$, by comparing these two relations we obtain the first assertion of the proposition.

The next two assertions may be obtained in the same way. The final two assertions follow from (38), together with (48), (49) and (51). ☐

It follows from Proposition 10 that the evaluation of the Jacobian elliptic functions for pure imaginary argument and parameter τ may be reduced to their evaluation for real argument and parameter $-1/\tau$.

From the definition (46) of the Jacobian elliptic functions and the duplication formulas for the theta functions we can also obtain formulas for the Jacobian functions when the parameter τ is doubled ('Landen's transformation'):

PROPOSITION 11 *For all $u \in \mathbb{C}$ and $\tau \in \mathfrak{H}$,*

$$\text{sn } (u''; 2\tau) = [1 + (1 - \lambda(\tau))^{1/2}] \text{ sn } (u;\tau) \text{ cn } (u;\tau)/\text{dn } (u;\tau),$$
$$\text{cn } (u''; 2\tau) = \{1 - [1 + (1 - \lambda(\tau))^{1/2}] \text{ sn}^2 (u;\tau)\}/\text{dn } (u;\tau),$$
$$\text{dn } (u''; 2\tau) = \{1 - [1 - (1 - \lambda(\tau))^{1/2}] \text{ sn}^2 (u;\tau)\}/\text{dn } (u;\tau),$$

where $u'' = [1 + (1 - \lambda(\tau))^{1/2}] u$ and $(1 - \lambda(\tau))^{1/2} = \theta_{01}^2(0;\tau)/\theta_{00}^2(0;\tau)$.

Furthermore,

$$\lambda(2\tau) = \lambda^2(\tau)/[1 + (1 - \lambda(\tau))^{1/2}]^4,$$
$$K(2\tau) = [1 + (1 - \lambda(\tau))^{1/2}] K(\tau)/2.$$

Proof If $u = \pi\theta_{00}^2(0;\tau)v$ and $u'' = \pi\theta_{00}^2(0;2\tau)2v$ then, by Proposition 5,

$$u'' = 2\theta_{00}^2(0;2\tau)\, u/\theta_{00}^2(0;\tau)$$
$$= [\theta_{00}^2(0;\tau) + \theta_{01}^2(0;\tau)]\, u/\theta_{00}^2(0;\tau).$$

Hence, by (49),

$$u'' = [1 + (1 - \lambda(\tau))^{1/2}]\, u.$$

By Proposition 5 also,

$$\text{sn } (u'';2\tau) = -i\, \theta_{00}(0;2\tau)\theta_{10}(v;\tau)\theta_{11}(v;\tau)/\theta_{10}(0;2\tau)\theta_{00}(v;\tau)\theta_{01}(v;\tau).$$

On the other hand,

$$\text{sn } (u;\tau)\, \text{cn } (u;\tau)/\text{dn } (u;\tau) = -i\, \theta_{00}^2(0;\tau)\theta_{10}(v;\tau)\theta_{11}(v;\tau)/\theta_{10}^2(0;\tau)\theta_{00}(v;\tau)\theta_{01}(v;\tau).$$

Since $2\theta_{00}(0;2\tau)\theta_{10}(0;2\tau) = \theta_{10}^2(0;\tau)$, it follows that

$$\text{sn } (u'';2\tau) = 2\theta_{00}^2(0;2\tau)\, \text{sn } (u;\tau)\, \text{cn } (u;\tau)/\theta_{00}^2(0;\tau)\, \text{dn } (u;\tau).$$

Since $2\theta_{00}^2(0;2\tau)/\theta_{00}^2(0;\tau) = u''/u$, this proves the first assertion of the proposition. The remaining assertions may be proved similarly. ◻

We show finally how the standard elliptic integrals of the second and third kinds, defined by (24) and (25), may be expressed in terms of theta functions. If we put

$$\Theta(u) = \theta_{01}(v), \tag{56}$$

where $u = \pi\theta_{00}^2(0)v$, then since

$$\lambda S(u) = \lambda\, \text{sn}^2 u = -\theta_{10}^2(0)\theta_{11}^2(v)/\theta_{00}^2(0)\theta_{01}^2(v),$$

we can rewrite (45) in the form

$$d\,\{\Theta'(u)/\Theta(u)\}/du = -\alpha + 1 - \lambda S(u),$$

where α is independent of u and the prime on the left denotes differentiation with respect to u. Since $\Theta'(0) = 0$, by integrating we obtain

$$E(u) = \Theta'(u)/\Theta(u) + \alpha u.$$

To determine α we take $u = \mathbf{K}$. Since $\theta_{01}'(1/2) = \theta_{00}'(1) = \theta_{00}'(0) = 0$, we obtain $\alpha = \mathbf{E/K}$, where

$$\mathbf{E} = \mathbf{E(K)} = \int_0^K \{1 - \lambda S(u)\}\, du = \int_0^1 (1 - \lambda x)\, dx/g_\lambda(x)^{1/2}$$

is a complete elliptic integral of the second kind. Thus

$$E(u) = \Theta'(u)/\Theta(u) + u\mathbf{E}/\mathbf{K}. \tag{57}$$

Substituting this expression for $E(u)$ in (27), we further obtain

$$\Pi(u,a) = u\,\Theta'(a)/\Theta(a) + (1/2)\log\{\Theta(u-a)/\Theta(u+a)\}. \tag{58}$$

6 The modular function

The function

$$\lambda(\tau):\,= \theta_{10}{}^4(0;\tau)/\theta_{00}{}^4(0;\tau),$$

which was introduced in §5, is known as the *modular function*. In this section we study its remarkable properties. (The term 'modular function', without the definite article, is also used in a more general sense, which we do not consider here.)

The modular function is holomorphic in the upper half-plane $\mathfrak{H}$. Furthermore, we have

PROPOSITION 12 *For any* $\tau \in \mathfrak{H}$,

$$\lambda(\tau + 1) = \lambda(\tau)/[\lambda(\tau) - 1],$$
$$\lambda(-1/\tau) = 1 - \lambda(\tau),$$
$$\lambda(-1/(\tau + 1)) = 1/[1 - \lambda(\tau)],$$
$$\lambda((\tau - 1)/\tau) = [\lambda(\tau) - 1]/\lambda(\tau),$$
$$\lambda(\tau/(\tau + 1)) = 1/\lambda(\tau).$$

Proof The first two relations have already been established in Propositions 9 and 10. If, as in §1, we put

$$U\lambda = 1 - \lambda, \quad V\lambda = 1/(1 - \lambda),$$

and if we also put $T\tau = \tau + 1$, $S\tau = -1/\tau$, then they may be written in the form

$$\lambda(T\tau) = UV\lambda(\tau), \quad \lambda(S\tau) = U\lambda(\tau).$$

It follows that

$$\lambda(-1/(\tau + 1)) = \lambda(ST\tau) = U\lambda(T\tau) = U^2V\lambda(\tau) = V\lambda(\tau) = 1/[1 - \lambda(\tau)].$$

Similarly,

$$\lambda((\tau - 1)/\tau) = \lambda(TS\tau) = V^2\lambda(\tau) = [\lambda(\tau) - 1]/\lambda(\tau),$$
$$\lambda(\tau/(\tau + 1)) = \lambda(TST\tau) = UV^2\lambda(\tau) = 1/\lambda(\tau). \quad \square$$

As we saw in Proposition IV.12, the transformations $S\tau = -1/\tau$ and $T\tau = \tau + 1$ generate the *modular group* Γ, consisting of all linear fractional transformations

$$\tau' = (a\tau + b)/(c\tau + d),$$

where $a,b,c,d \in \mathbb{Z}$ and $ad - bc = 1$. Consequently we can deduce the effect on $\lambda(\tau)$ of any modular transformation on τ. However, Proposition 12 contains the only cases which we require.

We will now study in some detail the behaviour of the modular function in the upper half-plane. We first observe that we need only consider the behaviour of $\lambda(\tau)$ in the right half of $\mathfrak{H}$. For, from the definitions of the theta functions as infinite series,

$$\overline{\theta_{00}(0;\tau)} = \theta_{00}(0;-\overline{\tau}), \quad \overline{\theta_{01}(0;\tau)} = \theta_{01}(0;-\overline{\tau}),$$

where the bar denotes complex conjugation, and hence

$$\lambda(-\overline{\tau}) = \overline{\lambda(\tau)}. \tag{59}$$

We next note that, by taking $\tau = i$ in the relation $\lambda(-1/\tau) = 1 - \lambda(\tau)$, we obtain $\lambda(i) = 1/2$. We have already seen in §5 that $\lambda(\tau)$ is real on the imaginary axis $\tau = iy$ ($y > 0$), and decreases from 1 to 0 as y increases from 0 to ∞. Since $\lambda(\tau + 1) = \lambda(\tau)/[\lambda(\tau) - 1]$, it follows that $\lambda(\tau)$ is real also on the half-line $\tau = 1 + iy$ ($y > 0$), and increases from $-\infty$ to 0 as y increases from 0 to ∞. Moreover, $\lambda(1 + i) = -1$.

The linear fractional map $\tau = (\tau' - 1)/\tau'$ maps the half-line $\mathfrak{R}\tau' = 1$, $\mathfrak{I}\tau' > 0$ onto the semi-circle $|\tau - 1/2| = 1/2$, $\mathfrak{I}\tau > 0$, and $\tau' = 1 + i$ is mapped to $\tau = (1 + i)/2$. Since

$$\lambda((\tau' - 1)/\tau') = [\lambda(\tau') - 1]/\lambda(\tau'),$$

it follows from what we have just proved that, as τ traverses this semi-circle from 0 to 1, $\lambda(\tau)$ is real and increases from 1 to ∞. Moreover, $\lambda((1 + i)/2) = 2$.

If $\mathfrak{R}\tau = 1/2$, then $\overline{\tau} = 1 - \tau$ and hence, by (59),

$$\overline{\lambda(\tau)} = \lambda(\tau - 1) = \lambda(\tau)/[\lambda(\tau) - 1],$$

which implies

$$|\lambda(\tau) - 1|^2 = 1.$$

Thus $w = \lambda(\tau)$ maps the half-line $\mathfrak{R}\tau = 1/2$, $\mathfrak{I}\tau > 0$ into the circle $|w - 1| = 1$. Furthermore, the map is injective. For if $\lambda(\tau_1) = \lambda(\tau_2)$, then $\lambda(2\tau_1) = \lambda(2\tau_2)$, by Proposition 11, and the map is injective on the half-line $\mathfrak{R}\tau = 1$, $\mathfrak{I}\tau > 0$. If $\tau = 1/2 + iy$, where $y \to +\infty$, then

$$\theta_{00}(0;\tau) \to 1, \quad \theta_{10}(0;\tau) \sim 2\, e^{\pi i \tau/4}$$

and hence

$$\lambda(\tau) \sim 16i\, e^{-\pi y}.$$

In particular, $\lambda(\tau) \in \mathfrak{H}$ and $\lambda(\tau) \to 0$. Since $\lambda((1 + i)/2) = 2$, it follows that $w = \lambda(\tau)$ maps the half-line $\tau = 1/2 + iy$ ($y > 1/2$) bijectively onto the semi-circle $|w - 1| = 1$, $\mathcal{I}w > 0$.

If $|\tau| = 1$, $\mathcal{I}\tau > 0$ and $\tau' = \tau/(1 + \tau)$, then $\mathcal{R}\tau' = 1/2$, $\mathcal{I}\tau' > 0$ and $\lambda(\tau') = 1/\lambda(\tau)$. Consequently, by what we have just proved, $w = \lambda(\tau)$ maps the semi-circle $|\tau| = 1$, $\mathcal{I}\tau > 0$ bijectively onto the half-line $\mathcal{R}w = 1/2$, $\mathcal{I}w > 0$.

The point $e^{\pi i/3} = (1 + i\sqrt{3})/2$ is in $\mathfrak{H}$ and lies on both the line $\mathcal{R}\tau = 1/2$ and the circle $|\tau| = 1$. Hence $\lambda(e^{\pi i/3})$ lies on both the semi-circle $|w - 1| = 1$, $\mathcal{I}w > 0$ and the line $\mathcal{R}w = 1/2$, which implies that

$$\lambda(e^{\pi i/3}) = e^{\pi i/3}.$$

Again, since $\lambda(\tau - 1) = \lambda(\tau)/[\lambda(\tau) - 1]$, $w = \lambda(\tau)$ maps the semi-circle $|\tau - 1| = 1$, $\mathcal{I}\tau > 0$ bijectively onto the semi-circle $|w| = 1$, $\mathcal{I}w > 0$.

In particular, we have the behaviour illustrated in Figure 2: $w = \lambda(\tau)$ maps the boundary of the (non-Euclidean) 'triangle' $\mathcal{T}$ with vertices $A = 0$, $B = (1 + i)/2$, $C = e^{\pi i/3}$ bijectively onto the boundary of the 'triangle' $\mathcal{T}'$ with vertices $A' = 1$, $B' = 2$, $C' = e^{\pi i/3}$. We are going to deduce from this that the region inside $\mathcal{T}$ is mapped bijectively onto the region inside $\mathcal{T}'$. The reasoning here does not depend on special properties of the function or the domain, but is quite general (the 'principle of the argument'). To emphasize this, we will temporarily denote the independent variable by z, instead of τ.

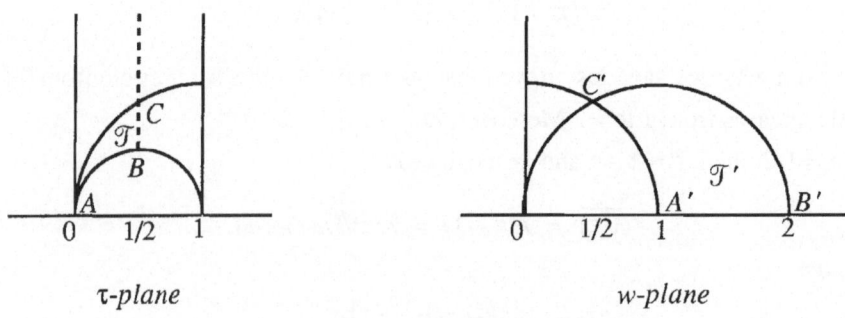

τ-plane w-plane

Figure 2: $w = \lambda(\tau)$ maps $\mathcal{T}$ onto $\mathcal{T}'$

Choose any $w_0 \in \mathbb{C}$ which is either inside or outside the 'triangle' $\mathcal{T}'$, and let Δ denote the change in the argument of $w - w_0$ as w traverses $\mathcal{T}'$ in the direction $A'B'C'$. Thus $\Delta = 2\pi$ or 0

according as w_0 is inside or outside $\mathcal{T}'$. But Δ is also the change in the argument of $\lambda(z) - w_0$ as z traverses $\mathcal{T}$ in the direction ABC. Since $\lambda(z)$ is a nonconstant holomorphic function, the number of times that it assumes the value w_0 inside $\mathcal{T}$ is either zero or a positive integer p.

Suppose the latter, and let $z = \zeta_1,...,\zeta_p$ be the points inside $\mathcal{T}$ for which $\lambda(z) = w_0$. In the neighbourhood of ζ_j we have, for some positive integer m_j and some $a_{0j} \neq 0$,

$$\lambda(z) - w_0 = a_{0j}(z - \zeta_j)^{m_j} + a_{1j}(z - \zeta_j)^{m_j+1} + ...$$

and

$$\lambda'(z) = m_j a_{0j}(z - \zeta_j)^{m_j-1} + (m_j + 1)a_{1j}(z - \zeta_j)^{m_j} + ...$$

Hence

$$\lambda'(z)/[\lambda(z) - w_0] = m_j/(z - \zeta_j) + f_j(z),$$

where $f_j(z)$ is holomorphic at ζ_j. Consequently

$$f(z): = \lambda'(z)/[\lambda(z) - w_0] - \sum_{j=1}^{p} m_j/(z - \zeta_j)$$

is holomorphic at every point z inside $\mathcal{T}$. Hence, by Cauchy's theorem,

$$\int_{\mathcal{T}} f(z)\, dz = 0.$$

But, since $\log \lambda(z) = \log |\lambda(z)| + i \arg \lambda(z)$,

$$\int_{\mathcal{T}} \lambda'(z)\, dz/[\lambda(z) - w_0] = i\Delta.$$

Similarly, since ζ_j is inside $\mathcal{T}$,

$$\int_{\mathcal{T}} dz/(z - \zeta_j) = 2\pi i.$$

It follows that

$$\Delta = 2\pi \sum_{j=1}^{p} m_j.$$

If w_0 is outside $\mathcal{T}'$, then $\Delta = 0$ and we have a contradiction. Hence $\lambda(z)$ is never outside $\mathcal{T}'$ if z is inside $\mathcal{T}$. If w_0 is inside $\mathcal{T}'$, then $\Delta = 2\pi$. Hence $\lambda(z)$ assumes each value inside $\mathcal{T}'$ at exactly one point z inside $\mathcal{T}$, and at this point $\lambda'(z) \neq 0$.

Finally, if $\lambda(z)$ assumed a value w_0 *on* $\mathcal{T}'$ at a point z_0 inside $\mathcal{T}$, then it would assume all values near w_0 in the neighbourhood of z_0. In particular, it would assume values outside $\mathcal{T}'$, which we have shown to be impossible. It follows that $w = \lambda(z)$ maps the region inside $\mathcal{T}$ bijectively onto the region inside $\mathcal{T}'$, and $\lambda'(z) \neq 0$ for all z inside $\mathcal{T}$.

We must also have $\lambda'(z) \neq 0$ for all $z \neq 0$ on $\mathcal{T}$. Otherwise, if $\lambda(z_0) = w_0$ and $\lambda'(z_0) = 0$ for some $z_0 \in \mathcal{T} \cap \tilde{\mathfrak{H}}$ then, for some $m > 1$ and $c \neq 0$,

$$\lambda(z) - w_0 \sim c(z - z_0)^m \quad \text{as } z \to z_0.$$

But this implies that $\lambda(z)$ takes values outside $\mathcal{T}'$ for some z near z_0 inside $\mathcal{T}$.

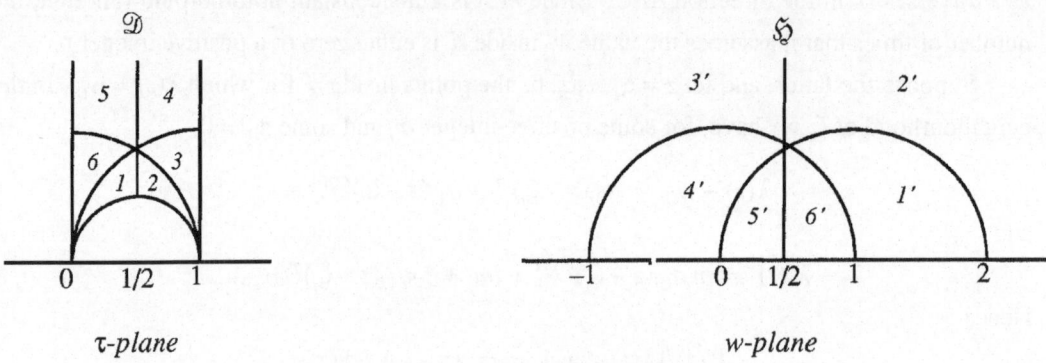

Figure 3: $w = \lambda(\tau)$ *maps* $\mathcal{D}$ *onto* $\mathfrak{H}$

By putting together the preceding results we see that $w = \lambda(\tau)$ maps the domain

$$\mathcal{D} = \{\tau \in \mathfrak{H}: 0 < \Re\tau < 1, |\tau - 1/2| > 1/2\}$$

bijectively onto the upper half-plane $\mathfrak{H}$, with the subdomain k of $\mathcal{D}$ mapped onto the subdomain k' of $\mathfrak{H}$ ($k = 1,...,6$), as illustrated in Figure 3. Moreover, the boundary in $\mathfrak{H}$ of $\mathcal{D}$ is mapped bijectively onto the real axis, with the points 0 and 1 omitted.

If we denote by $\overline{\mathcal{D}}$ the closure of $\mathcal{D}$ in $\mathfrak{H}$ and by $\mathcal{D}^*$ the reflection of $\mathcal{D}$ in the imaginary axis, then it follows from (59) that $w = \lambda(\tau)$ maps the region

$$\overline{\mathcal{D}} \cup \mathcal{D}^* = \{\tau \in \mathfrak{H}: 0 \le \Re\tau \le 1, |\tau - 1/2| \ge 1/2\} \cup \{\tau \in \mathfrak{H}: -1 < \Re\tau < 0, |\tau + 1/2| > 1/2\}$$

bijectively onto the whole complex plane $\mathbb{C}$, with the points 0 and 1 omitted. *This answers the question raised in* §5.

There remains the practical problem, for a given $w \in \mathbb{C}$, of determining $\tau \in \mathfrak{H}$ such that $\lambda(\tau) = w$. If $0 < w < 1$, we can calculate τ by the *AGM* algorithm, using the formula (4), since $\tau = iK(1 - w)/K(w)$. For complex w we can use an extension of the *AGM* algorithm, or proceed in the following way.

Since

$$(1 - \lambda(\tau))^{1/4} = \theta_{01}(0;\tau)/\theta_{00}(0;\tau)$$

and

$$\theta_{00}(0;\tau) = 1 + 2\sum_{n=1}^{n=\infty} q^{n^2}, \quad \theta_{01}(0;\tau) = 1 + 2\sum_{n=1}^{n=\infty} (-1)^n q^{n^2},$$

we have

$$[1 - (1 - \lambda(\tau))^{1/4}]/[1 + (1 - \lambda(\tau))^{1/4}] = [\theta_{00}(0;\tau) - \theta_{01}(0;\tau)]/[\theta_{00}(0;\tau) + \theta_{01}(0;\tau)]$$

$$= 2(q + q^9 + q^{25} + \dots)/(1 + 2q^4 + 2q^{16} + \dots).$$

Thus if we put

$$\ell := [1 - (1 - w)^{1/4}]/[1 + (1 - w)^{1/4}],$$

we have to solve for q the equation

$$\ell/2 = (q + q^9 + q^{25} + \dots)/(1 + 2q^4 + 2q^{16} + \dots).$$

Expanding the right side as a power series in q and inverting the relationship, we obtain

$$q = \ell/2 + 2(\ell/2)^5 + 15(\ell/2)^9 + 150(\ell/2)^{13} + O(\ell/2)^{17}.$$

To ensure rapid convergence we may suppose that, in Figure 3, w lies in the region $5'$ or on its boundary, since the general case may be reduced to this by a linear fractional transformation. It is not difficult to show that in this region $|\ell|$ takes its maximum value when $w = e^{\pi i/3}$, and then

$$\ell = (1 - e^{-\pi i/12})/(1 + e^{-\pi i/12}) = i \tan \pi/24.$$

Thus $|\ell| \le \tan \pi/24 < 2/15$ and $|\ell/2|^4 < 2 \times 10^{-5}$. Since $\mathcal{I}\tau \ge \sqrt{3}/2$ for τ in the region 5, for the solution q we have

$$|q| \le e^{-\pi\sqrt{3}/2} < 1/15.$$

Having determined q, we may calculate $\mathbf{K}(\tau)$, sn u, ... from their representations by theta functions.

7 Further remarks

Numerous references to the older literature on elliptic integrals and elliptic functions are given by Fricke [12]. The more important original contributions are readily available in Euler [10], Lagrange [21], Legendre [22], Gauss [13], Abel [1] and Jacobi [16], which includes his lecture course of 1838.

It was shown by Landen (1775) that the length of arc of a hyperbola could be expressed as the difference of the lengths of two elliptic arcs. The change of variables involved is equivalent to that used by Lagrange (1784/5) in his application of the *AGM* algorithm. However, Lagrange used the transformation in much greater generality, and it was his idea that elliptic integrals could be calculated numerically by iterating the transformation. The connection with the result of Landen was made explicit by Legendre (1786).

By bringing together his own results and those of others the treatise of Legendre [22], and his earlier *Exercices de calcul integral* (1811/19), contributed substantially to the discoveries of Abel and Jacobi. The supplementary third volume of his treatise, published in 1828 when he was 76, contains the first account of their work in book form.

The most important contribution of Abel (1827) was not the replacement of elliptic integrals by elliptic functions, but the study of the latter in the complex domain. In this way he established their double periodicity, determined their zeros and poles and (besides much else) showed that they could be represented as quotients of infinite products.

The triple product formula of Jacobi (1829) identified these infinite products with infinite series, whose rapid convergence made them well suited for numerical computation. Infinite series of this type had in fact already appeared in the *Théorie analytique de la Chaleur* of Fourier (1822), and Proposition 3 had essentially been proved by Poisson (1827). Remarkable generalizations of the Jacobi triple product formula to affine Lie algebras have recently been obtained by Macdonald [23] and Kac and Peterson [17]. For an introductory account, see Neher [24].

It is difficult to understand the glee with which some authors attribute to Gauss results on elliptic functions, since the world owes its knowledge of these results not to him, but to others. Gauss's work was undoubtedly independent and in most cases earlier, although not in the case of the arithmetic-geometric mean. The remark, in §335 of his *Disquisitiones Arithmeticae* (1801), that his results on the division of the circle into n equal parts applied also to the lemniscate, was one of the motivations for Abel, who carried out this extension. (For a modern account, see Rosen [25].) However, Gauss's claim in a letter to Schumacher of 30 May 1828, quoted in Krazer [20], that Abel had anticipated about a third of his own research is quite unjustified, and not only because of his inability to bring his work to a form in which it could be presented to the world.

It was *proved* by Liouville (1834) that elliptic integrals of the first and second kinds are always 'nonelementary'. For an introductory account of Liouville's theory, see Kasper [18].

The three kinds of elliptic integral may also be characterized function-theoretically. On the Riemann surface of the algebraic function $w^2 = g(z)$, where g is a cubic without repeated roots, the differential dz/w is everywhere holomorphic, the differential $z\,dz/w$ is holomorphic except for a double pole at ∞ with zero residue, and the differential $[w(z) + w(a)]\,dz/2(z - a)w(z)$ is holomorphic except for two simple poles at a and ∞ with residues 1 and -1 respectively.

Many integrals which are not visibly elliptic may be reduced to elliptic integrals by a change of variables. A compilation is given by Byrd and Friedman [8], pp. 254-271.

The arithmetic-geometric mean may also be defined for pairs of complex numbers; a thorough discussion is given by Cox [9]. For the application of the *AGM* algorithm to integrals which are not strictly elliptic, see Bartky [4].

The differential equation (6) is a special case of the hypergeometric differential equation. In fact, if $|\lambda| < 1$, then by expanding $(1 - \lambda x)^{-1/2}$, resp. $(1 - \lambda x)^{1/2}$, by the binomial theorem and integrating term by term, the complete elliptic integrals

$$K(\lambda) = \int_0^1 [4x(1-x)(1-\lambda x)]^{-1/2}\, dx, \quad E(\lambda) = \int_0^1 [(1-\lambda x)/4x(1-x)]^{1/2}\, dx,$$

may be identified with the hypergeometric functions

$$(\pi/2)\, F(1/2,1/2;1;\lambda), \quad (\pi/2)\, F(-1/2,1/2;1;\lambda),$$

where

$$F(\alpha,\beta;\gamma;z) = 1 + \alpha\beta z/1\cdot\gamma + \alpha(\alpha + 1)\beta(\beta + 1)z^2/1\cdot 2\cdot\gamma(\gamma + 1) + \dots .$$

Many transformation formulas for the complete elliptic integrals may be regarded as special cases of more general transformation formulas for the hypergeometric function.

The proof in §3 that $K(1 - \lambda)/K(\lambda)$ has positive real part is due to Falk [11].

It follows from (12)-(13) by induction that $S(nu)$ and $S'(nu)/S'(u)$ are rational functions of $S(u)$ for every integer n. The elliptic function $S(u)$ is said to admit *complex multiplication* if $S(\mu u)$ is a rational function of $S(u)$ for some complex number μ which is not an integer. It may be shown that $S(u)$ admits complex multiplication if and only if $\lambda \neq 0,1$ and the period ratio $iK(1 - \lambda)/K(\lambda)$ is a quadratic irrational, in the sense of Chapter IV. This condition is obviously satisfied if $\lambda = 1/2$, the case of the lemniscate.

A function $f(u)$ is said to possess an *algebraic addition theorem* if there is a polynomial $p(x,y,z)$, not identically zero and with coefficients independent of u,v, such that

$$p(f(u + v),f(u),f(v)) = 0 \quad \text{for all } u,v.$$

It may be shown that a function f, which is meromorphic in the whole complex plane, has an algebraic addition theorem if and only if it is either a rational function or, when the independent variable is scaled by a constant factor, a rational function of $S(u,\lambda)$ and its derivative $S'(u,\lambda)$ for some $\lambda \in \mathbb{C}$. This result (in different notation) is due to Weierstrass and is proved in Akhiezer [3], for example. A generalization of Weierstrass' theorem, due to Myrberg, is proved in Belavin and Drinfeld [6].

The term 'elliptic function' is often used to denote any function which is meromorphic in the whole complex plane and has two periods whose ratio is not real. It may be shown that, if the independent variable is scaled by a constant factor, an elliptic function in this general sense is a rational function of $S(u,\lambda)$ and $S'(u,\lambda)$ for some $\lambda \neq 0,1$.

The functions $f(v)$ which are holomorphic in the whole complex plane $\mathbb{C}$ and satisfy the functional equations

$$f(v + 1) = f(v), \quad f(v + \tau) = e^{-n\pi i(2v+\tau)} f(v),$$

where $n \in \mathbb{N}$ and $\tau \in \mathfrak{H}$, form an n-dimensional complex vector space. It was shown by Hermite (1862) that this may be used to derive many relations between theta functions, such as Proposition 6.

Proposition 11 can be extended to give transformation formulas for the Jacobian functions when the parameter τ is multiplied by any positive integer n. See, for example, Tannery and Molk [27], vol. II.

The modular function was used by Picard (1879) to prove that a function $f(z)$, which is holomorphic for all $z \in \mathbb{C}$ and not a constant, assumes every complex value except perhaps one. The exponential function $\exp z$, which does not assume the value 0, illustrates that an exceptional value may exist. A careful proof of Picard's theorem is given in Ahlfors [2].

It was already observed by Lagrange (1813) that there is a correspondence between addition formulas for elliptic functions and the formulas of spherical trigonometry. This correspondence has been most intensively investigated by Study [26].

There is an n-dimensional generalization of theta functions, which has a useful application to the lattices studied in Chapter VIII. The theta function of an *integral lattice* Λ in $\mathbb{R}^n$ is defined by

$$\theta_\Lambda(\tau) = \sum_{u \in \Lambda} q^{(u,u)} = 1 + \sum_{m \geq 1} N_m q^m,$$

where $q = e^{\pi i \tau}$ and N_m is the number of vectors in Λ with square-norm m. If $n = 1$ and $\Lambda = \mathbb{Z}$, then

$$\theta_\mathbb{Z}(\tau) = 1 + 2q + 2q^4 + 2q^9 + \ldots = \theta(0;\tau).$$

It is easily seen that $\theta_\Lambda(\tau)$ is a holomorphic function of τ in the half-plane $\mathcal{I}\tau > 0$. It follows from Poisson's summation formula that the theta function of the *dual lattice* Λ^* is given by

$$\theta_{\Lambda^*}(\tau) = d(\Lambda) (i/\tau)^{n/2} \theta_\Lambda(-1/\tau) \quad \text{for } \mathcal{I}\tau > 0.$$

Many geometrical properties of a lattice are reflected in its theta function. However, a lattice is not uniquely determined by its theta function, since there are lattices in $\mathbb{R}^4$ (and in higher dimensions) which are not isometric but have the same theta function.

For applications of elliptic functions and theta functions to classical mechanics, conformal mapping, geometry, theoretical chemistry, statistical mechanics and approximation theory, see Halphen [15] (vol. 2), Kober [19], Bos *et al.* [7], Glasser and Zucker [14], Baxter [5] and Todd [28]. Applications to number theory will be considered in the next chapter.

8 Selected references

[1] N.H. Abel, *Oeuvres complètes,* Tome 1, 2nd ed. (ed. L. Sylow et S. Lie), Grondahl, Christiania, 1881. [Reprinted J. Gabay, Sceaux, 1992]

[2] L.V. Ahlfors, *Complex analysis*, 3rd ed., McGraw-Hill, New York, 1979.

[3] N.I. Akhiezer, *Elements of the theory of elliptic functions*, American Mathematical Society, Providence, R.I., 1990. [English transl. of 2nd Russian edition, 1970]

[4] W. Bartky, Numerical calculation of a generalized complete elliptic integral, *Rev. Modern Phys.* **10** (1938), 264-269.

[5] R.J. Baxter, *Exactly solved models in statistical mechanics*, Academic Press, London, 1982. [Reprinted, 1989]

[6] A.A. Belavin and V.G. Drinfeld, Triangle equations and simple Lie algebras, *Soviet Sci. Rev. Sect. C: Math. Phys.* **4** (1984), 93-165. [Reprinted, Harwood, Amsterdam, 1998]

[7] H.J.M. Bos, C. Kers, F. Oort and D.W. Raven, Poncelet's closure theorem, *Exposition. Math.* **5** (1987), 289-364.

[8] P.F. Byrd and M.D. Friedman, *Handbook of elliptic integrals for engineers and scientists*, 2nd ed., Springer, Berlin, 1971.

[9] D.A. Cox, The arithmetic-geometric mean of Gauss, *Enseign. Math.* **30** (1984), 275-330.

[10] L. Euler, *Opera omnia*, Ser. I, Vol. XX (ed. A. Krazer), Leipzig, 1912.

[11] M. Falk, Beweis eines Satzes aus der Theorie der elliptischen Functionen, *Acta Math.* **7** (1885/6), 197-200.

[12] R. Fricke, *Elliptische Funktionen*, Encyklopädie der Mathematischen Wissenschaften, Band II, Teil 2, pp. 177-348, Teubner, Leipzig, 1921.

[13] C.F. Gauss, *Werke*, Band III, Göttingen, 1866. [Reprinted G. Olms, Hildesheim, 1973]

[14] M.L. Glasser and I.J. Zucker, Lattice sums, *Theoretical chemistry: Advances and perspectives* **5** (1980), 67-139.

[15] G.H. Halphen, *Traité des fonctions elliptiques et de leurs applications,* 3 vols., Gauthier-Villars, Paris, 1886-1891.

[16] C.G.J. Jacobi, *Gesammelte Werke,* Band I (ed. C.W. Borchardt), Berlin, 1881. [Reprinted Chelsea, New York, 1969]

[17] V.G. Kac and D.H. Peterson, Infinite-dimensional Lie algebras, theta functions and modular forms, *Adv. in Math.* **53** (1984), 125-264.

[18] T. Kasper, Integration in finite terms: the Liouville theory, *Math. Mag.* **53** (1980), 195-201.

[19] H. Kober, *Dictionary of conformal representations*, Dover, New York, 1952.

[20] A. Krazer, Zur Geschichte des Umkehrproblems der Integral, *Jahresber. Deutsch. Math.-Verein.* **18** (1909), 44-75.

[21] J.L. Lagrange, *Oeuvres,* t. 2 (ed. J.-A. Serret), Gauthier-Villars, Paris, 1868. [Reprinted G. Olms, Hildesheim, 1973]

[22] A.M. Legendre, *Traité des fonctions elliptiques et des intégrales Eulériennes, avec des tables pour en faciliter le calcul numérique*, Paris, t.1 (1825), t.2 (1826), t.3 (1828). [Microform, Readex Microprint Corporation, New York, 1970]

[23] I.G. Macdonald, Affine root systems and Dedekind's η-function, *Invent. Math.* **15** (1972), 91-143.

[24] E. Neher, Jacobis Tripelprodukt-Identität und η-Identitäten in der Theorie affiner Lie-Algebren, *Jahresber. Deutsch. Math.-Verein.* **87** (1985), 164-181.

[25] M. Rosen, Abel's theorem on the lemniscate, *Amer. Math. Monthly* **88** (1981), 387-395.

[26] E. Study, *Sphärische Trigonometrie, orthogonale Substitutionen und elliptische Funktionen*, Leipzig, 1893.

[27] J. Tannery and J. Molk, *Éléments de la théorie des fonctions elliptiques*, 4 vols., Gauthier-Villars, Paris, 1893-1902. [Reprinted Chelsea, New York, 1972]

[28] J. Todd, Applications of transformation theory: a legacy from Zolotarev (1847-1878), *Approximation theory and spline functions* (ed. S.P. Singh et al.), pp. 207-245, Reidel, Dordrecht, 1984.

[29] E.T. Whittaker and G.N. Watson, *A course of modern analysis*, 4th ed., Cambridge University Press, 1927. [Reprinted, 1996]

XIII
Connections with number theory

1 Sums of squares

In Proposition II.40 we proved Lagrange's theorem that every positive integer can be represented as a sum of 4 squares. Jacobi (1829), at the end of his *Fundamenta Nova*, gave a completely different proof of this theorem with the aid of theta functions. Moreover, his proof provided a formula for the number of different representations. Hurwitz (1896), by developing further the arithmetic of quaternions which was used in Chapter II, also derived this formula. Here we give Jacobi's argument preference since, although it is less elementary, it is more powerful.

PROPOSITION 1 *The number of representations of a positive integer m as a sum of* 4 *squares of integers is equal to* 8 *times the sum of those positive divisors of m which are not divisible by* 4.

Proof From the series expansion

$$\theta_{00}(0) = \Sigma_{n \in \mathbb{Z}} \, q^{n^2}$$

we obtain

$$\theta_{00}{}^4(0) = \Sigma_{n_1,...,n_4 \in \mathbb{Z}} \, q^{n_1^2 + ..+ n_4^2} = 1 + \Sigma_{m \geq 1} \, r_4(m) q^m,$$

where $r_4(m)$ is the number of solutions in integers $n_1,...,n_4$ of the equation

$$n_1^2 + ... + n_4^2 = m.$$

We will prove the result by comparing this with another expression for $\theta_{00}{}^4(0)$.

We can write equation (43) of Chapter XII in the form

$$\theta_{10}{}'(v)/\theta_{10}(v) - \theta_{01}{}'(v)/\theta_{01}(v) = \pi i \, \theta_{00}{}^2(0)\theta_{00}(v)\theta_{11}(v)/\theta_{01}(v)\theta_{10}(v).$$

Differentiating with respect to v and then putting $v = 0$, we obtain

$$\theta_{10}{}''(0)/\theta_{10}(0) - \theta_{01}{}''(0)/\theta_{01}(0) = \pi i \, \theta_{00}{}^3(0)\theta_{11}{}'(0)/\theta_{01}(0)\theta_{10}(0) = -\pi^2\theta_{00}{}^4(0),$$

by (36) of Chapter XII. Since the theta functions are all solutions of the partial differential equation

$$\partial^2 y/\partial v^2 \;=\; -\,4\pi^2 q\;\partial y/\partial q,$$

the last relation can be written in the form

$$4q\,\partial\,/\partial q\,\log\{\theta_{10}(0)/\theta_{01}(0)\} \;=\; \theta_{00}{}^4(0).$$

On the other hand, the product expansions of the theta functions show that

$$\theta_{10}(0)/\theta_{01}(0) \;=\; 2q^{1/4}\,\prod_{n\geq 1}\,(1+q^{2n})^2\,/\,\prod_{n\geq 1}\,(1-q^{2n-1})^2$$

$$=\; 2q^{1/4}\,\prod_{n\geq 1}\,(1-q^{4n})^2\,/\,\prod_{n\geq 1}\,(1-q^{2n})^2(1-q^{2n-1})^2$$

$$=\; 2q^{1/4}\,\prod_{n\geq 1}\,(1-q^{4n})^2(1-q^{n})^{-2}.$$

Differentiating logarithmically, we obtain

$$\theta_{00}{}^4(0) \;=\; 4q\,\partial\,/\partial q\,\log\{\theta_{10}(0)/\theta_{01}(0)\}$$

$$=\; 1+8\,\sum_{n\geq 1}\,nq^n/(1-q^n)-8\,\sum_{n\geq 1}\,4nq^{4n}/(1-q^{4n})$$

$$=\; 1+8\,\sum_{n\geq 1}\,\sum_{k\geq 1}\,(nq^{kn}-4nq^{4kn})$$

$$=\; 1+8\,\sum_{m\geq 1}\,\{\sigma(m)-\sigma'(m)\}q^{m},$$

where $\sigma(m)$ is the sum of all positive divisors of m and $\sigma'(m)$ is the sum of all positive divisors of m which are divisible by 4. Since the coefficients in a power series expansion are uniquely determined, it follows that

$$r_4(m) \;=\; 8\{\sigma(m)-\sigma'(m)\}. \quad \square$$

Proposition 1 may also be restated in the form: the number of representations of m as a sum of 4 squares is equal to 8 times the sum of the odd positive divisors of m if m is odd, and 24 times this sum if m is even. For example,

$$r_4(10) = 24(1+5) = 144.$$

Since any positive integer has the odd positive divisor 1, Proposition 1 provides a new proof of Proposition II.40.

The number of representations of a positive integer as a sum of 2 squares may be treated in the same way, as Jacobi also showed (or, alternatively, by developing further the arithmetic of Gaussian integers):

PROPOSITION 2 *The number of representations of a positive integer m as a sum of 2 squares of integers is equal to 4 times the excess of the number of positive divisors of m of the form 4h + 1 over the number of positive divisors of the form 4h + 3.*

Proof We have

$$\theta_{00}^2(0) = \sum_{n_1,n_2 \in \mathbb{Z}} q^{n_1^2 + n_2^2} = 1 + \sum_{m \geq 1} r_2(m)q^m,$$

where $r_2(m)$ is the number of solutions in integers n_1, n_2 of the equation

$$n_1^2 + n_2^2 = m.$$

To obtain another expression for $\theta_{00}^2(0)$ we use again the relation

$$\theta_{10}'(v)/\theta_{10}(v) - \theta_{01}'(v)/\theta_{01}(v) = \pi i\, \theta_{00}^2(0)\theta_{00}(v)\theta_{11}(v)/\theta_{01}(v)\theta_{10}(v),$$

but this time we simply take $v = 1/4$. Since

$$\theta_{01}(1/4) = \sum_{n \in \mathbb{Z}} (-i)^n q^{n^2} = \sum_{n \in \mathbb{Z}} i^{-n} q^{n^2} = \theta_{00}(1/4),$$

and similarly $\theta_{11}(1/4) = i\, \theta_{10}(1/4)$, we obtain

$$\pi\theta_{00}^2(0) = \theta_{01}'(1/4)/\theta_{01}(1/4) - \theta_{10}'(1/4)/\theta_{10}(1/4).$$

By differentiating logarithmically the product expansion for $\theta_{10}(v)$ and then putting $v = 1/4$, we get

$$\theta_{10}'(1/4)/\theta_{10}(1/4) = -\pi - 4\pi \sum_{n \geq 1} q^{2n}/(1 + q^{4n}).$$

Similarly, by differentiating logarithmically the product expansion for $\theta_{01}(v)$ and then putting $v = 1/4$, we get

$$\theta_{01}'(1/4)/\theta_{01}(1/4) = 4\pi \sum_{n \geq 1} q^{2n-1}/(1 + q^{4n-2}).$$

Thus

$$\theta_{01}'(1/4)/\theta_{01}(1/4) - \theta_{10}'(1/4)/\theta_{10}(1/4) = \pi + 4\pi \sum_{n \geq 1} q^n/(1 + q^{2n})$$

and hence

$$\theta_{00}^2(0) = 1 + 4 \sum_{n \geq 1} q^n/(1 + q^{2n}).$$

Since

$$q^n/(1 + q^{2n}) = q^n(1 - q^{2n})/(1 - q^{4n}) = (q^n - q^{3n}) \sum_{k \geq 0} q^{4kn},$$

it follows that

$$\theta_{00}^2(0) = 1 + 4 \sum_{n \geq 1} \sum_{k \geq 0} \{q^{(4k+1)n} - q^{(4k+3)n}\}$$

$$= 1 + 4 \sum_{m \geq 1} \{d_1(m) - d_3(m)\}q^m,$$

where $d_1(m)$ and $d_3(m)$ are respectively the number of positive divisors of m congruent to 1 and 3 mod 4. Hence

$$r_2(m) = 4\{d_1(m) - d_3(m)\}. \quad \blacksquare$$

From Proposition 2 we immediately obtain again that any prime $p \equiv 1 \bmod 4$ may be represented as a sum of 2 squares and that the representation is essentially unique. Proposition II.39 may also be rederived.

The number $r_s(m)$ of representations of a positive integer m as a sum of s squares has been expressed by explicit formulas for many other values of s besides 2 and 4. Systematic ways of attacking the problem are provided by the theory of modular forms and the circle method of Hardy, Ramanujan and Littlewood.

2 Partitions

A *partition* of a positive integer n is a set of positive integers with sum n. For example, $\{2,1,1\}$ is a partition of 4. We denote the number of distinct partitions of n by $p(n)$. For example, $p(4) = 5$, since all partitions of 4 are given by

$$\{4\}, \{3,1\}, \{2,2\}, \{2,1,1\}, \{1,1,1,1\}.$$

It was shown by Euler (1748) that the sequence $p(n)$ has a simple *generating function*:

PROPOSITION 3 *If* $|x| < 1$, *then*

$$1/(1 - x)(1 - x^2)(1 - x^3)... = 1 + \sum_{n \geq 1} p(n)x^n.$$

Proof If $|x| < 1$, then the infinite product $\prod_{m \geq 1} (1 - x^m)$ converges and its reciprocal has a convergent power series expansion. To determine the coefficients of this expansion note that, since

$$(1 - x^m)^{-1} = \sum_{k \geq 0} x^{km},$$

the coefficient of x^n ($n \geq 1$) in the product $\prod_{m \geq 1} (1 - x^m)^{-1}$ is the number of representations of n in the form

$$n = 1k_1 + 2k_2 + ... ,$$

where the k_j are non-negative integers. But this number is precisely $p(n)$, since any partition is determined by the number of 1's, 2's, ... that it contains. $\quad \blacksquare$

For many purposes the discussion of convergence is superfluous and Proposition 3 may be regarded simply as a relation between formal products and formal power series.

Euler also obtained an interesting counterpart to Proposition 3, which we will derive from Jacobi's triple product formula.

PROPOSITION 4 *If* $|x| < 1$, *then*

$$(1 - x)(1 - x^2)(1 - x^3)\ldots = \sum_{m \in \mathbb{Z}} (-1)^m x^{m(3m+1)/2}.$$

Proof If we take $q = x^{3/2}$ and $z = -x^{1/2}$ in Proposition XII.2, we obtain at once the result, since

$$\prod_{n \geq 1} (1 - x^{3n})(1 - x^{3n-1})(1 - x^{3n-2}) = \prod_{k \geq 1} (1 - x^k). \quad \square$$

Proposition 4 also has a combinatorial interpretation. The coefficient of x^n ($n \geq 1$) in the power series expansion of $\prod_{k \geq 1} (1 - x^k)$ is

$$s_n = \sum (-1)^\nu,$$

where the sum is over all partitions of n into *unequal* parts and ν is the number of parts in the partition. In other words,

$$s_n = p_e{}^*(n) - p_o{}^*(n),$$

where $p_e{}^*(n)$, resp. $p_o{}^*(n)$, is the number of partitions of the positive integer n into an even, resp. odd, number of unequal parts. On the other hand,

$$\sum_{m \in \mathbb{Z}} (-1)^m x^{m(3m+1)/2} = 1 + \sum_{m \geq 1} (-1)^m \{x^{m(3m+1)/2} + x^{m(3m-1)/2}\}.$$

Thus Proposition 4 says that $p_e{}^*(n) = p_o{}^*(n)$ unless $n = m(3m \pm 1)/2$ for some $m \in \mathbb{N}$, in which case $p_e{}^*(n) - p_o{}^*(n) = (-1)^m$.

From Propositions 3 and 4 we obtain

$$[1 + \sum_{m \geq 1} (-1)^m \{x^{m(3m+1)/2} + x^{m(3m-1)/2}\}][1 + \sum_{k \geq 1} p(k)x^k] = 1.$$

Multiplying out on the left side and equating to zero the coefficient of x^n ($n \geq 1$), we obtain the recurrence relation:

$$\begin{aligned} p(n) = {} & p(n-1) + p(n-2) - p(n-5) - p(n-7) + \ldots \\ & + (-1)^{m-1}p(n-m(3m-1)/2) + (-1)^{m-1}p(n-m(3m+1)/2) + \ldots, \end{aligned}$$

where $p(0) = 1$ and $p(k) = 0$ for $k < 0$. This recurrence relation is quite an efficient way of calculating $p(n)$. It was used by MacMahon (1918) to calculate $p(n)$ for $n \leq 200$.

In the same way that we proved Proposition 3 we may show that, if $|x| < 1$, then

$$1/(1-x)(1-x^2)...(1-x^m) = 1 + \sum_{n\geq 1} p_m(n)x^n,$$

where $p_m(n)$ is the number of partitions of n into parts not exceeding m.

From the vast number of formulas involving partitions and their generating functions we select only one more pair, the celebrated *Rogers–Ramanujan identities*. The proof of these identities will be based on the following preliminary result:

PROPOSITION 5 *If* $|q| < 1$ *and* $|x| < |q|^{-1}$, *then*

$$1 + \sum_{n\geq 1} x^n q^{n^2}/(q)_n = \sum_{n\geq 0} (-1)^n x^{2n} q^{5n(n+1)/2-2n}\{1 - x^2 q^{2(2n+1)}\}/(q)_n(xq^{n+1})_\infty,$$

where $(a)_0 = 1$,

$$(a)_n = (1-a)(1-aq)...(1-aq^{n-1}) \text{ if } n \geq 1, \text{ and } (a)_\infty = (1-a)(1-aq)(1-aq^2)... .$$

Proof Consider the *q-difference equation*

$$f(x) = f(xq) + xq\, f(xq^2).$$

A formal power series $\sum_{n\geq 0} a_n x^n$ satisfies this equation if and only if

$$a_n(1-q^n) = a_{n-1}q^{2n-1} \quad (n \geq 1).$$

Thus the only formal power series solution with $a_0 = 1$ is

$$f(x) = 1 + xq/(1-q) + x^2 q^4/(1-q)(1-q^2) + x^3 q^9/(1-q)(1-q^2)(1-q^3) + ...$$

Moreover, if $|q| < 1$, this power series converges for all $x \in \mathbb{C}$.

If $|q| < 1$, the functions

$$F(x) = \sum_{n\geq 0} (-1)^n x^{2n} q^{5n(n+1)/2-2n}\{1 - x^2 q^{2(2n+1)}\}/(q)_n(xq^{n+1})_\infty,$$

$$G(x) = \sum_{n\geq 0} (-1)^n x^{2n} q^{5n(n+1)/2-n}\{1 - xq^{2n+1}\}/(q)_n(xq^{n+1})_\infty$$

are holomorphic for $|x| < |q|^{-1}$. We have

$$F(x) - G(x) = \sum_{n\geq 0} (-1)^n x^{2n} q^{5n(n+1)/2} \{q^{-2n} - x^2 q^{2(n+1)} - q^{-n} + xq^{n+1}\}/(q)_n(xq^{n+1})_\infty$$

$$= \sum_{n\geq 0} (-1)^n x^{2n} q^{5n(n+1)/2} \{q^{-2n}(1-q^n) + xq^{n+1}(1-xq^{n+1})\}/(q)_n(xq^{n+1})_\infty$$

$$= \sum_{n\geq 1} (-1)^n x^{2n} q^{5n(n+1)/2-2n}/(q)_{n-1}(xq^{n+1})_\infty + xq \sum_{n\geq 0} (-1)^n x^{2n} q^{5n(n+1)/2+n}/(q)_n(xq^{n+2})_\infty$$

$$= -x^2 q^3 \sum_{n\geq 0} (-1)^n x^{2n} q^{5n(n+1)/2+3n}/(q)_n(xq^{n+2})_\infty + xq\sum_{n\geq 0} (-1)^n x^{2n} q^{5n(n+1)/2+n}/(q)_n(xq^{n+2})_\infty$$

$$= xq \sum_{n \geq 0} (-1)^n (xq)^{2n} q^{5n(n+1)/2-n} \{1 - (xq)q^{2n+1}\}/(q)_n (xq^{n+2})_\infty$$

$$= xq \, G(xq).$$

Similarly,

$$G(x) = \sum_{n \geq 0} (-1)^n x^{2n} q^{5n(n+1)/2} \{q^{-n} - xq^{n+1}\}/(q)_n (xq^{n+1})_\infty$$

$$= \sum_{n \geq 0} (-1)^n x^{2n} q^{5n(n+1)/2} \{q^{-n}(1 - q^n) + 1 - xq^{n+1}\}/(q)_n (xq^{n+1})_\infty$$

$$= \sum_{n \geq 1} (-1)^n x^{2n} q^{5n(n+1)/2-n}/(q)_{n-1}(xq^{n+1})_\infty + \sum_{n \geq 0} (-1)^n x^{2n} q^{5n(n+1)/2}/(q)_n (xq^{n+2})_\infty$$

$$= \sum_{n \geq 0} (-1)^n (xq)^{2n} q^{5n(n+1)/2-2n} \{1 - (xq)^2 q^{2(2n+1)}\}/(q)_n (xq^{n+2})_\infty$$

$$= F(xq).$$

Combining this with the previous relation, we obtain

$$F(x) = F(xq) + xq \, F(xq^2).$$

But we have seen that this q-difference equation has a unique holomorphic solution $f(x)$ such that $f(0) = 1$. Hence $F(x) = f(x)$. ◻

The Rogers–Ramanujan identities may now be easily derived:

PROPOSITION 6 *If $|q| < 1$, then*

$$\sum_{n \geq 0} q^{n^2}/(1 - q)(1 - q^2)...(1 - q^n) = \prod_{m \geq 0} (1 - q^{5m+1})^{-1}(1 - q^{5m+4})^{-1},$$

$$\sum_{n \geq 0} q^{n(n+1)}/(1 - q)(1 - q^2)...(1 - q^n) = \prod_{m \geq 0} (1 - q^{5m+2})^{-1}(1 - q^{5m+3})^{-1}.$$

Proof Put $P = \prod_{k \geq 1} (1 - q^k)$. By Proposition 5 and its proof we have

$$\sum_{n \geq 0} q^{n^2}/(1 - q)(1 - q^2)...(1 - q^n) = F(1) = [1 + \sum_{n \geq 1} (-1)^n \{q^{n(5n+1)/2} + q^{n(5n-1)/2}\}]/P$$

and, since $F(q) = G(1)$,

$$\sum_{n \geq 0} q^{n(n+1)}/(1 - q)(1 - q^2)...(1 - q^n) = F(q) = [1 + \sum_{n \geq 1} (-1)^n \{q^{n(5n+3)/2} + q^{n(5n-3)/2}\}]/P.$$

On the other hand, by replacing q by $q^{5/2}$ and z by $-q^{1/2}$, resp. $-q^{3/2}$, in Jacobi's triple product formula (Proposition XII.2), we obtain

$$\sum_{n \in \mathbb{Z}} (-1)^n \, q^{n(5n+1)/2} = \prod_{m \geq 1} (1 - q^{5m})(1 - q^{5m-2})(1 - q^{5m-3})$$

$$= P/\prod_{m \geq 0} (1 - q^{5m+1})(1 - q^{5m+4})$$

and

$$\sum_{n\in\mathbb{Z}} (-1)^n \, q^{n(5n+3)/2} = \prod_{m\geq1} (1-q^{5m})(1-q^{5m-1})(1-q^{5m-4})$$

$$= P/\prod_{m\geq0} (1-q^{5m+2})(1-q^{5m+3}).$$

Combining these relations with the previous ones, we obtain the result. $\square$

The combinatorial interpretation of the Rogers–Ramanujan identities was pointed out by MacMahon (1916). The first identity says that the number of partitions of a positive integer n into parts congruent to ± 1 mod 5 is equal to the number of partitions of n into parts that differ by at least 2. The second identity says that the number of partitions of a positive integer n into parts congruent to ± 2 mod 5 is equal to the number of partitions of n into parts greater than 1 that differ by at least 2.

A remarkable application of the Rogers–Ramanujan identities to the hard hexagon model of statistical mechanics was found by Baxter (1981). Many other models in statistical mechanics have been exactly solved with the aid of theta functions. A unifying principle is provided by the vast theory of infinite-dimensional Lie algebras which has been developed over the past 25 years.

The number $p(n)$ of partitions of n increases rapidly with n. It was first shown by Hardy and Ramanujan (1918) that

$$p(n) \sim e^{\pi\sqrt{(2n/3)}}/4n\sqrt{3} \quad \text{as } n \to \infty.$$

They further obtained an asymptotic series for $p(n)$, which was modified by Rademacher (1937) into a convergent series, from which it is even possible to calculate $p(n)$ exactly. A key role in the difficult proof is played by the behaviour under transformations of the modular group of *Dedekind's eta function*

$$\eta(\tau) = q^{1/12} \prod_{k\geq1} (1-q^{2k}),$$

where $q = e^{\pi i \tau}$ and $\tau \in \mathfrak{H}$ (the upper half-plane).

The paper of Hardy and Ramanujan contained the first use of the 'circle method', which was subsequently applied by Hardy and Littlewood to a variety of problems in analytic number theory.

3 Cubic curves

We define an *affine plane curve* over a field K to be a polynomial $f(X,Y)$ in two indeterminates with coefficients from K, but we regard two polynomials $f(X,Y)$ and $f^*(X,Y)$ as defining the same affine curve if $f^* = \lambda f$ for some nonzero $\lambda \in K$. The *degree* of the curve is defined without ambiguity to be the degree of the polynomial f.

If

$$f(X,Y) = aX + bY + c$$

is a polynomial of degree 1, the curve is said to be an *affine line*. If

$$f(X,Y) = aX^2 + bXY + cY^2 + lX + mY + n$$

is a polynomial of degree 2, the curve is said to be an *affine conic*. If $f(X,Y)$ is a polynomial of degree 3, the curve is said to be an *affine cubic*. It is the cubic case in which we will be most interested.

Let $\mathscr{C}$ be an affine plane curve over the field K, defined by the polynomial $f(X,Y)$. We say that $(x,y) \in K^2$ is a *point* or, more precisely, a *K-point* of the affine curve $\mathscr{C}$ if $f(x,y) = 0$. The K-point (x,y) is said to be *non-singular* if there exist $a,b \in K$, not both zero, such that

$$f(x + X, y + Y) = aX + bY + ... ,$$

where all unwritten terms have degree > 1. Since a,b are uniquely determined by f, we can define the *tangent* to the affine curve $\mathscr{C}$ at the non-singular point (x,y) to be the affine line

$$\ell(X,Y) = aX + bY - (ax + by).$$

It is easily seen that these definitions do not depend on the choice of polynomial within an equivalence class $\{\lambda f: 0 \neq \lambda \in K\}$.

The study of the asymptotes of an affine plane curve leads one to consider also its 'points at infinity', the asymptotes being the tangents at these points. We will now make this precise.

If the polynomial $f(X,Y)$ has degree d, then

$$F(X,Y,Z) = Z^d f(X/Z,Y/Z)$$

is a homogeneous polynomial of degree d such that

$$f(X,Y) = F(X,Y,1).$$

Furthermore, if $\mathscr{F}(X,Y,Z)$ is any homogeneous polynomial such that $f(X,Y) = \mathscr{F}(X,Y,1)$, then $\mathscr{F}(X,Y,Z) = Z^m F(X,Y,Z)$ for some non-negative integer m.

We define a *projective plane curve* over a field K to be a homogeneous polynomial $F(X,Y,Z)$ of degree $d > 0$ in three indeterminates with coefficients from K, but we regard two homogeneous polynomials $F(X,Y,Z)$ and $F^*(X,Y,Z)$ as defining the same projective curve if

$F^* = \lambda F$ for some nonzero $\lambda \in K$. The projective curve is said to be a *projective line, conic* or *cubic* if F has degree 1,2 or 3 respectively.

If $\mathscr{C}$ is an affine plane curve, defined by a polynomial $f(X,Y)$ of degree $d > 0$, the projective plane curve $\overline{\mathscr{C}}$, defined by the homogeneous polynomial $Z^d f(X/Z,Y/Z)$ of the same degree, is called the *projective completion* of $\mathscr{C}$. Thus the projective completion of an affine line, conic or cubic is respectively a projective line, conic or cubic.

Let $\overline{\mathscr{C}}$ be a projective plane curve over the field K, defined by the homogeneous polynomial $F(X,Y,Z)$. We say that $(x,y,z) \in K^3$ is a *point*, or *K-point*, of $\overline{\mathscr{C}}$ if $(x,y,z) \neq (0,0,0)$ and $F(x,y,z) = 0$, but we regard two triples (x,y,z) and (x^*,y^*,z^*) as defining the same K-point if

$$x^* = \lambda x,\ y^* = \lambda y,\ z^* = \lambda z \text{ for some nonzero } \lambda \in K.$$

If $\overline{\mathscr{C}}$ is the projective completion of the affine plane curve $\mathscr{C}$, then a point (x,y,z) of $\overline{\mathscr{C}}$ with $z \neq 0$ corresponds to a point $(x/z,y/z)$ of $\mathscr{C}$, and a point $(x,y,0)$ of $\overline{\mathscr{C}}$ corresponds to a *point at infinity* of $\mathscr{C}$.

The K-point (x,y,z) of the projective plane curve defined by the homogeneous polynomial $F(X,Y,Z)$ is said to be *non-singular* if there exist $a,b,c \in K$, not all zero, such that

$$F(x + X, y + Y, z + Z)\ =\ aX + bY + cZ + \ldots,$$

where all unwritten terms have degree > 1. Since a,b,c are uniquely determined by F, we can define the *tangent* to the projective curve at the non-singular point (x,y,z) to be the projective line defined by $aX + bY + cZ$. It follows from Euler's theorem on homogeneous functions that (x,y,z) is itself a point of the tangent.

It is easily seen that if $\overline{\mathscr{C}}$ is the projective completion of an affine plane curve $\mathscr{C}$, and if $z \neq 0$, then (x,y,z) is a non-singular point of $\overline{\mathscr{C}}$ if and only if $(x/z,y/z)$ is a non-singular point of $\mathscr{C}$. Moreover, if the tangent to $\overline{\mathscr{C}}$ at (x,y,z) is the projective line

$$\overline{\ell}(X,Y,Z)\ =\ aX + bY + cZ,$$

then the tangent to $\mathscr{C}$ at $(x/z,y/z)$ is the affine line defined by

$$\ell(X,Y)\ =\ aX + bY + c.$$

Let $\mathscr{C}$ be an affine plane curve over the field K, defined by the polynomial $f(X,Y)$, and let (x,y) be a non-singular K-point of $\mathscr{C}$. Then we can write

$$f(x + X, y + Y)\ =\ aX + bY + f_2(X,Y) + \ldots,$$

where a,b are not both zero, $f_2(X,Y)$ is a homogeneous polynomial of degree 2, and all unwritten terms have degree > 2. The non-singular point (x,y) is said to be an *inflection point* or, more simply, a *flex* of $\mathscr{C}$ if $f_2(X,Y)$ is divisible by $aX + bY$.

Similarly we can define a flex for a projective plane curve. Let (x,y,z) be a non-singular point of the projective plane curve over the field K, defined by the homogeneous polynomial $F(X,Y,Z)$. Then we can write

$$F(x + X, y + Y, z + Z) = aX + bY + cZ + F_2(X,Y,Z) + \dots ,$$

where a,b,c are not all zero, $F_2(X,Y,Z)$ is a homogeneous polynomial of degree 2, and all unwritten terms have degree > 2. The non-singular point (x,y,z) is said to be a *flex* if $F_2(X,Y,Z)$ is divisible by $aX + bY + cZ$.

Two more definitions are required before we embark on our study of cubic curves. A projective curve over the field K, defined by the homogeneous polynomial $F(X,Y,Z)$ of degree $d > 0$, is said to be *reducible* over K if

$$F(X,Y,Z) = F_1(X,Y,Z)\, F_2(X,Y,Z),$$

where F_1 and F_2 are homogeneous polynomials of degree less than d with coefficients from K. The K-points of the curve defined by F are then just the K-points of the curve defined by F_1, together with the K-points of the curve defined by F_2. A curve is said to be *irreducible over K* if it is not reducible over K.

Two projective curves over the field K, defined by the homogeneous polynomials $F(X,Y,Z)$ and $G(X',Y',Z')$, are said to be *projectively equivalent* if there exists an invertible linear transformation

$$\begin{aligned}
X &= a_{11}X' + a_{12}Y' + a_{13}Z' \\
Y &= a_{21}X' + a_{22}Y' + a_{23}Z' \\
Z &= a_{31}X' + a_{32}Y' + a_{33}Z'
\end{aligned}$$

with coefficients $a_{ij} \in K$ such that

$$F(a_{11}X' + .. , a_{21}X' + .. , a_{31}X' + ..) = G(X',Y',Z').$$

It is clear that F and G necessarily have the same degree, and that projective equivalence is in fact an equivalence relation.

Consider now the affine cubic curve $\mathscr{C}$ defined by the polynomial

$$f(X,Y) = a_{30}X^3 + a_{21}X^2Y + a_{12}XY^2 + a_{03}Y^3 + a_{20}X^2 + a_{11}XY + a_{02}Y^2 + a_{10}X + a_{01}Y + a_{00}.$$

We assume that $\mathscr{C}$ has a non-singular K-point which is a flex. Without loss of generality, suppose that this is the origin. Then $a_{00} = 0$, a_{10} and a_{01} are not both zero, and

$$a_{20}X^2 + a_{11}XY + a_{02}Y^2 = (a_{10}X + a_{01}Y)(a_{10}'X + a_{01}'Y)$$

for some a_{10}', $a_{01}' \in K$. By an invertible linear change of variables we may suppose that $a_{10} = 0$, $a_{01} = 1$. Then f has the form

$$f(X,Y) = Y + a_1XY + a_3Y^2 - a_0X^3 - a_2X^2Y - a_4XY^2 - a_6Y^3.$$

If $a_0 = 0$, then f is divisible by Y and the corresponding projective curve is reducible. Thus we now assume $a_0 \neq 0$. In fact we may assume $a_0 = 1$, by replacing f by a constant multiple and then scaling Y. The projective completion $\overline{\mathscr{C}}$ of $\mathscr{C}$ is now defined by the homogeneous polynomial

$$YZ^2 + a_1XYZ + a_3Y^2Z - X^3 - a_2X^2Y - a_4XY^2 - a_6Y^3.$$

If we interchange Y and Z, the flex becomes the unique point at infinity of the affine cubic curve defined by the polynomial

$$Y^2 + a_1XY + a_3Y - (X^3 + a_2X^2 + a_4X + a_6).$$

This can be further simplified by making mild restrictions on the field K. If K has characteristic $\neq 2$, i.e. if $1 + 1 \neq 0$, then by replacing Y by $(Y - a_1X - a_3)/2$ we obtain the cubic curve defined by the polynomial

$$Y^2 - (4X^3 + b_2X^2 + 2b_4X + b_6).$$

If K also has characteristic $\neq 3$, i.e. if $1 + 1 + 1 \neq 0$, then by replacing X by $(X - 3b_2)/6^2$ and Y by $2Y/6^3$, we obtain the cubic curve defined by the polynomial $Y^2 - (X^3 + aX + b)$. Thus we have proved

PROPOSITION 7 *If a projective cubic curve over the field K is irreducible and has a non-singular K-point which is a flex, then it is projectively equivalent to the projective completion $\mathscr{W} = \mathscr{W}(a_1,...,a_6)$ of an affine curve of the form*

$$Y^2 + a_1XY + a_3Y - (X^3 + a_2X^2 + a_4X + a_6).$$

If K has characteristic $\neq 2,3$, then it is projectively equivalent to the projective completion $\mathscr{C} = \mathscr{C}_{a,b}$ of an affine curve of the form

$$Y^2 - (X^3 + aX + b). \qquad \square$$

It is easily seen that, conversely, for any choice of $a_1,...,a_6 \in K$ the curve $\mathcal{W}$, and in particular $\mathcal{C}_{a,b}$, is irreducible over K and that $\mathbf{0}$, the unique point at infinity, is a flex.

For any $u,r,s,t \in K$ with $u \neq 0$, the invertible linear change of variables

$$X = u^2X' + r, \quad Y = u^3Y' + su^2X' + t$$

replaces the curve $\mathcal{W} = \mathcal{W}(a_1,...,a_6)$ by a curve $\mathcal{W}' = \mathcal{W}'(a_1',...,a_6')$ of the same form. The numbering of the coefficients reflects the fact that if $r = s = t = 0$, then

$$a_1 = ua_1', \; a_2 = u^2a_2', \; a_3 = u^3a_3', \; a_4 = u^4a_4', \; a_6 = u^6a_6'.$$

In particular, for any nonzero $u \in K$, the invertible linear change of variables

$$X = u^2X', \quad Y = u^3Y'$$

replaces $\mathcal{C}_{a,b}$ by $\mathcal{C}_{a',b'}$, where

$$a = u^4a', \; b = u^6b'.$$

By replacing X by $x + X$ and Y by $y + Y$, we see that if a K-point (x,y) of $\mathcal{C}_{a,b}$ is singular, then

$$3x^2 + a = y = 0,$$

which implies $4a^3 + 27b^2 = 0$. Thus the curve $\mathcal{C}_{a,b}$ has no singular points if $4a^3 + 27b^2 \neq 0$.

We will call

$$d: = 4a^3 + 27b^2$$

the *discriminant* of the curve $\mathcal{C}_{a,b}$. It is not difficult to verify that if the cubic polynomial $X^3 + aX + b$ has roots e_1,e_2,e_3, then

$$d = -[(e_1 - e_2)(e_1 - e_3)(e_2 - e_3)]^2.$$

If $d = 0$, $a \neq 0$, then the polynomial $X^3 + aX + b$ has the repeated root $x_0 = -3b/2a$ and $P = (x_0,0)$ is the unique singular point. If $d = a = 0$, then $b = 0$ and $P = (0,0)$ is the unique singular point.

The different types of curve which arise when $K = \mathbb{R}$ is the field of real numbers are illustrated in Figure 1. The unique point at infinity $\mathbf{0}$ may be thought of as being at both ends of the y-axis. (In the case of a node, Figure 1 illustrates the situation for $x_0 > 0$. For $x_0 < 0$ the singular point is an isolated point of the curve.)

Suppose now that K is any field of characteristic $\neq 2,3$ and that the curve $\mathcal{C}_{a,b}$ has zero discriminant. Because of the geometrical interpretation when $K = \mathbb{R}$, the unique singular point

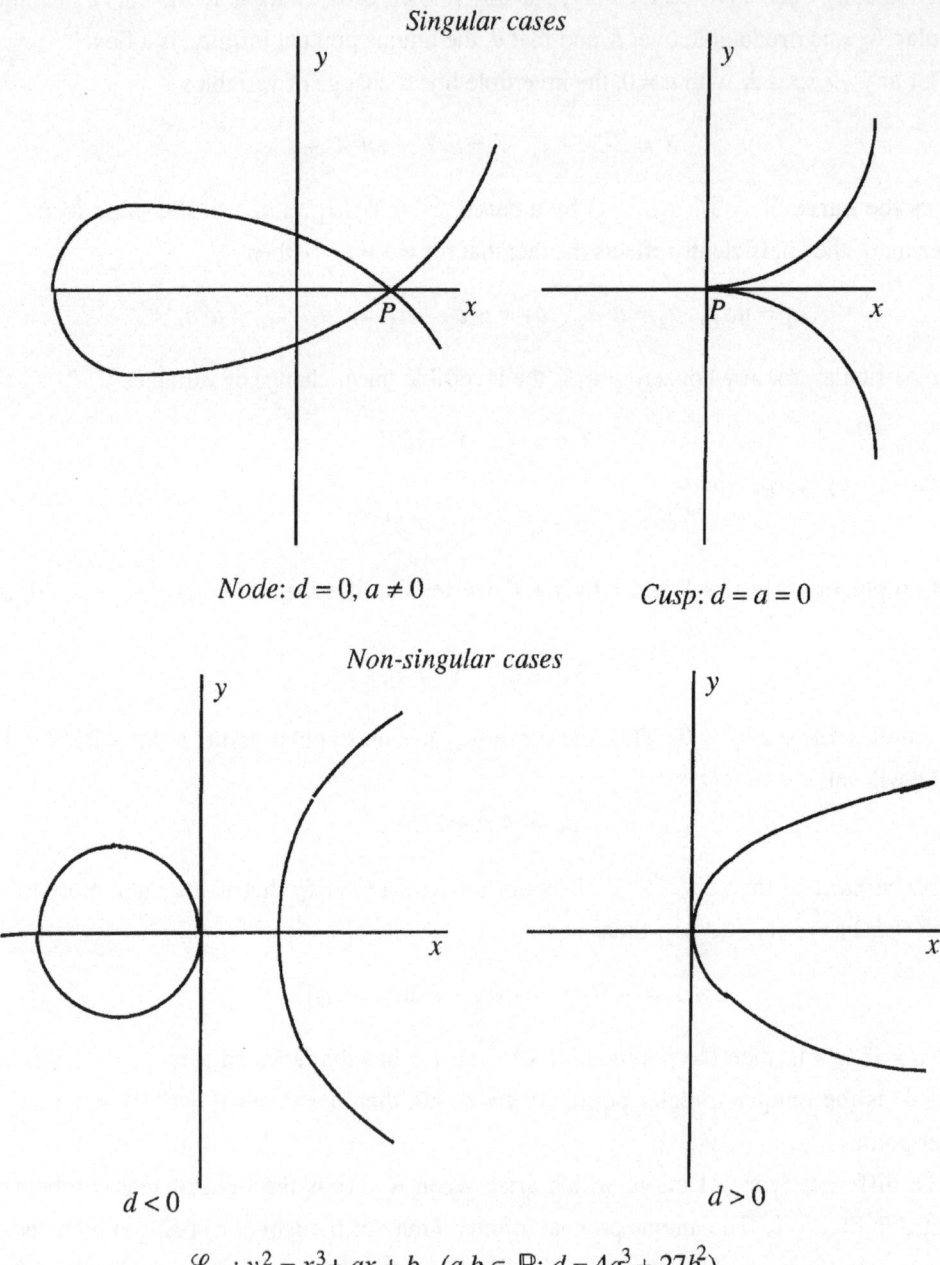

Figure 1: *Cubic curves over* $\mathbb{R}$

of the curve $\mathscr{C}_{a,b}$ is said to be a *node* if $a \neq 0$ and a *cusp* if $a = 0$. In the cusp case, if we put $T = Y/X$, then the cubic curve has the parametrization $X = T^2$, $Y = T^3$. In the node case, if we put $T = Y/(X + 3b/2a)$, then it has the parametrization

$$X = T^2 + 3b/a, \ Y = T^3 + 9bT/2a.$$

Thus in both cases the cubic curve is in fact elementary.

We now restrict attention to non-singular cubic curves, i.e. curves which do not have a singular point.

Two K-points of a projective cubic curve determine a projective line, which intersects the curve in a third K-point. This procedure for generating additional K-points was used implicitly by Diophantus and explicitly by Newton. There is also another procedure, which may be regarded as a limiting case: the tangent to a projective cubic curve at a K-point intersects the curve in another K-point. The combination of the two procedures is known as the 'chord and tangent' process. It will now be described analytically for the cubic curve $\mathscr{C}_{a,b}$.

If O is the unique point at infinity of the cubic curve $\mathscr{C}_{a,b}$ and if $P = (x,y)$ is any finite K-point, then the affine line determined by O and P is $X - x$ and its other point of intersection with $\mathscr{C}_{a,b}$ is $P^* = (x, -y)$.

Now let $P_1 = (x_1, y_1)$ and $P_2 = (x_2, y_2)$ be any two finite K-points. If $x_1 \neq x_2$, then the affine line determined by P_1 and P_2 is

$$Y - mX - c,$$

where

$$m = (y_2 - y_1)/(x_2 - x_1), \ c = (y_1 x_2 - y_2 x_1)/(x_2 - x_1),$$

and its third point of intersection with $\mathscr{C}_{a,b}$ is $P_3 = (x_3, y_3)$, where

$$x_3 = m^2 - x_1 - x_2, \ y_3 = mx_3 + c.$$

If $x_1 = x_2$, but $y_1 \neq y_2$, then the affine line determined by P_1 and P_2 is $X - x_1$ and its other point of intersection with $\mathscr{C}_{a,b}$ is O. Finally, if $P_1 = P_2$, it may be verified that the tangent to $\mathscr{C}_{a,b}$ at P_1 is the affine line

$$Y - mX - c,$$

where

$$m = (3x_1^2 + a)/2y_1, \ c = (-x_1^3 + ax_1 + 2b)/2y_1,$$

and its other point of intersection with $\mathscr{C}_{a,b}$ is the point $P_3 = (x_3, y_3)$, where x_3 and y_3 are given by the same formulas as before, but with the new values of m and c (and with $x_2 = x_1$).

It is rather remarkable that the K-points of a non-singular projective cubic curve can be given the structure of an abelian group. That this is possible is suggested by the addition theorem for elliptic functions.

Suppose that $K = \mathbb{C}$ is the field of complex numbers and that the cubic curve is the projective completion $\mathscr{C}_\lambda$ of the affine curve

$$Y^2 - g_\lambda(X),$$

where

$$g_\lambda(X) \ = \ 4\lambda X^3 - 4(1 + \lambda)X^2 + 4X$$

is Riemann's normal form and $\lambda \neq 0,1$. If $S(u)$ is the elliptic function defined in §3 of Chapter XII, then $P(u) = (S(u),S'(u))$ is a point of $\mathscr{C}_\lambda$ for any $u \in \mathbb{C}$. If we define the sum of $P(u)$ and $P(v)$ to be the point $P(u + v)$, then the set of all $\mathbb{C}$-points of $\mathscr{C}_\lambda$ becomes an abelian group, with $P(0) = (0,0)$ as identity element and with $P(-u) = (S(u), -S'(u))$ as the inverse of $P(u)$. In order to carry this construction over to the cubic curve $\mathscr{C}_{a,b}$ and to other fields than $\mathbb{C}$, we interpret it geometrically.

It was shown in (10) of Chapter XII that

$$S(u + v) \ = \ 4S(u)S(v)[S(v) - S(u)]^2/[S'(u)S(v) - S'(v)S(u)]^2.$$

The points $(x_1,y_1) = (S(u),S'(u))$ and $(x_2,y_2) = (S(v),S'(v))$ determine the affine line

$$Y - mX - c,$$

where

$$m = [S'(v) - S'(u)]/[S(v) - S(u)], \quad c = [S'(u)S(v) - S'(v)S(u)]/[S(v) - S(u)].$$

The third point of intersection of this line with the cubic $\mathscr{C}_\lambda$ is the point (x_3,y_3), where

$$x_3 \ = \ c^2/4\lambda x_1 x_2 \ = \ [S'(u)S(v) - S'(v)S(u)]^2/4\lambda S(u)S(v)[S(v) - S(u)]^2$$

$$= \ 1/\lambda S(u + v).$$

On the other hand, the points $(0,0) = (S(0),S'(0))$ and $(x_3{}^*,y_3{}^*) = (S(u+v),S'(u+v))$ determine the affine line $Y - (y_3{}^*/x_3{}^*)X$ and its third point of intersection with $\mathscr{C}_\lambda$ is the point (x_4,y_4), where $x_4 = 1/\lambda x_3{}^* = x_3$. Evidently $y_4{}^2 = y_3{}^2$, and it may be verified that actually $y_4 = y_3$. Thus $(x_3{}^*,y_3{}^*)$ is the third point of intersection with $\mathscr{C}_\lambda$ of the line determined by the points $(0,0)$ and (x_3,y_3).

The origin $(0,0)$ may not be a point of the cubic curve $\mathscr{C}_{a,b}$ but O, the point at infinity, certainly is. Consequently, as illustrated in Figure 2, we now define the *sum* $P_1 + P_2$ of two

K-points P_1, P_2 of $\mathcal{C}_{a,b}$ to be the K-point $P_3{}^*$, where P_3 is the third point of $\mathcal{C}_{a,b}$ on the line determined by P_1, P_2 and $P_3{}^*$ is the third point of $\mathcal{C}_{a,b}$ on the line determined by O, P_3. If $P_1 = P_2$, the line determined by P_1, P_2 is understood to mean the tangent to $\mathcal{C}_{a,b}$ at P_1.

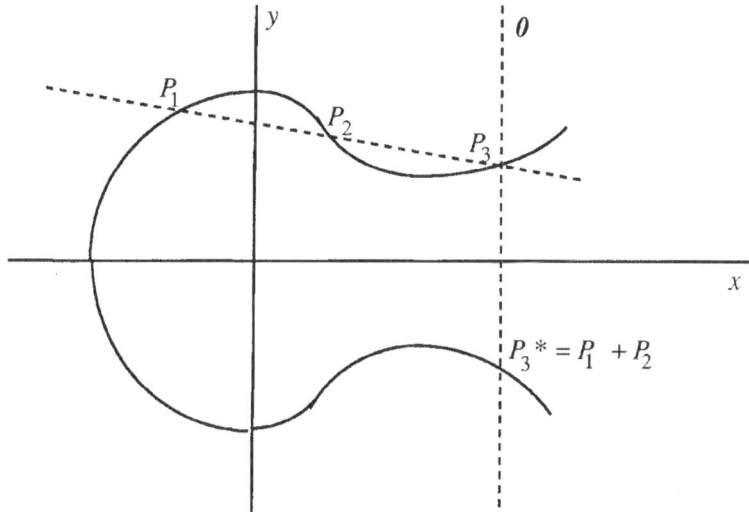

Figure 2: Addition on $\mathcal{C}_{a,b}$

It is simply a matter of elementary algebra to deduce from the formulas given earlier that, when addition is defined in this way, the set of all K-points of $\mathcal{C}_{a,b}$ becomes an abelian group, with O as identity element and with $-P = (x, -y)$ as the inverse of $P = (x,y)$. Since $-P = P$ if and only if $y = 0$, the elements of order 2 in this group are the points $(x_0, 0)$, where x_0 is a root of the polynomial $X^3 + aX + b$ (if it has any roots in K).

Throughout the preceding discussion of cubic curves we restricted attention to those with a flex. It will now be shown that in a sense this is no restriction.

Let $\mathcal{C}$ be a projective cubic curve over the field K, defined by the homogeneous polynomial $F_1(X,Y,Z)$, and suppose that $\mathcal{C}$ has a non-singular K-point P. Without loss of generality we assume that $P = (1,0,0)$ and that the tangent at P is the projective line Z. Then F_1 has no term in X^3 or in X^2Y:

$$F_1(X,Y,Z) = aY^3 + bY^2Z + cYZ^2 + dZ^3 + eX^2Z + gXY^2 + hXZ^2.$$

Here $e \neq 0$, since P is non-singular, and we may suppose $g \neq 0$, since otherwise P is a flex. If we replace $gX + aY$ by X, this assumes the form

$$F_2(X,Y,Z) = XY^2 + bY^2Z + cYZ^2 + dZ^3 + eX^2Z + gXYZ + hXZ^2,$$

with new values for the coefficients. If we now replace $X + bZ$ by X, this assumes the form

$$F_3(X,Y,Z) \;=\; XY^2 + cYZ^2 + dZ^3 + eX^2Z + gXYZ + hXZ^2,$$

again with new values for the coefficients. The projective cubic curve $\mathcal{D}$ over the field K, defined by the homogeneous polynomial

$$F_4(U,V,W) \;=\; VW^2 + cV^2W + dUV^2 + eU^3 + gUVW + hU^2V,$$

has a flex at the point $(0,0,1)$. Moreover,

$$F_3(U^2,VW,UV) \;=\; U^2V\,F_4(U,V,W),$$

$$F_4(XZ,Z^2,XY) \;=\; XZ^2\,F_3(X,Y,Z).$$

This shows that *any projective cubic curve over the field K with a non-singular K-point is* birationally equivalent *to one with a flex.*

Birational equivalence may be defined in the following way. A *rational transformation* of the projective plane with points $X = (X_1,X_2,X_3)$ is a map $X \to Y = \varphi(X)$, where

$$\varphi(X) \;=\; (\varphi_1(X),\varphi_2(X),\varphi_3(X))$$

and $\varphi_1,\varphi_2,\varphi_3$ are homogeneous polynomials without common factor of the same degree m, say. (In the corresponding affine plane the coordinates are transformed by *rational* functions.) The transformation is *birational* if there exists an inverse map $Y \to X = \psi(Y)$, where

$$\psi(Y) \;=\; (\psi_1(Y),\psi_2(Y),\psi_3(Y))$$

and ψ_1,ψ_2,ψ_3 are homogeneous polynomials without common factor of the same degree n, say, such that

$$\psi[\varphi(X)] = \omega(X)\,X, \quad \varphi[\psi(Y)] = \theta(Y)\,Y$$

for some scalar polynomials $\omega(X),\theta(Y)$. Two irreducible projective plane curves $\mathcal{C}$ and $\mathcal{D}$ over the field K, defined respectively by the homogeneous polynomials $F(X)$ and $G(Y)$ (not necessarily of the same degree), are *birationally equivalent* if there exists a birational transformation $Y = \varphi(X)$ with inverse $X = \psi(Y)$ such that $G[\varphi(X)]$ is divisible by $F(X)$ and $F[\psi(Y)]$ is divisible by $G[(Y)]$.

It is clear that birational equivalence is indeed an equivalence relation, and that irreducible projective curves which are projectively equivalent are also birationally equivalent. Birational transformations are often used to simplify the singular points of a curve. Indeed the theorem on *resolution of singularities* says that any irreducible curve is birationally equivalent to a non-singular curve, although it may be a curve in a higher-dimensional space rather than in the plane. The algebraic geometry of curves may be regarded as the study of those properties which are invariant under birational equivalence.

It was shown by Poincaré (1901) that any non-singular curve of *genus* 1 defined over the field $\mathbb{Q}$ of rational numbers and with at least one rational point is birationally equivalent over $\mathbb{Q}$ to a cubic curve. Such a curve is now said to be an *elliptic curve* (for the somewhat inadequate reason that it may be parametrized by elliptic functions over the field of complex numbers.) However, for our purposes it is sufficient to define an elliptic curve to be a non-singular cubic curve of the form $\mathcal{W}$, over a field K of arbitrary characteristic, or of the form $\mathcal{C}_{a,b}$, over a field K of characteristic $\neq 2,3$.

4 Mordell's theorem

We showed in the previous section that, for any field K of characteristic $\neq 2,3$, the K-points of the elliptic curve $\mathcal{C}_{a,b}$ defined by the polynomial

$$Y^2 - X^3 - aX - b,$$

where $a,b \in K$ and $d := 4a^3 + 27b^2 \neq 0$, form an abelian group, $E(K)$ say. We now restrict our attention to the case when $K = \mathbb{Q}$ is the field of rational numbers, and we write simply $E := E(\mathbb{Q})$. This section is devoted to the basic theorem of Mordell (1922), which says that *the abelian group E is finitely generated*.

By replacing X by X/c^2 and Y by Y/c^3 for some nonzero $c \in \mathbb{Q}$, we may (and will) assume that a and b are both integers. Let $P = (x,y)$ be any finite rational point of $\mathcal{C}_{a,b}$ and write $x = p/q$, where p and q are coprime integers. The *height* $h(P)$ of P is uniquely defined by

$$h(P) = \log \max(|p|,|q|).$$

We also set $h(O) = 0$, where O is the unique point at infinity of $\mathcal{C}_{a,b}$.

Evidently $h(P) \geq 0$. Furthermore, $h(-P) = h(P)$, since $P = (x,y)$ implies $-P = (x,-y)$. Also, for any $r > 0$, there exist only finitely many elements $P = (x,y)$ of E with $h(P) \leq r$, since x determines y up to sign.

PROPOSITION 8 *There exists a constant $C = C(a,b) > 0$ such that*

$$\left| h(2P) - 4h(P) \right| \leq C \quad \text{for all } P \in E.$$

Proof By the formulas given in §3, if $P = (x,y)$, then $2P = (x',y')$, where

$$x' = m^2 - 2x, \quad m = (3x^2 + a)/2y.$$

Since $y^2 = x^3 + ax + b$, it follows that

$$x' = (x^4 - 2ax^2 - 8bx + a^2)/4(x^3 + ax + b).$$

If $x = p/q$, where p and q are coprime integers, then $x' = p'/q'$, where

$$p' = p^4 - 2ap^2q^2 - 8bpq^3 + a^2q^4,$$
$$q' = 4q(p^3 + apq^2 + bq^3).$$

Evidently p' and q' are also integers, but they need not be coprime. However, since

$$p' = ep'', \quad q' = eq'',$$

where e,p'',q'' are integers and p'',q'' are coprime, we have

$$h(2P) = \log \max(|p''|,|q''|) \le \log \max(|p'|,|q'|).$$

Since

$$\max(|p'|,|q'|) \le \max(|p|,|q|)^4 \max\{1 + 2|a| + 8|b| + a^2, 4(1 + |a| + |b|)\},$$

it follows that

$$h(2P) \le 4h(P) + C'$$

for some constant $C' = C'(a,b) > 0$.

The Euclidean algorithm may be used to derive the polynomial identity

$$(3X^2 + 4a)(X^4 - 2aX^2 - 8bX + a^2) - (3X^3 - 5aX - 27b)(X^3 + aX + b) = d,$$

where once again $d = 4a^3 + 27b^2$. Substituting p/q for X, we obtain

$$4dq^7 = 4(3p^2q + 4aq^3)p' - (3p^3 - 5apq^2 - 27bq^3)q'.$$

Similarly, the Euclidean algorithm may be used to derive the polynomial identity

$$f(X)(1 - 2aX^2 - 8bX^3 + a^2X^4) + g(X)X(1 + aX^2 + bX^3) = d,$$

where

$$f(X) = 4a^3 + 27b^2 - a^2bX + a(3a^3 + 22b^2)X^2 + 3b(a^3 + 8b^2)X^3,$$
$$g(X) = a^2b + a(5a^3 + 32b^2)X + 2b(13a^3 + 96b^2)X^2 - 3a^2(a^3 + 8b^2)X^3.$$

Substituting q/p for X, we obtain

$$4dp^7 = 4\{(4a^3 + 27b^2)p^3 - a^2bp^2q + (3a^4 + 22ab^2)pq^2 + 3(a^3b + 8b^3)q^3\}p'$$

$$+ \{a^2bp^3 + (5a^4 + 32ab^2)p^2q + (26a^3b + 192b^3)pq^2 - 3(a^5 + 8a^2b^2)q^3\}q'.$$

Since $d \neq 0$, it follows from these two relations that

$$\max(|p|,|q|)^7 \leq C_1 \max(|p|,|q|)^3 \max(|p'|,|q'|)$$

and hence

$$\max(|p|,|q|)^4 \leq C_1 \max(|p'|,|q'|).$$

But the two relations also show that the greatest common divisor e of p' and q' divides both $4dq^7$ and $4dp^7$, and hence also $4d$, since p and q are coprime. Consequently

$$\max(|p'|,|q'|) \leq 4|d| \max(|p''|,|q''|).$$

Combining this with the previous inequality, we obtain

$$4h(P) \leq h(2P) + C''$$

for some constant $C'' = C''(a,b) > 0$.

This proves the result, with $C = \max(C',C'')$. $\square$

PROPOSITION 9 *There exists a unique function* $\hat{h}: E \to \mathbb{R}$ *such that*

(i) $\hat{h} - h$ *is bounded,*
(ii) $\hat{h}(2P) = 4\hat{h}(P)$ *for every* $P \in E$.

Furthermore, it is given by the formula $\hat{h}(P) = \lim_{n \to \infty} h(2^nP)/4^n$.

Proof Suppose $\hat{h}$ has the properties (i),(ii). Then, by (ii), $4^n \hat{h}(P) = \hat{h}(2^nP)$ and hence, by (i), $4^n \hat{h}(P) - h(2^nP)$ is bounded. Dividing by 4^n, we see that $h(2^nP)/4^n \to \hat{h}(P)$ as $n \to \infty$. This proves uniqueness.

To prove existence, choose C as in the statement of Proposition 8. Then, for any integers m,n with $n > m > 0$,

$$|4^{-n} h(2^nP) - 4^{-m} h(2^mP)| = |\textstyle\sum_{j=m}^{n-1} \{4^{-j-1} h(2^{j+1}P) - 4^{-j} h(2^jP)\}|$$

$$\leq \textstyle\sum_{j=m}^{n-1} 4^{-j-1} |h(2^{j+1}P) - 4h(2^jP)\}|$$

$$\leq \textstyle\sum_{j=m}^{n-1} 4^{-j-1} C < 4^{-m} C/3.$$

Thus the sequence $\{4^{-n} h(2^n P)\}$ is a fundamental sequence and consequently convergent. If we denote its limit by $\hat{h}(P)$, then clearly $\hat{h}(2P) = 4\hat{h}(P)$. On the other hand, by taking $m = 0$ and letting $n \to \infty$ in the preceding inequality we obtain

$$|\hat{h}(P) - h(P)| \le C/3.$$

Thus $\hat{h}$ has both the required properties. ☐

The value $\hat{h}(P)$ is called the *canonical height* of the rational point P. The formula for $\hat{h}(P)$ shows that, for all $P \in E$,

$$\hat{h}(-P) = \hat{h}(P) \ge 0.$$

Moreover, by Proposition 9(i), for any $r > 0$ there exist only finitely many elements P of E with $\hat{h}(P) \le r$.

It will now be shown that the canonical height satisfies the *parallelogram law*:

PROPOSITION 10 *For all* $P_1, P_2 \in E$,

$$\hat{h}(P_1 + P_2) + \hat{h}(P_1 - P_2) = 2\hat{h}(P_1) + 2\hat{h}(P_2).$$

Proof It is sufficient to show that there exists a constant $C' > 0$ such that, for all $P_1, P_2 \in E$,

$$h(P_1 + P_2) + h(P_1 - P_2) \le 2h(P_1) + 2h(P_2) + C'. \tag{*}$$

For it then follows from the formula in Proposition 9 that, for all $P_1, P_2 \in E$,

$$\hat{h}(P_1 + P_2) + \hat{h}(P_1 - P_2) \le 2\hat{h}(P_1) + 2\hat{h}(P_2).$$

But, replacing P_1 by $P_1 + P_2$ and P_2 by $P_1 - P_2$, we also have

$$\hat{h}(2P_1) + \hat{h}(2P_2) \le 2\hat{h}(P_1 + P_2) + 2\hat{h}(P_1 - P_2)$$

and hence, by Proposition 9(ii),

$$2\hat{h}(P_1) + 2\hat{h}(P_2) \le \hat{h}(P_1 + P_2) + \hat{h}(P_1 - P_2).$$

To prove (*) we may evidently assume that $P_1 = (x_1, y_1)$ and $P_2 = (x_2, y_2)$ are both finite. Moreover, by Proposition 8, we may assume that $P_1 \ne P_2$. Then, by the formulas of §3,

$$P_1 + P_2 = (x_3, y_3), \quad P_1 - P_2 = (x_4, y_4),$$

where

$$x_3 = (y_2 - y_1)^2/(x_2 - x_1)^2 - (x_1 + x_2),$$
$$x_4 = (y_2 + y_1)^2/(x_2 - x_1)^2 - (x_1 + x_2).$$

Hence

$$x_3 + x_4 = 2[y_2^2 + y_1^2 - (x_2 - x_1)(x_2^2 - x_1^2)]/(x_2 - x_1)^2$$

and

$$x_3 x_4 = (y_2^2 - y_1^2)^2/(x_2 - x_1)^4 - 2(x_1 + x_2)(y_1^2 + y_2^2)/(x_2 - x_1)^2 + (x_1 + x_2)^2.$$

Since $y_j^2 = x_j^3 + a x_j + b$ $(j = 1,2)$, these relations simplify to

$$x_3 + x_4 = 2[x_1 x_2(x_1 + x_2) + a(x_1 + x_2) + 2b]/(x_2 - x_1)^2$$

and

$$x_3 x_4 = N/(x_2 - x_1)^2,$$

where

$$N = (x_2^2 + x_1 x_2 + x_1^2 + a)^2 - 2(x_1 + x_2)^2(x_2^2 - x_1 x_2 + x_1^2 + a) - 4b(x_1 + x_2) + (x_2^2 - x_1^2)^2$$

$$= (x_1 x_2 - a)^2 - 4b(x_1 + x_2).$$

Put $x_j = p_j/q_j$, where $(p_j,q_j) = 1$ $(1 \leq j \leq 4)$. Then x_3,x_4 are the roots of the quadratic polynomial

$$AX^2 + BX + C$$

with integer coefficients

$$A = (p_2 q_1 - p_1 q_2)^2,$$
$$B = (p_1 p_2 + a q_1 q_2)(p_1 q_2 + p_2 q_1) + 2b q_1^2 q_2^2,$$
$$C = (p_1 p_2 - a q_1 q_2)^2 - 4b q_1 q_2(p_1 q_2 + p_2 q_1).$$

Consequently

$$A p_3 p_4 = C q_3 q_4,$$
$$A(p_3 q_4 + p_4 q_3) = B q_3 q_4.$$

By Proposition II.16, q_3 and q_4 each divide A, and so their product divides A^2. Hence, for some integer $D \neq 0$,

$$A^2 = D q_3 q_4, \quad AC = D p_3 p_4, \quad AB = D(p_3 q_4 + p_4 q_3).$$

But it is easily seen that $q_3 q_4$, $p_3 p_4$ and $p_3 q_4 + p_4 q_3$ have no common prime divisor. It follows that A divides D.

Hence, if we put

$$\rho_j = \max(|p_j|,|q_j|) \quad (1 \leq j \leq 4),$$

then

$$|q_3 q_4| \leq |A| \leq 4\rho_1^2 \rho_2^2,$$

$$|p_3 p_4| \leq |C| \leq [(1 + |a|)^2 + 8|b|]\rho_1^2 \rho_2^2,$$

$$|p_3 q_4 + p_4 q_3| \leq |B| \leq 2(1 + |a| + |b|)\rho_1^2 \rho_2^2.$$

But

$$\max(|p_3|, |q_3|) \max(|p_4|, |q_4|) \leq \max(|p_3 p_4|, |q_3 q_4| + |p_3 q_4 + p_4 q_3|),$$

since if $|q_3| \leq |p_3|$ and $|p_4| \leq |q_4|$, for example, then

$$|p_3 q_4| \leq |p_4 q_3| + |p_3 q_4 + p_4 q_3| \leq |q_3 q_4| + |p_3 q_4 + p_4 q_3|.$$

It follows that there exists a constant $C'' > 0$ such that

$$\rho_3 \rho_4 \leq C'' \rho_1^2 \rho_2^2,$$

which is equivalent to (*) with $C' = \log C''$. $\square$

COROLLARY 11 *For any $P \in E$ and any integer n,*

$$\hat{h}(nP) = n^2 \hat{h}(P).$$

Proof Since $\hat{h}(-P) = \hat{h}(P)$, we may assume $n > 0$. We may actually assume $n > 2$, since the result is trivial for $n = 1$ and it holds for $n = 2$ by Proposition 9. By Proposition 10 we have

$$\hat{h}(nP) + \hat{h}((n-2)P) = 2\hat{h}((n-1)P) + 2\hat{h}(P),$$

from which the general case follows by induction. $\square$

It follows from Corollary 11 that if an element P of the group E has finite order, then $\hat{h}(P) = 0$. The converse is also true. In fact, by Proposition 10, the set of all $P \in E$ such that $\hat{h}(P) = 0$ is a subgroup of E, and this subgroup is finite since there are only finitely many points P such that $\hat{h}(P) < 1$.

We now deduce from Proposition 10 that a non-negative quadratic form can be constructed from the canonical height. If we put

$$(P,Q) = \hat{h}(P + Q) - \hat{h}(P) - \hat{h}(Q),$$

then evidently

$$(P,Q) = (Q,P), \quad (P,P) = 2\hat{h}(P) \geq 0.$$

It remains to show that

$$(P,Q + R) = (P,Q) + (P,R),$$

and we do this by proving that

$$\hat{h}(P + Q + R) = \hat{h}(P + Q) + \hat{h}(P + R) + \hat{h}(Q + R) - \hat{h}(P) - \hat{h}(Q) - \hat{h}(R).$$

But, by the parallelogram law,

$$\hat{h}(P + Q + R + P) + \hat{h}(Q + R) = \hat{h}(P + Q + R + P) + \hat{h}(P + Q + R - P)$$

$$= 2\hat{h}(P + Q + R) + 2\hat{h}(P)$$

and

$$\hat{h}(P + Q + R + P) + \hat{h}(Q - R) = \hat{h}(P + Q + R + P) + \hat{h}(P + Q - P - R)$$

$$= 2\hat{h}(P + Q) + 2\hat{h}(P + R).$$

Subtracting the second relation from the first, we obtain

$$\hat{h}(Q + R) - \hat{h}(Q - R) = 2\hat{h}(P + Q + R) + 2\hat{h}(P) - 2\hat{h}(P + Q) - 2\hat{h}(P + R).$$

Since, by the parallelogram law again,

$$\hat{h}(Q + R) + \hat{h}(Q - R) = 2\hat{h}(Q) + 2\hat{h}(R),$$

this is equivalent to what we wished to prove.

PROPOSITION 12 *The abelian group E is finitely generated if, for some integer $m > 1$, the factor group E/mE is finite.*

Proof Let S be a set of representatives of the cosets of the subgroup mE. Since S is finite, by hypothesis, we can choose $C > 0$ so that $\hat{h}(Q) \le C$ for all $Q \in S$. The set

$$S' = \{Q' \in E: \hat{h}(Q') \le C\}$$

contains S and is also finite. We will show that it generates E.

Let E' be the subgroup of E generated by the elements of S'. If $E' \ne E$, choose $P \in E \setminus E'$ so that $\hat{h}(P)$ is minimal. Then

$$P = mP_1 + Q_1 \text{ for some } P_1 \in E \text{ and } Q_1 \in S.$$

Since

$$\hat{h}(P + Q_1) + \hat{h}(P - Q_1) = 2\,\hat{h}(P) + 2\,\hat{h}(Q_1),$$

it follows that

$$\hat{h}(mP_1) = \hat{h}(P - Q_1) \leq 2\,\hat{h}(P) + 2C$$

and hence

$$\hat{h}(P_1) \leq 2[\,\hat{h}(P) + C]/m^2 \leq [\,\hat{h}(P) + C]/2.$$

But $P_1 \notin E'$, since $P \notin E'$, and hence $\hat{h}(P_1) \geq \hat{h}(P)$. It follows that $\hat{h}(P) \leq C$, which is a contradiction. Hence $E' = E$. $\square$

Proposition 12 shows that to complete the proof of Mordell's theorem it is enough to show that the factor group $E/2E$ is finite. *We will prove this only for the case when E contains an element of order* 2. A similar proof may be given for the general case, but it requires some knowledge of algebraic number theory.

The assumption that E contains an element of order 2 means that there is a rational point $(x_0,0)$, where x_0 is a root of the polynomial $X^3 + aX + b$. Since a and b are taken to be integers, and the polynomial has highest coefficient 1, x_0 must also be an integer. By changing variable from X to $x_0 + X$, we replace the cubic $\mathscr{C}_{a,b}$ by a cubic $C_{A,B}$ defined by a polynomial

$$Y^2 - (X^3 + AX^2 + BX),$$

where $A,B \in \mathbb{Z}$. The non-singularity condition $d := 4a^3 + 27b^2 \neq 0$ becomes

$$D := B^2(4B - A^2) \neq 0,$$

but this is the only restriction on A,B. The chord joining two rational points of $C_{A,B}$ is given by the same formulas as for $\mathscr{C}_{a,b}$ in §3, but the tangent to $C_{A,B}$ at the finite point $P_1 = (x_1,y_1)$ is now the affine line

$$Y - mX - c,$$

where

$$m = (3x_1{}^2 + 2Ax_1 + B)/2y_1, \quad c = -x_1(x_1{}^2 - B)/2y_1.$$

The geometrical interpretation of the group law remains the same as before. We will now denote by E the group of all rational points of $C_{A,B}$. Our change of variable has made the point $N = (0,0)$ an element of E of order 2.

Let $P = (x,y)$ be a rational point of $C_{A,B}$ with $x \neq 0$. We are going to show that, in a sense which will become clear, there are only finitely many rational square classes to which x can belong.

Write $x = m/n$, $y = p/q$, where m,n,p,q are integers with $n,q > 0$ and $(m,n) = (p,q) = 1$. Then

$$p^2 n^3 = (m^3 + Am^2 n + Bmn^2)q^2,$$

which implies both $q^2|n^3$ and $n^3|q^2$. Thus $n^3 = q^2$. From $n^2|q^2$ we obtain $n|q$. Hence $q = en$ for some integer e, and it follows that $n = e^2$, $q = e^3$. Thus

$$x = m/e^2, \quad y = p/e^3, \text{ where } e > 0 \text{ and } (m,e) = (p,e) = 1.$$

Moreover,

$$p^2 = m(m^2 + Ame^2 + Be^4).$$

This shows that each prime which divides m, but not $m^2 + Ame^2 + Be^4$, must occur to an even power in m. On the other hand, each prime which divides both m and $m^2 + Ame^2 + Be^4$ must also divide B, since $(m,e) = 1$. Consequently we can write

$$x = \pm p_1^{\varepsilon_1} \cdots p_k^{\varepsilon_k} (u/e)^2,$$

where $u \in \mathbb{N}$, $p_1,\ldots,p_k$ are the distinct primes dividing B and $\varepsilon_j \in \{0,1\}$ $(1 \le j \le k)$. Hence there are at most 2^{k+1} rational square classes to which x can belong.

Suppose now that $P_1 = (x_1,y_1)$ and $P_2 = (x_2,y_2)$ are distinct rational points of $C_{A,B}$ for which $x_1 x_2$ is a nonzero rational square, and let $P_3 = (x_3,y_3)$ be the third point of intersection with $C_{A,B}$ of the line through P_1 and P_2. Then x_1,x_2,x_3 are the three roots of a cubic equation

$$(mX + c)^2 = X^3 + AX^2 + BX.$$

From the constant term we see that $x_1 x_2 x_3 = c^2$. It follows that x_3 is a nonzero rational square if $c \ne 0$. If $c = 0$, then $P_3 = N$ and $x_1 x_2 = B$.

Suppose next that $P = (x,y)$ is any rational point of $C_{A,B}$ with $x \ne 0$, and let $2P = (\bar{x},-\bar{y})$. Then $\bar{P} = (\bar{x},\bar{y})$ is the other point of intersection with $C_{A,B}$ of the tangent to $C_{A,B}$ at P. By the same argument as before, $x^2 \bar{x} = c^2$. Hence $\bar{x}$ is a nonzero rational square if $c \ne 0$. If $c = 0$, then $2P = N$ and $x^2 = B$.

To deduce that $E/2E$ is finite from these observations we will use an arithmetic analogue of Landen's transformation. We saw in Chapter XII that, over the field $\mathbb{C}$ of complex numbers, the cubic curve $\mathcal{C}_\lambda$ defined by the polynomial $Y^2 - g_\lambda(X)$, where $g_\lambda(X) = 4X(1-X)(1-\lambda X)$, admits the parametrization

$$X = S(u,\lambda), \quad Y = S'(u,\lambda).$$

It follows from Proposition XII.11 that the cubic curve $\mathscr{C}_{\lambda'}$, where $\lambda' = \lambda^2/[1 + (1 - \lambda)^{1/2}]^4$, admits the parametrization

$$X' = [1 + (1 - \lambda)^{1/2}]X(1 - X)/(1 - \lambda X),$$

$$Y' = [1 + (1 - \lambda)^{1/2}]Y(1 - 2X + \lambda X^2)/(1 - \lambda X)^2,$$

where again $X = S(u,\lambda)$, $Y = S'(u,\lambda)$ and where $(1 - 2X + \lambda X^2)/(1 - \lambda X)^2$ is the derivative with respect to X of $X(1 - X)/(1 - \lambda X)$. Since also $X' = S(u',\lambda')$, where $u' = [1 + (1 - \lambda)^{1/2}]u$, the map $(X,Y) \rightarrow (X',Y')$ defines a homomorphism of the group of complex points of $\mathscr{C}_\lambda$ into the group of complex points of $\mathscr{C}_{\lambda'}$.

We will simply state analogous results for the cubic curve $C_{A,B}$ over the field $\mathbb{Q}$ of rational numbers, since their verification is elementary. If (x,y) is a rational point of $C_{A,B}$ with $x \neq 0$ and if

$$x' = (x^2 + Ax + B)/x, \quad y' = y(x^2 - B)/x^2,$$

then (x',y') is a rational point of $C_{A',B'}$, where

$$A' = -2A, \quad B' = A^2 - 4B.$$

Moreover, if we define a map φ of the group E of all rational points of $C_{A,B}$ into the group E' of all rational points of $C_{A',B'}$ by putting

$$\varphi(x,y) = (x',y') \text{ if } x \neq 0, \quad \varphi(N) = \varphi(O) = O,$$

then φ is a homomorphism, i.e.

$$\varphi(P + Q) = \varphi(P) + \varphi(Q), \quad \varphi(-P) = -\varphi(P).$$

The range $\varphi(E)$ may not be the whole of E'. In fact, since

$$x' = (x^3 + Ax^2 + Bx)/x^2 = (y/x)^2,$$

the first coordinate of any finite point of $\varphi(E)$ must be a rational square. Furthermore, if $N = (0,0)$ is a point of $\varphi(E)$, the integer $B' = A^2 - 4B$ must be a square. We will show that these conditions completely characterize $\varphi(E)$.

Evidently if $A^2 - 4B$ is a square, then the quadratic polynomial $X^2 + AX + B$ has a rational root $x_0 \neq 0$ and $\varphi(x_0,0) = N$. Suppose now that (x',y') is a rational point of $C_{A',B'}$ and that $x' = t^2$ is a nonzero rational square. We will show that if

$$x_1 = (t^2 - A + y'/t)/2, \quad y_1 = tx_1,$$
$$x_2 = (t^2 - A - y'/t)/2, \quad y_2 = -tx_2,$$

then $(x_j, y_j) \in E$ and $\varphi(x_j, y_j) = (x', y')$ $(j = 1, 2)$. It is easily seen that $(x_j, y_j) \in E$ if and only if

$$t^2 = x_j + A + B/x_j.$$

But

$$x_1 x_2 = [(t^2 - A)^2 - y'^2/t^2]/4$$

$$= [(x' - A)^2 - y'^2/x']/4$$

$$= (x'^3 - 2Ax'^2 + A^2 x' - y'^2)/4x'.$$

Since

$$y'^2 = x'^3 - 2Ax'^2 + (A^2 - 4B)x',$$

it follows that $x_1 x_2 = B$. Hence (x_1, y_1) and (x_2, y_2) are both in E if $t^2 = x_1 + A + x_2$, and this condition is certainly satisfied by the definitions of x_1 and x_2.

In addition to

$$x_j + A + B/x_j = t^2 = x' \quad (j = 1, 2),$$

we have

$$y_1(x_1^2 - B)/x_1^2 = t(x_1^2 - x_1 x_2)/x_1 = t(x_1 - x_2) = y',$$

and similarly $y_2(x_2^2 - B)/x_2^2 = y'$. It follows that

$$\varphi(x_1, y_1) = \varphi(x_2, y_2) = (x', y').$$

Since φ is a homomorphism, the range $\varphi(E)$ is a subgroup of E'. We are going to show that this subgroup is of finite index in E'. By what we have already proved for E, there exists a finite (or empty) set $P_1' = (x_1', y_1'), \ldots, P_s' = (x_s', y_s')$ of points of E' such that x_i' is not a rational square $(1 \le i \le s)$ and such that, if $P' = (x', y')$ is any other point of E' for which x' is not a rational square, then $x' x_j'$ is a nonzero rational square for a unique $j \in \{1, \ldots, s\}$. Let $P'' = (x'', y'')$ be the third point of intersection with $C_{A',B'}$ of the line through P' and P_j', so that

$$P' + P_j' + P'' = O.$$

By what we have already proved, either x'' is a nonzero rational square or $P'' = N$ and $x' x_j' = B'$ is a square. In either case, $P'' \in \varphi(E)$. Furthermore, if $2P_j' = (\bar{x}, -\bar{y})$, then either $\bar{x}$ is a nonzero rational square or $2P_j' = N$ and $x_j'^2 = B'$. In either case again, $2P_j' \in \varphi(E)$. Since

$$P' = P_j' - (2P_j' + P''),$$

it follows that P' and P_j' are in the same coset of $\varphi(E)$. Consequently $P_1',...,P_s'$, together with O, and also N if B' is not a square, form a complete set of representatives of the cosets of $\varphi(E)$ in E'.

The preceding discussion can be repeated with $C_{A',B'}$ in the place of $C_{A,B}$. It yields a homomorphism φ' of the group E' of all rational points of $C_{A',B'}$ into the group E'' of all rational points of $C_{A'',B''}$, where

$$A'' = -2A' = 4A, \quad B'' = A'^2 - 4B' = 16B.$$

But the simple transformation $(X,Y) \to (X/4,Y/8)$ replaces $C_{A'',B''}$ by $C_{A,B}$ and defines an isomorphism χ of E'' with E. Hence the composite map $\psi = \chi \circ \varphi'$ is a homomorphism of E' into E, and $\psi \circ \varphi$ is a homomorphism of E into itself.

We now show that the homomorphism $P \to \psi \circ \varphi(P)$ is just the doubling map $P \to 2P$. Since this is obvious if $P = O$ or N, we need only verify it for $P = (x,y)$ with $x \neq 0$.

For $P'' = \varphi' \circ \varphi(P)$ we have

$$x'' = (y'/x')^2 = [y(1 - B/x^2) \cdot x^2/y^2]^2 = (x^2 - B)^2/y^2$$

and

$$y'' = y'(1 - B'/x'^2) = y(1 - B/x^2)[1 - (A^2 - 4B)x^4/y^4]$$

$$= (x^2 - B)[y^4 - (A^2 - 4B)x^4]/x^2y^3$$

$$= (x^2 - B)[(x^2 + Ax + B)^2 - (A^2 - 4B)x^2]/y^3.$$

Hence for $\psi \circ \varphi(P) = P^* = (x^*,y^*)$ we have

$$x^* = (x^2 - B)^2/4y^2,$$
$$y^* = (x^2 - B)[(x^2 + Ax + B)^2 - (A^2 - 4B)x^2]/8y^3.$$

On the other hand, if the tangent to $C_{A,B}$ at P intersects $C_{A,B}$ again at $(\bar{x},\bar{y})$, then $2P = (\bar{x},-\bar{y})$. The cubic equation

$$(mx + c)^2 = X^3 + AX^2 + BX$$

has x as a double root and $\bar{x}$ as its third root. Hence $\bar{x} = (c/x)^2$. Using the formula for c given previously, we obtain

$$\bar{x} = (x^2 - B)^2/4y^2 = x^*.$$

Furthermore, using the formula for m given previously,

$$\bar{y} = m\bar{x} + c = [(3x^2 + 2Ax + B)\bar{x} - x(x^2 - B)]/2y$$

$$= (x^2 - B)[(3x^2 + 2Ax + B)(x^2 - B) - 4xy^2)]/8y^3.$$

Substituting $x^3 + Ax^2 + Bx$ for y^2, we obtain $\bar{y} = -y^*$. Thus $\psi \circ \varphi(P) = 2P$, as claimed.

Since $\varphi(E)$ has finite index in E', and likewise $\psi(E')$ has finite index in E, it follows that $2E = \psi \circ \varphi(E)$ has finite index in E. (The proof shows that the index is at most $2^{\alpha+\beta+2}$, where α is the number of distinct prime divisors of B and β is the number of distinct prime divisors of $A^2 - 4B$.)

By the remarks after the proof of Proposition 12, Mordell's theorem has now been completely proved in the case where E contains an element of order 2.

5 Further results and conjectures

Let $\mathcal{C}_{a,b}$ be the elliptic curve defined by the polynomial

$$Y^2 - (X^3 + aX + b),$$

where $a, b \in \mathbb{Z}$ and $d := 4a^3 + 27b^2 \neq 0$. By Mordell's theorem, the abelian group $E = E_{a,b}(\mathbb{Q})$ of all rational points of $\mathcal{C}_{a,b}$ is finitely generated. It follows from the structure theorem for finitely generated abelian groups (Chapter III, §4) that E is the direct sum of a finite abelian group E^t and a 'free' abelian group E^f, which is the direct sum of $r \geq 0$ infinite cyclic subgroups. The non-negative integer r is called the *rank* of the elliptic curve and E^t its *torsion group*.

The torsion group can, in principle, be determined by a finite amount of computation. A theorem of Nagell (1935) and Lutz (1937) says that if $P = (x, y)$ is a point of E of finite order, then x and y are integers and either $y = 0$ or y^2 divides d. Thus there are only finitely many possibilities to check.

A deep theorem of Mazur (1977) says that the torsion group must be one of the following:

(i)   a cyclic group of order n ($1 \leq n \leq 10$ or $n = 12$),

(ii) the direct sum of a cyclic group of order 2 and a cyclic group of order $2n$ ($1 \leq n \leq 4$).

It was already known that each of these possibilities occurs. It is easy to check if the torsion group is of type (i) or type (ii), since in the latter case there are three elements of order 2, whereas in the former case there is at most one. Mazur's result shows that an element has infinite order, if it does not have order ≤ 12.

It is conjectured that there exist elliptic curves over $\mathbb{Q}$ with arbitrarily large rank. (Examples are known of elliptic curves with rank ≥ 22.) At present no infallible algorithm is known for determining the rank of an elliptic curve, let alone a basis for the torsion-free group $E\!f$. However, Manin (1971) devised a conditional algorithm, based on the strong conjecture of Birch and Swinnerton-Dyer which will be mentioned later. This conjecture is still unproved, but is supported by much numerical evidence.

An important way of obtaining arithmetic information about an elliptic curve is by reduction modulo a prime p. We regard the coefficients not as integers, but as integers mod p, and we look not for $\mathbb{Q}$-points, but for $\mathbb{F}_p$-points. Since the normal form $\mathscr{C}_{a,b}$ was obtained by assuming that the field had characteristic $\neq 2,3$, we now adopt a more general normal form.

Let $\mathscr{W} = \mathscr{W}(a_1,...,a_6)$ be the projective completion of the affine cubic curve defined by the polynomial

$$Y^2 + a_1XY + a_3Y - (X^3 + a_2X^2 + a_4X + a_6),$$

where $a_j \in \mathbb{Q}$ $(j = 1,2,3,4,6)$. It may be shown that $\mathscr{W}$ is non-singular if and only if the *discriminant* $\Delta \neq 0$, where

$$\Delta = -b_2^2 b_8 - 8b_4^3 - 27b_6^2 + 9b_2 b_4 b_6$$

and

$$b_2 = a_1^2 + 4a_2,$$
$$b_4 = a_1 a_3 + 2a_4,$$
$$b_6 = a_3^2 + 4a_6,$$
$$b_8 = a_1^2 a_6 - a_1 a_3 a_4 + 4a_2 a_6 + a_2 a_3^2 - a_4^2.$$

(We retain the name 'discriminant', although $\Delta = -16d$ for $\mathscr{W} = \mathscr{C}_{a,b}$.) The definition of addition on $\mathscr{W}$ has the same geometrical interpretation as on $\mathscr{C}_{a,b}$, although the corresponding algebraic formulas are different. They are written out in §7.

For any $u,r,s,t \in \mathbb{Q}$ with $u \neq 0$, the invertible linear change of variables

$$X = u^2 X' + r, \quad Y = u^3 Y' + su^2 X' + t$$

replaces $\mathscr{W}$ by a curve $\mathscr{W}'$ of the same form with discriminant $\Delta' = u^{-12}\Delta$. By means of such a transformation we may assume that the coefficients a_j are integers and that Δ, which is now an integer, has minimal absolute value. (It has been proved by Tate that we then have $|\Delta| > 1$.) The discussion which follows presupposes that $\mathscr{W}$ is chosen in this way so that, in particular, discriminant means 'minimal discriminant'. We say that such a $\mathscr{W}$ is a *minimal model* for the elliptic curve.

For any prime p, let $\mathcal{W}_p$ be the cubic curve defined over the finite field $\mathbb{F}_p$ by the polynomial

$$Y^2 + \tilde{a}_1 XY + \tilde{a}_3 Y - (X^3 + \tilde{a}_2 X^2 + \tilde{a}_4 X + \tilde{a}_6),$$

where $\tilde{a}_j \in a_j + p\mathbb{Z}$. If $p \nmid \Delta$ the cubic curve $\mathcal{W}_p$ is non-singular, and thus an elliptic curve, but if $p | \Delta$ then $\mathcal{W}_p$ has a unique singular point. The singular point (x_0, y_0) of $\mathcal{W}_p$ is a *cusp* if, on replacing X and Y by $x_0 + X$ and $y_0 + Y$, we obtain a polynomial of the form

$$c(aX + bY)^2 + \dots ,$$

where $a, b, c \in \mathbb{F}_p$ and the unwritten terms are of degree > 2. Otherwise, the singular point is a *node*.

For any prime p, let N_p denote the number of $\mathbb{F}_p$-points of $\mathcal{W}_p$, including the point at infinity O, and put

$$c_p = p + 1 - N_p.$$

It was conjectured by Artin (1924), and proved by Hasse (1934), that

$$|c_p| \le 2p^{1/2} \text{ if } p \nmid \Delta.$$

Since $2p^{1/2}$ is not an integer, this inequality says that the quadratic polynomial

$$1 - c_p T + pT^2$$

has conjugate complex roots $\gamma_p, \bar{\gamma}_p$ of absolute value $p^{-1/2}$ or, if we put $T = p^{-s}$, that the zeros of

$$1 - c_p p^{-s} + p^{1-2s}$$

lie on the line $\Re s = 1/2$. Thus it is an analogue of the Riemann hypothesis on the zeros of $\zeta(s)$, but differs from it by having been proved. (As mentioned in §5 of Chapter IX, Hasse's result was considerably generalized by Weil (1948) and Deligne (1974).)

The *L-function* of the original elliptic curve $\mathcal{W}$ is defined by

$$L(s) = L(s, \mathcal{W}) := \prod_{p | \Delta} (1 - c_p p^{-s})^{-1} \prod_{p \nmid \Delta} (1 - c_p p^{-s} + p^{1-2s})^{-1}.$$

The first product on the right side has only finitely many factors. The infinite second product is convergent for $\Re s > 3/2$, since

$$1 - c_p p^{-s} + p^{1-2s} = (p^{1/2-s} - p^{1/2}\gamma_p)(p^{1/2-s} - p^{1/2}\bar{\gamma}_p)$$

and $|\gamma_p| = |\overline{\gamma}_p| = p^{-1/2}$. Multiplying out the products, we obtain for $\Re s > 3/2$ an absolutely convergent Dirichlet series

$$L(s) = \sum_{n \geq 1} c_n n^{-s}$$

with integer coefficients c_n. (If $n = p$ is prime, then c_n is the previously defined c_p.)

The *conductor* $N = N(W)$ of the elliptic curve W is defined by the singular reductions W_p of W:

$$N = \prod_{p \mid \Delta} p^{f_p},$$

where $f_p = 1$ if W_p has a node, whereas $f_p = 2$ if $p > 3$ and W_p has a cusp. We will not define f_p if $p \in \{2,3\}$ and W_p has a cusp, but we mention that f_p is then an integer ≥ 2 which can be calculated by an algorithm due to Tate (1975). (It may be shown that $f_2 \leq 8$ and $f_3 \leq 5$.)

The elliptic curve W is said to be *semi-stable* if W_p has a node for every $p \mid \Delta$. Thus, for a semi-stable elliptic curve, the conductor N is precisely the product of the distinct primes dividing the discriminant Δ. (The semi-stable case is the only one in which the conductor is square-free.)

Three important conjectures about elliptic curves, involving their L-functions and conductors, will now be described.

It was conjectured by Hasse (1954) that the function

$$\zeta(s,W): = \zeta(s)\zeta(s-1)/L(s,W)$$

may be analytically continued to a function which is meromorphic in the whole complex plane and that $\zeta(2-s,W)$ is connected with $\zeta(s,W)$ by a functional equation similar to that satisfied by the Riemann zeta-function $\zeta(s)$. In terms of L-functions, Hasse's conjecture was given the following precise form by Weil (1967):

HW-Conjecture: *If the elliptic curve W has L-function $L(s)$ and conductor N, then $L(s)$ may be analytically continued, so that the function*

$$\Lambda(s) = (2\pi)^{-s}\,\Gamma(s)L(s),$$

where $\Gamma(s)$ denotes Euler's gamma-function, is holomorphic throughout the whole complex plane and satisfies the functional equation

$$\Lambda(s) = \pm N^{1-s}\,\Lambda(2-s).$$

(In fact it is the functional equation which determines the precise definition of the conductor.)

The second conjecture, due to Birch and Swinnerton-Dyer (1965), connects the L-function with the group of rational points:

BSD-Conjecture: *The L-function L(s) of the elliptic curve $\mathcal{W}$ has a zero at s = 1 of order exactly equal to the rank r ≥ 0 of the group E = E($\mathcal{W}$, $\mathbb{Q}$) of all rational points of $\mathcal{W}$.*

This is sometimes called the 'weak' conjecture of Birch and Swinnerton-Dyer, since they also gave a 'strong' version, in which the nonzero constant C such that

$$L(s) \sim C(s-1)^r \text{ for } s \to 1$$

is expressed by other arithmetic invariants of $\mathcal{W}$. The strong conjecture may be regarded as an analogue for elliptic curves of a known formula for the Dedekind zeta-function of an algebraic number field. An interesting reformulation of the strong form has been given by Bloch (1980).

The statement of the third conjecture requires some preparation. For any positive integer N, let $\Gamma_0(N)$ denote the multiplicative group of all matrices

$$A = \begin{pmatrix} a & b \\ c & d \end{pmatrix},$$

where a,b,c,d are integers such that $ad - bc = 1$ and $c \equiv 0 \bmod N$. A function $f(\tau)$ which is holomorphic for $\tau \in \mathfrak{H}$ (the upper half-plane) is said to be a *modular form of weight 2 for* $\Gamma_0(N)$ if, for every such A,

$$f((a\tau + b)/(c\tau + d)) = (c\tau + d)^2 f(\tau).$$

An elliptic curve $\mathcal{W}$, with L-function

$$L(s) = \sum_{n \geq 1} c_n n^{-s}$$

and conductor N, is said to be *modular* if the function

$$f(\tau) = \sum_{n \geq 1} c_n e^{2\pi i n \tau},$$

which is certainly holomorphic in $\mathfrak{H}$, is a modular form of weight 2 for $\Gamma_0(N)$. This actually implies that f is a 'cusp form' and satisfies a functional equation

$$f(-1/N\tau) = \mp N\tau^2 f(\tau).$$

It follows that the *Mellin transform*

$$\Lambda(s) = \int_0^\infty f(iy) y^{s-1} dy$$

may be analytically continued for all $s \in \mathbb{C}$ and satisfies the functional equation

$$\Lambda(s) = \pm N^{1-s} \Lambda(2-s).$$

(Note the reversal of sign.) But

$$\Lambda(s) = (2\pi)^{-s} \Gamma(s) L(s),$$

since, by (9) of Chapter IX,

$$\int_0^\infty e^{-2\pi ny} y^{s-1} \, dy = (2\pi n)^{-s} \Gamma(s).$$

Hence any modular elliptic curve satisfies the *HW*-conjecture.

It was shown by Weil (1967) that, conversely, an elliptic curve is modular if not only its *L*-function $L(s) = \sum_{n\geq 1} c_n n^{-s}$ has the properties required in the *HW*-conjecture but also, for sufficiently many Dirichlet characters χ, the 'twisted' *L*-functions

$$L(s,\chi) = \sum_{n\geq 1} \chi(n) c_n n^{-s}$$

have analogous properties.

The definition of modular elliptic curve can be given a more intuitive form: the elliptic curve $\mathscr{C}_{a,b}$ is modular if there exist non-constant functions $X = f(\tau)$, $Y = g(\tau)$ which are holomorphic in the upper half-plane, which are invariant under $\Gamma_0(N)$, i.e.

$$f((a\tau + b)/(c\tau + d)) = f(\tau), \; g((a\tau + b)/(c\tau + d)) = g(\tau)$$

for every

$$A = \begin{pmatrix} a & b \\ c & d \end{pmatrix} \in \Gamma_0(N),$$

and which parametrize $\mathscr{C}_{a,b}$:

$$g^2(\tau) = f^3(\tau) + af(\tau) + b.$$

The significance of modular elliptic curves is that one can apply to them the extensive analytic theory of modular forms. For example, through the work of Kolyvagin (1990), together with results of Gross and Zagier (1986) and others, it is known that (as the *BSD*-conjecture predicts) a modular elliptic curve has rank 0 if its *L*-function does not vanish at $s = 1$, and has rank 1 if its *L*-function has a simple zero at $s = 1$.

The third conjecture, stated rather roughly by Taniyama (1955) and more precisely by Weil (1967), is simply this:

TW-Conjecture: *Every elliptic curve over the field $\mathbb{Q}$ of rational numbers is modular.*

The name of Shimura is often also attached to this conjecture, since he certainly contributed to its ultimate formulation. Shimura (1971) further showed that any elliptic curve which admits complex multiplication is modular. A big step forward was made by Wiles (1995) who, with assistance from Taylor, showed that any semi-stable elliptic curve is modular. A complete proof of the *TW*-conjecture, due to Diamond and others, has recently been announced by Darmon (1999). Thus all the results which had previously been established for modular elliptic curves actually hold for all elliptic curves over $\mathbb{Q}$.

It should be mentioned that there is also a 'Riemann hypothesis' for elliptic curves over $\mathbb{Q}$, namely that all zeros of the *L*-function in the critical strip $1/2 < \Re s < 3/2$ lie on the line $\Re s = 1$.

Mordell's theorem was extended from elliptic curves over $\mathbb{Q}$ to abelian varieties over any algebraic number field by Weil (1928). Many other results in the arithmetic of elliptic curves have been similarly extended. The topic is too vast to be considered here, but it should be said that our exposition for the prototype case is not always in the most appropriate form for such generalizations.

In the same paper in which he proved his theorem, Mordell (1922) conjectured that if a non-singular irreducible projective curve, defined by a homogeneous polynomial $F(x,y,z)$ with rational coefficients, has infinitely many rational points, then it is birationally equivalent to a line, a conic or a cubic. Mordell's conjecture was first proved by Faltings (1983). Actually Falting's result was not restricted to *plane* algebraic curves, and on the way he proved two other important conjectures of Tate and Shafarevich.

Falting's result implies that the Fermat equation $x^n + y^n = z^n$ has at most finitely many solutions in integers if $n > 3$. In the next section we will see that Wiles' result that semi-stable elliptic curves are modular implies that there are *no* solutions in nonzero integers.

6 Some applications

The arithmetic of elliptic curves has an interesting application to the ancient problem of congruent numbers. A positive integer n is (confusingly) said to be *congruent* if it is the area of a right-angled triangle whose sides all have rational length, i.e. if there exist positive rational numbers u,v,w such that $u^2 + v^2 = w^2$, $uv = 2n$. For example, 6 is congruent, since it is the area of the right-angled triangle with sides of length 3,4,5. Similarly, 5 is congruent, since it is the area of the right-angled triangle with sides of length 3/2, 20/3, 41/6.

In the margin of his copy of Diophantus' *Arithmetica* Fermat (c. 1640) gave a complete proof that 1 is not congruent. The following is a paraphrase of his argument. Assume that 1 is congruent. Then there exist positive rational numbers u,v,w such that

$$u^2 + v^2 = w^2, \quad uv = 2.$$

Since an integer is a rational square only if it is an integral square, on clearing denominators it follows that there exist positive integers a,b,c,d such that

$$a^2 + b^2 = c^2, \quad 2ab = d^2.$$

Choose such a quadruple a,b,c,d for which c is minimal. Then $(a,b) = 1$. Since d is even, exactly one of a,b is even and we may suppose it to be a. Then

$$a = 2g^2, \quad b = h^2$$

for some positive integers g,h. Since b and c are both odd and $(b,c) = 1$,

$$(c - b, c + b) = 2.$$

Since

$$(c - b)(c + b) = a^2 = 4g^4,$$

it follows that

$$c + b = 2c_1^4, \quad c - b = 2d_1^4,$$

for some relatively prime positive integers c_1, d_1. Then

$$(c_1^2 - d_1^2)(c_1^2 + d_1^2) = c_1^4 - d_1^4 = b = h^2.$$

But

$$(c_1^2 - d_1^2, c_1^2 + d_1^2) = 1,$$

since $(c_1^2, d_1^2) = 1$ and b is odd. Hence

$$c_1^2 - d_1^2 = p^2, \quad c_1^2 + d_1^2 = q^2,$$

for some odd positive integers p,q. Thus

$$a_1 = (q + p)/2, \quad b_1 = (q - p)/2$$

are positive integers and

$$a_1^2 + b_1^2 = (q^2 + p^2)/2 = c_1^2,$$

$$2a_1 b_1 = (q^2 - p^2)/2 = d_1^2.$$

Since $c_1 \leq c_1{}^4 < c$, this contradicts the minimality of c.

It follows that the Fermat equation

$$x^4 + y^4 = z^4$$

has no solutions in nonzero integers x, y, z. For if a solution existed and if we put

$$u = 2|yz|/x^2, \quad v = x^2/|yz|, \quad w = (y^4 + z^4)/x^2|yz|,$$

we would have $u^2 + v^2 = w^2$, $uv = 2$.

It is easily seen that a positive integer n is congruent if and only if there exists a rational number x such that x, $x + n$ and $x - n$ are all rational squares. For suppose

$$x = r^2, \quad x + n = s^2, \quad x - n = t^2,$$

and put

$$u = s - t, \quad v = s + t, \quad w = 2r.$$

Then

$$uv = s^2 - t^2 = 2n$$

and

$$u^2 + v^2 = 2(s^2 + t^2) = 4x = w^2.$$

Conversely, if u, v, w are rational numbers such that $uv = 2n$ and $u^2 + v^2 = w^2$, then

$$(u + v)^2 = w^2 + 4n, \quad (u - v)^2 = w^2 - 4n.$$

Thus, if we put $x = (w/2)^2$, then x, $x + n$ and $x - n$ are all rational squares.

It may be noted that if x is a rational number such that x, $x + n$ and $x - n$ are all rational squares, then $x \neq -n, 0, n$, since $n > 0$ and 2 is not a rational square.

The problem of determining which positive integers are congruent was considered by Arab mathematicians of the 10th century AD, and later by Fibonacci (1225) in his *Liber Quadratorum*. The connection with elliptic curves will now be revealed:

PROPOSITION 13 *A positive integer n is congruent if and only if the cubic curve C_n defined by the polynomial*

$$Y^2 - (X^3 - n^2X)$$

has a rational point $P = (x, y)$ with $y \neq 0$.

Proof Suppose first that n is congruent. Then there exists a rational number x such that x, $x + n$ and $x - n$ are all rational squares. Hence their product

$$x^3 - n^2x = x(x - n)(x + n)$$

is also a rational square. Since $x \neq -n, 0, n$, it follows that $x^3 - n^2x = y^2$, where y is a nonzero rational number.

Suppose now that $P = (x,y)$ is any rational point of the curve C_n with $y \neq 0$. If we put

$$u = |(x^2 - n^2)/y|, \quad v = |2nx/y|, \quad w = |(x^2 + n^2)/y|,$$

then u, v, w are positive rational numbers such that

$$u^2 + v^2 = w^2, \quad uv = 2n. \quad \square$$

It is readily verified that $\lambda = 1/2$ in the Riemann normal form for C_n.

We show next that if x is an *integer* such that $x, x + n$ and $x - n$ are all rational squares, then n is divisible by 4. We have $x = r^2$, $x + n = s^2$, $x - n = t^2$ for some positive *integers* r, s, t. Thus

$$2n = s^2 - t^2 = (s - t)(s + t).$$

If n were odd, exactly one of $s - t$ and $s + t$ would be even, which is impossible. Hence n is even. Since $n = s^2 - r^2$ and any integral square is congruent to 0 or 1 mod 4, we cannot have $n \equiv 2 \bmod 4$. Hence $n \equiv 0 \bmod 4$.

We now show that, for any positive integer n, the torsion group of C_n has order 4, consisting of the identity element O, and the three elements $(0,0)$, $(n,0)$, $(-n,0)$ of order 2. Let $P = (x,y)$ be any rational point of C_n with $y \neq 0$. We wish to show that P is of infinite order. Assume on the contrary that P is of finite order. Then $2P = (x',y')$ is also a rational point of finite order. The formula for the other point of intersection with C_n of the tangent to C_n at P shows that

$$x' = [(x^2 + n^2)/2y]^2.$$

It follows that

$$x' + n = [(x^2 - n^2 + 2nx)/2y]^2,$$
$$x' - n = [(x^2 - n^2 - 2nx)/2y]^2.$$

Thus $x', x' + n$ and $x' - n$ are all rational squares. Since $2P$ is of finite order, the theorem of Nagell and Lutz mentioned in §5 implies that x' is an integer. Consequently n is divisible by 4. But, by replacing X by $2^{2\alpha}X$ and Y by $2^{3\alpha}Y$ for some integer $\alpha \geq 1$, we replace C_n by C_ν, where $\nu = 2^{-2\alpha}n$, and we may choose α so that ν is an integer not divisible by 4. Since $(\xi, \eta) = (2^{-2\alpha}x, 2^{-3\alpha}y)$ is a rational point of C_ν of finite order with $\eta \neq 0$, this is a contradiction.

If n is congruent, then so also is $m^2 n$ for any positive integer m. Thus it is enough to determine which square-free positive integers are congruent. By what we have just proved and Proposition 13, a square-free positive integer n is congruent if and only if the elliptic curve C_n has positive rank. Since C_n admits complex multiplication, a result of Coates and Wiles (1977) shows that if C_n has positive rank, then its L-function vanishes at $s = 1$. (According to the *BSD*-conjecture, C_n has positive rank if and only if its L-function vanishes at $s = 1$.)

By means of the theory of modular forms, Tunnell (1983) has obtained a practical necessary and sufficient condition for the L-function $L(s, C_n)$ of C_n to vanish at $s = 1$: if n is a square-free positive integer, then $L(1, C_n) = 0$ if and only if $A_+(n) = A_-(n)$, where $A_+(n)$, resp. $A_-(n)$, is the number of triples $(x, y, z) \in \mathbb{Z}^3$ with z even, resp. z odd, such that

$$x^2 + 2y^2 + 8z^2 = n \text{ if } n \text{ is odd, or } 2x^2 + 2y^2 + 16z^2 = n \text{ if } n \text{ is even.}$$

It is not difficult to show that $A_+(n) = A_-(n)$ when $n \equiv 5, 6$ or $7 \bmod 8$, but there seems to be no such simple criterion in other cases. With the aid of a computer it has been verified that, for every $n < 10000$, n is congruent if and only if $A_+(n) = A_-(n)$.

The arithmetic of elliptic curves also has a useful application to the class number problem of Gauss. For any square-free integer $d < 0$, let $h(d)$ be the *class number* of the quadratic field $\mathbb{Q}(\sqrt{d})$. As mentioned in §8 of Chapter IV, it was conjectured by Gauss (1801), and proved by Heilbronn (1934), that $h(d) \to \infty$ as $d \to -\infty$. However, the proof does not provide a method of determining an upper bound for the values of d for which the class number $h(d)$ has a given value. As mentioned in Chapter II, Stark (1967) showed that there are no other negative values of d for which $h(d) = 1$ besides the nine values already known to Gauss. Using methods developed by Baker (1966) for the theory of transcendental numbers, it was shown by Baker (1971) and Stark (1971) that there are exactly 18 negative values of d for which $h(d) = 2$. A simpler and more powerful method for attacking the problem was found by Goldfeld (1976). He obtained an effective lower bound for $h(d)$, provided that there exists an elliptic curve over $\mathbb{Q}$ whose L-function has a triple zero at $s = 1$. Gross and Zagier (1986) showed that such an elliptic curve does indeed exist. However, they needed to know that this elliptic curve was modular, and this required a considerable amount of computation. The proof of the *TW*-conjecture makes any computation unnecessary.

The most celebrated application of the arithmetic of elliptic curves has been the recent proof of Fermat's last theorem. In his copy of the translation by Bachet of Diophantus' *Arithmetica* Fermat also wrote "It is impossible to separate a cube into two cubes, or a fourth power into two fourth powers or, in general, any power higher than the second into two like powers. I have discovered a truly marvellous proof of this, which this margin is too narrow to contain."

In other words, Fermat asserted that, if $n > 2$, the equation

$$x^n + y^n = z^n$$

has no solutions in nonzero integers x,y,z. In §2 of Chapter III we pointed out that it was sufficient to prove his assertion when $n = 4$ and when $n = p$ is an odd prime, and we gave a proof there for $n = 3$.

A nice application to cubic curves of the case $n = 3$ was made by Kronecker (1859). If we make the change of variables

$$x = 2a/(3b - 1), \quad y = (3b + 1)/(3b - 1),$$

with inverse

$$a = x/(y - 1), \quad b = (y + 1)/3(y - 1),$$

then

$$x^3 + y^3 - 1 = 2(4a^3 + 27b^2 + 1)/(3b - 1)^3.$$

Since the equation $x^3 + y^3 = 1$ has no solution in nonzero rational numbers, the only solutions in rational numbers of the equation

$$4a^3 + 27b^2 = -1$$

are $a = -1$, $b = \pm 1/3$. Consequently the only cubic curves $\mathscr{C}_{a,b}$ with rational coefficients a,b and discriminant $d = -1$ are $Y^2 - X^3 + X \pm 1/3$.

We return now to Fermat's assertion. In the present section we have already given Fermat's own proof for $n = 4$. Suppose now that $p \geq 5$ is prime and assume, contrary to Fermat's assertion, that the equation

$$a^p + b^p + c^p = 0$$

does have a solution in nonzero integers a,b,c. By removing any common factor we may assume that $(a,b) = 1$, and then also $(a,c) = (b,c) = 1$. Since a,b,c cannot all be odd, we may assume that b is even. Then a and c are odd, and we may assume that $a \equiv -1 \bmod 4$.

We now consider the projective cubic curve $\mathscr{C}_{A,B}$ defined by the polynomial

$$Y^2 - X(X - A)(X + B),$$

where $A = a^p$ and $B = b^p$. By construction, $(A,B) = 1$ and

$$A \equiv -1 \bmod 4, \quad B \equiv 0 \bmod 32.$$

Moreover, if we put $C = -(A + B)$, then $C \neq 0$ and $(A,C) = (B,C) = 1$. The linear change of variables

$$X \to 4X, \ Y \to 8Y + 4X$$

replaces $\mathcal{E}_{A,B}$ by the elliptic curve $\mathcal{W}_{A,B}$ defined by

$$Y^2 + XY - \{X^3 + (B - A - 1)X^2/4 - ABX/16\},$$

which has discriminant

$$\Delta = (ABC)^2/2^8.$$

Our hypotheses ensure that the coefficients of $\mathcal{W}_{A,B}$ are integers and that Δ is a nonzero integer. It may be shown that $\mathcal{W}_{A,B}$ is actually a minimal model for $\mathcal{E}_{A,B}$. Moreover, when we reduce modulo any prime ℓ which divides Δ, the singular point which arises is a node. Thus $\mathcal{W}_{A,B}$ is semi-stable and its conductor N is the product of the distinct primes dividing ABC.

Fermat's last theorem will be proved, for any prime $p \geq 5$, if we show that such an elliptic curve cannot exist if A,B,C are all p-th powers. If p is large, one reason for suspecting that such an elliptic curve cannot exist is that the discriminant is then very large compared with the conductor. Another reason, which does not depend on the size of p, was suggested by Frey (1986). Frey gave a heuristic argument that $\mathcal{E}_{A,B}$ could not then be modular, which would contradict the *TW*-conjecture.

Frey's intuition was made more precise by Serre (1987). Let G be the group of all automorphisms of the field of all algebraic numbers. With any modular form for $\Gamma_0(N)$ one can associate a 2-dimensional representation of G over a finite field. Serre showed that Fermat's last theorem would follow from the *TW*-conjecture, together with a conjecture about lowering the level of such 'Galois representations' associated with modular forms. The latter conjecture was called Serre's ε-conjecture, because it was a special case of a much more general conjecture which Serre made.

Serre's ε-conjecture was proved by Ribet (1990), although the proof might be described as being of order ε^{-1}. Now, for the first time, the falsity of Fermat's last theorem would have a significant consequence: the falsity of the *TW*-conjecture. Since $\mathcal{E}_{A,B}$ is semi-stable with the normalizations made above, to prove Fermat's last theorem it was actually enough to show that any semi-stable elliptic curve was modular. As stated in §5, this was accomplished by Wiles (1995) and Taylor and Wiles (1995). We will not attempt to describe the proof since, besides Fermat's classic excuse, it is beyond the scope of this work.

Fermat's last theorem contributed greatly to the development of mathematics, but Fermat was perhaps lucky that his assertion turned out to be correct. After proving Fermat's assertion for $n = 3$, that the cube of a positive integer could not be the sum of two cubes of positive integers, Euler asserted that, also for any $n \geq 4$, an n-th power of a positive integer could not be expressed as a sum of $n - 1$ n-th powers of positive integers. A counterexample to Euler's conjecture was first found, for $n = 5$, by Lander and Parkin (1966):

$$27^5 + 84^5 + 110^5 + 133^5 = 144^5.$$

Elkies (1988) used the arithmetic of elliptic curves to find infinitely many counterexamples for $n = 4$, the simplest being

$$95800^4 + 217519^4 + 414560^4 = 422481^4.$$

A prize has been offered by Beal (1997) for a proof or disproof of his conjecture that the equation

$$x^l + y^m = z^n$$

has no solution in coprime positive integers x,y,z if l,m,n are integers > 2. (The exponent 2 must be excluded since, for example, $2^5 + 7^2 = 3^4$ and $2^7 + 17^3 = 71^2$.) Will Beal's conjecture turn out to be like Fermat's or like Euler's?

7 Further remarks

For sums of squares, see Grosswald [31], Rademacher [46], and Volume II, Chapter IX of Dickson [23]. A recent contribution is Milne [42].

A general reference for the theory of partitions is Andrews [2]. Proposition 4 is often referred to as *Euler's pentagonal number theorem*, since the numbers $m(3m - 1)/2$ $(m > 1)$ represent the number of dots needed to construct successively larger and larger pentagons. A direct proof of the combinatorial interpretation of Proposition 4 was given by Franklin (1881). It is reproduced in Andrews [2] and in van Lint and Wilson [41]. The replacement of proofs using generating functions by purely combinatorial proofs has become quite an industry; see, for example, Bressoud and Zeilberger [13], [14].

Besides the q-difference equations used in the proof of Proposition 5, there are also q-*integrals*:

$$\int_0^a f(x) \, d_q x := \sum_{n\geq 0} f(aq^n)(aq^n - aq^{n+1}).$$

The *q-binomial coefficients* (mentioned in §2 of Chapter II)

$$\begin{bmatrix} n \\ m \end{bmatrix} = \begin{bmatrix} n \\ m \end{bmatrix}_q := (q)_n/(q)_m(q)_{n-m} \quad (0 \leq m < n),$$

where $(a)_0 = 1$ and

$$(a)_n = (1-a)(1-aq)...(1-aq^{n-1}) \quad (n \geq 1),$$

have recurrence properties similar to those of ordinary binomial coefficients:

$$\begin{bmatrix} n \\ m \end{bmatrix} = \begin{bmatrix} n-1 \\ m-1 \end{bmatrix} + q^m \begin{bmatrix} n-1 \\ m \end{bmatrix} = \begin{bmatrix} n-1 \\ m \end{bmatrix} + q^{n-m} \begin{bmatrix} n-1 \\ m-1 \end{bmatrix} \quad (0 < m < n).$$

The *q-hypergeometric series*

$$\sum_{n \geq 0} (a)_n (b)_n x^n / (c)_n (q)_n$$

was already studied by Heine (1847). There is indeed a whole world of q-analysis, which may be regarded as having the same relation to classical analysis as quantum mechanics has to classical mechanics. (The choice of the letter 'q' nearly a century before the advent of quantum mechanics showed remarkable foresight.) There are introductions to this world in Andrews *et al.* [4] and Vilenkin and Klimyk [58]. For Macdonald's conjectures concerning q-analogues of orthogonal polynomials, see Kirillov [36].

Although q-analysis always had its devotees, it remained outside the mainstream of mathematics until recently. Now it arises naturally in the study of *quantum groups*, which are not groups but q-deformations of the universal enveloping algebra of a Lie algebra.

The Rogers–Ramanujan identities were discovered independently by Rogers (1894), Ramanujan (1913) and Schur (1917). Their romantic history is retold in Andrews [2], which contains also generalizations. For the applications of the identities in statistical mechanics, see Baxter's article (pp. 69-84) in Andrews *et al.* [3]. (The same volume contains other interesting articles on mathematical developments arising from Ramanujan's work.)

The Jacobi triple product formula was derived in Chapter XII as the limit of a formula for polynomials. Andrews [1] has given a similar derivation of the Rogers–Ramanujan identities. This approach has found applications and generalizations in conformal field theory, the two sides of the polynomial identity corresponding to fermionic and bosonic bases for Fock space; see Berkovich and McCoy [9].

These connections go much further than the Rogers–Ramanujan identities. There is now a vast interacting area which involves, besides the theory of partitions, solvable models of statistical mechanics, conformal field theory, integrable systems in classical and quantum mechanics, infinite-dimensional Lie algebras, quantum groups, knot theory and operator algebras. For introductory accounts, see [45], [10] and various articles in [24] and [27]. More detailed treatments of particular aspects are given in Baxter [8], Faddeev and Takhtajan [26], Jantzen [33], Jones [34], Kac [35] and Korepin *et al.* [38].

For the Hardy–Ramanujan–Rademacher expansion for $p(n)$, see Rademacher [46] and Andrews [2]. An interesting proof by means of probability theory for the first term of the expansion has been given by Báez-Duarte [5].

The definition of birational equivalence in §3 is adequate for our purposes, but has been superseded by a more general definition in the language of 'schemes', which is applicable to algebraic varieties of arbitrary dimension without any given embedding in a projective space. For the evolution of the modern concept, see Čižmár [18].

The history of the discovery of the group law on a cubic curve is described by Schappacher [48].

Several good accounts of the arithmetic of elliptic curves are now available; e.g., Knapp [37] and the trilogy [52],[50],[51]. Although the subject has been transformed in the past 25 years, the survey articles by Cassels [16], Tate [55] and Gelbart [28] are still of use. Tate gives a helpful introduction, Cassels has many references to the older literature, and Gelbart explains the connection with the Langlands program, for which see also Gelbart [29].

For reference, we give here the formulas for addition on an elliptic curve in the so-called Weierstrass's normal form. If $P_1 = (x_1, y_1)$ and $P_2 = (x_2, y_2)$ are points of the curve

$$Y^2 + a_1 XY + a_3 Y - (X^3 + a_2 X^2 + a_4 X + a_6),$$

then

$$-P_1 = (x_1, -y_1 - a_1 x_1 - a_3), \quad P_1 + P_2 = P_3^* = (x_3, -y_3),$$

where

$$x_3 = \lambda(\lambda + a_1) - a_2 - x_1 - x_2, \quad y_3 = (\lambda + a_1)x_3 + \mu + a_3,$$

and

$$\lambda = (y_2 - y_1)/(x_2 - x_1), \quad \mu = (y_1 x_2 - y_2 x_1)/(x_2 - x_1) \quad \text{if } x_1 \neq x_2;$$

$$\lambda = (3x_1^2 + 2a_2 x_1 + a_4 - a_1 y_1)/N, \quad \mu = (-x_1^3 + a_4 x_1 + 2a_6 - a_3 y_1)/N \text{ if } x_1 = x_2, P_2 \neq -P_1,$$

with $N = 2y_1 + a_1 x_1 + a_3$.

An algorithm for obtaining a minimal model of an elliptic curve is described in Laska [40]. Other algorithms connected with elliptic curves are given in Cremona [21].

The original conjecture of Birch and Swinnerton-Dyer was generalized by Tate [54] and Bloch [11]. For a first introduction to the theory of modular forms see Serre [49], and for a second see Lang [39].

Hasse actually showed that, if $\mathscr{E}$ is an elliptic curve over any finite field $\mathbb{F}_q$ containing q elements, then the number N_q of $\mathbb{F}_q$-points on $\mathscr{E}$ (including the point at infinity) satisfies the inequality

$$|N_q - (q + 1)| \le 2q^{1/2}.$$

For an elementary proof, see Chahal [17]. Hasse's result is the special case, when the genus $g = 1$, of the Riemann hypothesis for function fields, which was mentioned in Chapter IX, §5.

It follows from the result of Siegel (1929), mentioned in §9 of Chapter IV, and even from the earlier work of Thue (1909), that an elliptic curve with integral coefficients has at most finitely many *integral* points. However, their method is not constructive. Baker [6], using the results on linear forms in the logarithms of algebraic numbers which he developed for the theory of transcendental numbers, obtained an explicit upper bound for the magnitude of any integral point in terms of an upper bound for the absolute values of all coefficients. Sharper bounds have since been obtained, e.g. by Bugeaud [15]. (For modern proofs of Baker's theorem on the linear independence of logarithms of algebraic numbers, see Waldschmidt [59]. The history of Baker's method is described in Baker [7].)

For information about the proof of Mordell's conjecture we refer to Bloch [12], Szpiro [53], and Cornell and Silverman [19]. The last includes an English translation of Faltings' original article. As mentioned in §9 of Chapter IV, Vojta (1991) has given a proof of the Mordell conjecture which is completely different from that of Faltings. There is an exposition of this proof, with simplifications due to Bombieri (1990), in Hindry and Silverman [32].

For congruent numbers, see Volume II, Chapter XVI of Dickson [23], Tunnell [57], and Noda and Wada [43]. The survey articles of Goldfeld [30] and Oesterlé [44] deal with Gauss's class number problem.

References for earlier work on Fermat's last theorem were given in Chapter III. Ribet [47] and Cornell *et al.* [20] provide some preparation for the original papers of Wiles [60] and Taylor and Wiles [56]. For the *TW*-conjecture, see also Darmon [22]. For Euler's conjecture, see Elkies [25].

8 Selected references

[1] G.E. Andrews, A polynomial identity which implies the Rogers–Ramanujan identities, *Scripta Math.* **28** (1970), 297-305.

[2] G.E. Andrews, *The theory of partitions*, Addison-Wesley, Reading, Mass., 1976. [Paperback edition, Cambridge University Press, 1998]

[3]   G.E. Andrews, R.A. Askey, B.C. Berndt, K.G. Ramanathan and R.A. Rankin (ed.), *Ramanujan revisited*, Academic Press, London, 1988.

[4]   G.E. Andrews, R. Askey and R. Roy, *Special functions*, Cambridge University Press, 1999.

[5]   L. Báez-Duarte, Hardy–Ramanujan's asymptotic formula for partitions and the central limit theorem, *Adv. in Math.* **125** (1997), 114-120.

[6]   A. Baker, The diophantine equation $y^2 = ax^3 + bx^2 + cx + d$, *J. London Math. Soc.* **43** (1968), 1-9.

[7]   A. Baker, The theory of linear forms in logarithms, *Transcendence theory: advances and applications* (ed. A. Baker and D.W. Masser), pp. 1-27, Academic Press, London, 1977.

[8]   R.J. Baxter, *Exactly solved models in statistical mechanics*, Academic Press, London, 1982. [Reprinted, 1989]

[9]   A. Berkovich and B.M. McCoy, Rogers–Ramanujan identities: a century of progress from mathematics to physics, *Proceedings of the International Congress of Mathematicians: Berlin 1998*, Vol. III, pp. 163-172, Documenta Mathematica, Bielefeld, 1998.

[10]   J.S. Birman, New points of view in knot theory, *Bull. Amer. Math. Soc. (N.S.)* **28** (1993), 253-287.

[11]   S. Bloch, A note on height pairings, Tamagawa numbers, and the Birch and Swinnerton-Dyer conjecture, *Invent. Math.* **58** (1980), 65-76.

[12]   S. Bloch, The proof of the Mordell conjecture, *Math. Intelligencer* **6** (1984), no. 2, 41-47.

[13]   D.M. Bressoud and D. Zeilberger, A short Rogers–Ramanujan bijection, *Discrete Math.* **38** (1982), 313-315.

[14]   D.M. Bressoud and D. Zeilberger, Bijecting Euler's partitions-recurrence, *Amer. Math. Monthly* **92** (1985), 54-55.

[15]   Y. Bugeaud, On the size of integer solutions of elliptic equations, *Bull. Austral. Math. Soc.* **57** (1998), 199-206.

[16] J.W.S. Cassels, Diophantine equations with special reference to elliptic curves, *J. London Math. Soc.* **41** (1966), 193-291.

[17] J.S. Chahal, Manin's proof of the Hasse inequality revisited, *Nieuw Arch. Wisk.* (4) **13** (1995), 219-232.

[18] J. Čižmár, Birationale Transformationen (Ein historischer Überblick), *Period. Polytech. Mech. Engrg.* **39** (1995), 9-24.

[19] G. Cornell and J.H. Silverman (ed.), *Arithmetic geometry*, Springer-Verlag, New York, 1986.

[20] G. Cornell, J.H. Silverman and G. Stevens (ed.), *Modular forms and Fermat's last theorem*, Springer, New York, 1997.

[21] J.E. Cremona, *Algorithms for modular elliptic curves*, 2nd ed., Cambridge University Press, 1997.

[22] H. Darmon, A proof of the full Shimura–Taniyama–Weil conjecture is announced, *Notices Amer. Math. Soc.* **46** (1999), 1397-1401.

[23] L.E. Dickson, *History of the theory of numbers*, 3 vols., Carnegie Institute, Washington, D.C., 1919-1923. [Reprinted Chelsea, New York, 1992]

[24] L. Ehrenpreis and R.C. Gunning (ed.), *Theta functions: Bowdoin* 1987, Proc. Symp. Pure Math. **49**, Amer. Math. Soc., Providence, R.I., 1989.

[25] N.D. Elkies, On $A^4 + B^4 + C^4 = D^4$, *Math. Comp.* **51** (1988), 825-835.

[26] L.D. Faddeev and L.A. Takhtajan, *Hamiltonian methods in soliton theory*, Springer-Verlag, Berlin, 1987.

[27] A.S. Fokas and V.E. Zakharov (ed.), *Important developments in soliton theory*, Springer-Verlag, Berlin, 1993.

[28] S. Gelbart, Elliptic curves and automorphic representations, *Adv. in Math.* **21** (1976), 235-292.

[29] S. Gelbart, An elementary introduction to the Langlands program, *Bull. Amer. Math. Soc.* (*N.S.*) **10** (1984), 177-219.

[30] D. Goldfeld, Gauss' class number problem for imaginary quadratic fields, *Bull. Amer. Math. Soc. (N.S.)* **13** (1985), 23-37.

[31] E. Grosswald, *Representations of integers as sums of squares*, Springer-Verlag, New York, 1985.

[32] M. Hindry and J.H. Silverman, *Diophantine geometry*, Springer, New York, 2000.

[33] J.C. Jantzen, *Lectures on quantum groups*, American Mathematical Society, Providence, R.I., 1996.

[34] V.F.R. Jones, *Subfactors and knots*, CBMS Regional Conference Series in Mathematics **80**, Amer. Math. Soc., Providence, R.I., 1991.

[35] V.G. Kac, *Infinite-dimensional Lie algebras*, 3rd ed., Cambridge University Press, 1990.

[36] A.A. Kirillov, Jr., Lectures on affine Hecke algebras and Macdonald's conjectures, *Bull. Amer. Math. Soc. (N.S.)* **34** (1997), 251-292.

[37] A.W. Knapp, *Elliptic curves*, Princeton University Press, Princeton, N.J., 1992.

[38] V.E. Korepin, N.M. Bogoliubov and A.G. Izergin, *Quantum inverse scattering method and correlation functions*, Cambridge University Press, 1993.

[39] S. Lang, *Introduction to modular forms*, Springer-Verlag, Berlin, corr. reprint, 1995.

[40] M. Laska, An algorithm for finding a minimal Weierstrass equation for an elliptic curve, *Math. Comp.* **38** (1982), 257-260.

[41] J.H. van Lint and R.M. Wilson, *A course in combinatorics*, Cambridge University Press, 1992.

[42] S.C. Milne, New infinite families of exact sums of squares formulas, Jacobi elliptic functions and Ramanujan's tau function, *Proc. Nat. Acad. Sci. U.S.A.* **93** (1996), 15004-15008.

[43] K. Noda and H. Wada, All congruent numbers less than 10000, *Proc. Japan Acad. Ser. A Math. Sci.* **69** (1993), 175-178.

[44] J. Oesterlé, Le problème de Gauss sur le nombre de classes, *Enseign. Math.* **34** (1988), 43-67.

[45] M. Okado, M. Jimbo and T. Miwa, Solvable lattice models in two dimensions and modular functions, *Sugaku Exp.* **2** (1989), 29-54.

[46] H. Rademacher, *Topics in analytic number theory*, Springer-Verlag, Berlin, 1973.

[47] K.A. Ribet, Galois representations and modular forms, *Bull. Amer. Math. Soc. (N.S.)* **32** (1995), 375-402.

[48] N. Schappacher, Développement de la loi de groupe sur une cubique, *Séminaire de Théorie des Nombres, Paris* 1988-89 (ed. C. Goldstein), pp. 159-184, Birkhäuser, Boston, 1990.

[49] J.-P. Serre, *A course in arithmetic*, Springer-Verlag, New York, 1973.

[50] J.H. Silverman, *The arithmetic of elliptic curves*, Springer-Verlag, New York, 1986.

[51] J.H. Silverman, *Advanced topics in the arithmetic of elliptic curves*, Springer-Verlag, New York, 1994.

[52] J.H. Silverman and J. Tate, *Rational points on elliptic curves*, Springer-Verlag, New York, 1992.

[53] L. Szpiro, La conjecture de Mordell [d'après G. Faltings], *Astérisque* **121-122** (1985), 83-103.

[54] J.T. Tate, On the conjectures of Birch and Swinnerton-Dyer and a geometric analog, *Séminaire Bourbaki: Vol. 1965/1966, Exposé no. 306*, Benjamin, New York, 1966.

[55] J.T. Tate, The arithmetic of elliptic curves, *Invent. Math.* **23** (1974), 179-206.

[56] R.L. Taylor and A. Wiles, Ring theoretic properties of certain Hecke algebras, *Ann. of Math.* **141** (1995), 553-572.

[57] J.B. Tunnell, A classical Diophantine problem and modular forms of weight 3/2, *Invent. Math.* **72** (1983), 323-334.

[58] N. Ja. Vilenkin and A.V. Klimyk, *Representation of Lie groups and special functions*, 4 vols., Kluwer, Dordrecht, 1991-1995.

[59] M. Waldschmidt, *Diophantine approximation on linear algebraic groups*, Springer, Berlin, 2000.

[60] A. Wiles, Modular elliptic curves and Fermat's last theorem, *Ann. of Math.* **141** (1995), 443-551.

Notations

R/S 73

$R \oplus R'$, αv, $v + w$ 74

D^n, $\mathscr{C}(I)$, O 75

$\mathscr{C}'(I)$, $U_1 + U_2$, $U_1 \oplus U_2$, $<S>$ 76

$\dim V$, $[E{:}F]$, $e_1,...,e_n$ 77

Tv, TS 78

$S + T$, $GL(V)$, $M_n(F)$ 79

$M_n(D)$ 80

$<u,v>$ 82

$\|v\|$ 83

$\mathscr{B}$ 86

ℓ^2, $L^2(I)$ 87

$*\mathbb{R}$ 88

$b|a$, $b \nmid a$, $\times$, (a,b) 97

$[a,b]$ 98

$a \wedge b$, $a \vee b$ 99

$(a_1,a_2,...,a_n)$, $[a_1,a_2,...,a_n]$ 100

$K[t]$ 102, 112

$K(t)$ 102, 306

$^m C_n$ 108, 129

$\partial(f)$, $|f|$, $R[t]$, $R[[t]]$ 113

$K[t,t^{-1}]$ 115

$c(f)$ 117

$R[t_1,...,t_m]$ 118

$\Phi_p(x)$ 119

f' 120

$\delta(a)$ 121

$\mathbb{Q}(\sqrt{d})$, $\mathbb{O}_d$ 123

$a \equiv b \bmod m$, $\not\equiv$, $\mathbb{Z}_{(m)}$ 124

$\mathbb{Z}_{(m)}{}^\times$, $\mathbb{F}_p$, $\varphi(m)$ 127

$\Phi_n(x)$ 129, 130

$\bar{f}(x)$ 130

$\mathbb{F}_p{}^\times$ 133

$N(\gamma)$, $\mathscr{G}$ 139

$\mathscr{H}$, $\bar{\gamma}$ 140

$N(\gamma)$, $(\alpha,\beta)_r$ 141

$\lfloor x \rfloor$, $g(k)$, $w(k)$, $G(k)$ 143

$K[[t_1,...,t_m]]$ 145

$\mathbb{F}_q$ 146

(a/n), $\mathrm{sgn}\,(\pi_a)$ 152

(a/p) 156

$G(m,n)$ 159

$\mathbb{Q}(\sqrt{d})$, α', $N(\alpha)$ 163

$\mathbb{O}_d$, ω, $\mathscr{E}$, $\mathscr{G}$ 164

$(a_1,...,a_m)$ 168

AB 169

A' 170

$h(d)$, $\mathbb{O}(K)$ 175

$f*g$ 176

$\delta(n)$, $\mathscr{A}$, $|f|$ 177

$i(n)$, $j(n)$ 178

$\tau(n)$, $\sigma(n)$ 179

$\mu(n)$, $\hat{f}(n)$ 180

M_p 182

γ 183, 443

F_n 184

$GL_n(\mathbb{Z})$ 186

$A \oplus B$ 187, 352

$M_1 \cap M_2$, $M_1 + M_2$ 191

Δ_k 197

$|a|$ 199

(f/g) 201

$\lfloor \xi \rfloor$, ξ_n, τ 209

$[a_0,a_1,a_2,...\,]$ 209, 213, 246, 247

p_n, q_n 210

$[a_0,a_1,...,a_N]$ 212, 213

p_n/q_n 212, 247

The *Landau order symbols* are defined in the following way: if $I = [t_0,\infty)$ is a half-line and if $f, g: I \to \mathbb{R}$ are real-valued functions with $g(t) > 0$ for all $t \in I$, we write

$f = O(g)$ if there exists a constant $C > 0$ such that $|f(t)|/g(t) \le C$ for all $t \in I$;

$f = o(g)$ if $f(t)/g(t) \to 0$ as $t \to \infty$;

$f \sim g$ if $f(t)/g(t) \to 1$ as $t \to \infty$.

The *end of a proof* is denoted by $\square$.

Axioms

Index